SIGNAL PROCESSING, II

OTHER IEEE PRESS BOOKS

SIGNAL PROCESSING, II

Edited by
Sun-Yuan Kung
University of Southern California

Robert E. Owen
Bipolar Integrated Technology

J. Greg Nash
Hughes Research Laboratories

The chapters in this book are based on presentations given at the IEEE Acoustics, Speech, and Signal Processing Society Workshop on VLSI Signal Processing held November 5–7, 1986 at the University of Southern California, Los Angeles. The Program Committee consisted of

Robert W. Brodersen, University of California
Peter R. Cappello, University of California
Chi-Foon Chan, Intel
Gwyn P. Edwards, Racal-Vadic
James H. Hesson, GE—Intersil
Leah H. Jamieson, Purdue University
Yih C. Jenq, Tektronix
Richard F. Lyon, Schlumberger
Rikio Maruta, NEC
Earl E. Schwartzlander, TRW
Paul M. Toldalagi, Analog Devices

**IEEE
PRESS**

**Published under the sponsorship of the
IEEE Acoustics, Speech, and Signal Processing Society.**

The Institute of Electrical and Electronics Engineers, Inc., New York

RECID = 10151 - 1

Copyright © 1986 by
THE INSTITUTE OF ELECTRICAL AND ELECTRONICS ENGINEERS, INC.
345 East 47th Street, New York, NY 10017-2394
All rights reserved.

PRINTED IN THE UNITED STATES OF AMERICA

IEEE Order Number: PC02022

Library of Congress Cataloging-in-Publication Data

VLSI signal processing, II.

"The chapters in this book are based on presentations at the IEEE Acoustics, Speech, and Signal Processing Society Workshop on VLSI Signal Processing held November 5–7, 1986 at the University of Southern California, Los Angeles."
Published under the sponsorship of the IEEE Acoustics, Speech, and Signal Processing Society."
Includes index.
1. Signal processing—Digital techniques—Congresses.
2. Integrated circuits—Very large scale integration—Congresses.   I. Kung, S. Y. (Sun-Yuan)   II. Owen, Robert E.   III. Nash, J. Greg.   IV. Workshop on VLSI Signal Processing (1986: University of Southern California)   V. IEEE Acoustics, Speech, and Signal Processing Society.
TK5102.5.V563   1986   621.38'043   86-20865

ISBN 0-87942-210-6

iv

TABLE OF CONTENTS

Part III: Implementation Examples

Part IV: Application Requirements

Note: Due to time constraints, the presentation on radar digital signal processing requirements by T. W. Miller could not be included in this book.

Preface

The chapters of this book are based on presentations at the 1986 IEEE Acoustics, Speech, and Signal Processing Society Workshop on VLSI Signal Processing, which was held in the Davidson Conference Center at the University of Southern California in Los Angeles on November 5–7, 1986. This was the third such biennial workshop sponsored and organized by the VLSI Technical Committee of the IEEE ASSP. The first, in 1982, had a focus on algorithms and architectures for DSP made possible with VLSI. The next workshop, held in 1984,[1] broadened this scope to include VLSI implementations.

Discussions following the previous workshops yielded two major observations in this embryonic field: (1) the design of VLSI circuits for signal processing is highly interdisciplinary and (2) the technology transfer between disciplines and the progress within these have been very uneven, particularly with respect to actual circuits used in operational systems. Therefore, the organizing committee of the third workshop sought, as primary objectives, to encourage discussion between major contributors in the various areas, and to further an understanding of what forces were shaping the overall progress of new results. Invited and submitted papers were selected to promote these objectives, rather than just to present new results.

For this workshop, four core disciplines of research and practice were defined: Integrated Circuit Technology, Algorithm/Architecture, Implementation Examples, and Application Requirements. These disciplines comprise the four parts of this book.

Part I: Integrated Circuit Technology

The chapters of Part I are based on presentations at the first two workshop sessions: one on integrated circuit design technology, and the other on analog integrated circuit technology. These sessions were organized by R. Brodersen of the University of California, Y. C. Jenq of Tektronix, R. Lyon of Schlumberger, and E. Swartzlander of TRW. This part stresses that an understanding of the basic capabilities of the fabrication technologies is essential in any VLSI circuit pursuit. It emphasizes both design tools and methods specialized for DSP

---

[1] The presentations in the 1984 workshop were published in the IEEE PRESS book *VLSI Signal Processing* (Order Number PC 01800).

and the particular technology issues involved in the vital analog-to-digital conversion interface.

Part II: VLSI Algorithms and Architectures

The chapters of Part II are based on presentations at three sessions: one on mapping algorithms to arrays, and the other two on special purpose architectures. These sessions were organized by P. R. Cappello of the University of California, G. Edwards of Racal-Vadic, L. Jamieson of Purdue University, and S. Y. Kung of the University of Southern California. The chapters describe systematic procedures for finding a systolic implementation for a given algorithm; array processor design for solving recursive algorithms; Viterbi's algorithm, rank order, and median filtering; and systolic design tool development. They also discuss new designs for a wavefront array that handles independent data flow and has applications to Kalman filtering; a content addressable array with applications to sparse matrix operations; and a parallel processing system that performs most Boolean and geometric operations. Also included are a VLSI FIR filter design with sound area efficiency; a decomposition-based architecture for convolution and DFT with a very high processor-level parallelism; and a VLSI processor, which is based on a quadratic residue number system and offers a very good function-level parallelism.

Part III: Implementation Examples

The chapters of Part III are based on presentations at two sessions: one on programmable processor implementations and the other on application-specific implementations. These sessions were organized by J. Hesson of GE-Intersil, R. Maruta of NEC, and J. G. Nash of Hughes Research Laboratories. The chapters describe digital signal processing implementation examples with various levels of problem specificity. They are illustrative of a growing trend toward the use of application-specific VLSI hardware for solutions of demanding digital signal processing problems. Of particular interest are the different techniques (e.g., algorithmic, architectural or technological) used to achieve the desired application goals. In all cases the solutions are described in terms of a custom or semi-custom VLSI circuit representing a unique compromise between such factors as desired performance, chip control complexity or associated software support, cost, dynamic range/precision requirements, available technology, and generality of usage.

Part IV: Application Requirements

The chapters of Part IV are based on presentations at two workshop sessions on application requirements. These sessions were organized by C. F. Chan of Intel, P. Toldalagi of Analog Devices, and R. E. Owen of Bipolar Integrated

Technology. Ultimately, the basis for any VLSI approach to digital signal processing must be justified in terms of its utility in solving a specific application problem or class of problems. The chapters describe the problem domains in such areas as image processing, telecommunications, and sound generation. In addition, there are discussions of the various architectural compromises necessitated by different application requirements, with emphasis on custom VLSI approaches to implementing these architectures.

The developments in the above four disciplines are seen as driven by basic application requirements or economic forces. These forces were discussed by a panel of representatives from industry, government, and universities. The panel included people from three segments of industry, each with a different outlook: commercial IC manufacturers, commercial system producers, and military system producers. The workshop also featured a Plenary Opening Address by Professor Jonathan Allen of MIT, as well as a banquet keynote speech entitled "Feedback or Learning? Signal Processing for a Competitive World," by Dr. Peter Cannon of Rockwell International Corporation. (The panel discussion and keynote speech are not included in this book.)

The preparation of the workshop was a rewarding experience. This was partly because "VLSI Signal Processing" is such an inspiring subject to its workers. A more important factor was the great pleasure we found in collaborating with a group of very spirited colleagues. These individuals should be credited in full for the success of the workshop. In particular, we are indebted to the members of the Workshop Program Committee and the VLSI Technical Committee of ASSP for their initial definition of the workshop program. Thanks are also due to the authors who prepared camera-ready manuscripts in a very short time. The production of this book has benefited significantly from the very professional management of Hans Leander of the IEEE PRESS. Finally, we wish to express our gratitude to Barbara Cory of Hughes Research Laboratory and Linda Varilla of USC who took care of every detail regarding the workshop with their professional expertise and natural cheerfulness.

S. Y. Kung, R. E. Owen, J. G. Nash

Part I

INTEGRATED CIRCUIT TECHNOLOGY

CHAPTER 1

SYSTEM COMPILERS

Peter B. Denyer
University of Edinburgh
UK

Abstract

There is a wide gap between our capability to develop and model data processing algorithms and our ability to cast these algorithms as real-time electronic systems, especially in integrated (VLSI) form. Existing CAD technologies ("silicon compilers") have failed to bridge this gap. This paper reviews current capabilities and explores possible approaches to full system compilation for real-time applications.

1. Introduction

The interface to many commercial and military systems is characterised by requirements for real-time acquisition and processing of environmental signals. Primary examples are speech, vision, radar, sonar and touch. A great deal of current research is targetted at the development of algorithms to process and extract information from these signals. Real-time execution of these algorithms demands prodigious signal and symbolic processing, such that real-time hardware solutions are characteristically expensive to develop, and to produce. However, the true commercial potential of today's algorithmic research will come only through solutions that may be developed within reasonable costs and timescales and produced for tens or hundreds of dollars. This joint requirement of exceptional computational throughput and low production cost can be met with integrated VLSI solutions.

Unfortunately, no methodology yet exists by which we may rapidly and reliably map these systems from algorithmic to integrated form. Existing silicon compilers certainly do not offer a complete answer. These tools are more accurately silicon assemblers, offering a set of MSI and LSI functions, but no unifying architectural methodology (or at best an excessively restrictive form). Nor do standard or programmable parts offer the best solution; for any particular application the resultant systems consume many times more volume and power than a custom VLSI solution can offer.

We require a new generation of compiler technology that should incorporate architectural knowledge above the macro-function level, and be capable of mapping high-level specifications to a corpus of architectural forms. The structure and regularity of many DSP algorithms makes them particularly amenable to this approach.

We review below the current status of silicon compiler technology and outline a route to full DSP system compilation.

2. Silicon vs Software Compilation

A transformation took place some time ago in the field of computer programming. As the power and memory size of computers grew, programmers could produce larger and more complex programs, whose design and debug times increased accordingly. The solution to that complexity problem was the high-level compiler, which today is used for practically all programming. The slight reduction in efficiency of resulting code is more than compensated for by the enormous reduction in design effort. Furthermore, there is no need to verify that the machine code correctly implements the source specification - the compiler, being known good, always produces code that is correct by construction. Any malfunction of the object code must result from errors of specification originating from the designer.

2.1. Similarities

Silicon compilers must ultimately offer the same advantages to the digital systems designer as do software compilers to the programmer. The designer must be able to specify his system by intent or behaviour, rather than as a mass of arcane geometrical shapes, and thus should far more easily keep pace with progress in his design. Low-level details such as mask layers used, the physical location of functional elements, and design rules are hidden from him.

2.2. Differences

Significant differences exist between the two compiler types. The output of a software compiler has only one dimension, whilst the silicon compiler must produce a two-dimensional pattern of mask geometries. Resolving the options offered by this extra dimension is not a trivial task, since execution times are generally sensitive to placement.

A further difference lies in the cost of communication. A software jump is easy to implement, and the cost is independent of distance. The cost of communication on an IC rises with the distance, in terms of area, time and power consumption.

A final difference is the issue of concurrency. The majority of software compilers are targetted at single-processor von-Neumann machine architectures. Silicon offers the opportunity of concurrent operation. Indeed, the majority of VLSI applications derive a primary advantage from this factor. Solutions are currently ad-hoc and may be addressed by purely structural tools. True system compilers will have to break new territory in concurrent compilation.

The goal of system compilation may be met by solving these problems through the development of appropriate *architectures*. The sub goals of compilability, dimensionality, communication and concurrency must direct research in this area.

3. VLSI CAD Review

We begin with a review of current VLSI CAD. This is relevant since we may presume that, through market forces, these tools represent a good compromise between user requirements and state-of-the-art CAD technology.

3.1. "Macrocells" and Structured Arrays

A contemporary approach to the development of systems on silicon is the macro- or mega-cell. These are generally micro-architectural functional blocks of LSI complexity. They are intended to support a purely structural approach to VLSI design, mimicing the standard-part design style familiar to system designers. Typical micro-architectural functions are RAMs, ROMs, device controllers, bus interface blocks, etc.

The use of megacells in custom silicon enables a greater degree of flexibility (over the use of standard LSI parts), because they may be *parameterised*. Thus the user may for example specify precisely the width and depth of a RAM block. This leads to crucial gains in efficiency (silicon area, power consumption, cost) over a fixed-block approach.

Frequently, function libraries are tied to proprietory fabrication processes. At least one process-independent library is commercially available, although this generality comes at the expense of some layout inefficiency.

A similar theme can be traced in Gate Array technologies. However, since gate array wafers are prediffused, function blocks must be predefined in convenient combinations on a range of product arrays. These are referred to as 'structured arrays'. Typical structured arrays contain combinations of ROM and RAM blocks and possibly also computational functions, such as multipliers and ALU's. In principle these functions may be committed within a conventional array of gates, but they are normally highly structured and benefit greatly from custom implementation.

The existence of parameterised function library generally prompts vendors to coin the term 'silicon compiler', although such tools are not. Putting semantics aside however, these systems represent a significant advance to the system designer; placing him on familiar architectural ground. The advent of hardware multiply functions in particular is a boost for DSP applications.

The software capability needed to support parameterisable cells often takes the form of an embedded language; a set of extensions within a standard high-level language which support the concepts of geometric placement. Recognising this requirement, some vendors now offer independent language products to support those wishing to develop their own parameterised cell libraries.

3.2. Silicon Assemblers

The next level in our product taxonomy is to support a set of parameterised macro-functions within a consistent *architectural framework*. This allows the development of a formal system design methodology; a significant step towards ideal design automation.

The most prevalent examples of this form address the synthesis of *datapath* architectures. Effectively, the user is invited to develop a custom von-Neumann processor. A choice of common ALU functions, and of parameterised register sets, again offers substantial implementation efficiencies. These architectures already appear to offer satisfactory solutions for low-bandwidth real-time applications in speech, modems and control.

Since the user is specifying a machine structure and microprogramming it, the term "silicon assembler" seems most appropriate. This level of CAD is evidenced by function libraries which have been developed to fit a common architectural methodology, probably incorporating some communication standard through which macro-functions can be linked. Typically this might be a common bus standard, or possibly a serial link protocol.

4. Towards System Compilation

4.1. Architectural Methodologies

It is not surprising that each of the tools identified above is driven by currently perceived market requirements for system design on silicon. Since they all offer a wholly structural approach we may identify a new generation of *behavioural* compilers yet to come. These will support specific architectural methodologies, and it is to some University-based research programmes that we turn for indication of future products in this field.

The author is responsible for the FIRST programme at Edinburgh University that is centred around bit-serial architectures [6]. The target architecture is of a pipeline of arithmetic primitives, dedicated to one algorithm. This leads to a high (close to optimal) computational throughout for fixed algorithms. The user is (currently) offered a single, high-level interface to FIRST in the form of a structural description language.

The structural nature of FIRST networks is extremely simple, and this makes it amenable to behavioural compilation. In essence, FIRST-generated functional pipelines are isomorphic with inner-loop code (for example, the arithmetic evaluation of an FFT butterfly algorithm). The controlled execution of this function is a matter of interpreting the loop control statements and from these generating a suitable state memory and controller. An investigation of such a behavioural input capability is underway.

Whilst FIRST constructs high-throughput, fixed-function processors, a complementary approach is pursued by Brodersen et. al. in the development of "Lager" [16]. Lager constructs systems as a collection of small datapath processors. Each processor implements an autonomous task by executing a dedicated programme held in ROM. The processes are linked (bit-serially), and several of them may be contained on one chip.

The processors contain simple shift/add-based ALU's (tuned to efficiently execute multiplication), with RAM and ROM space tailored to the process to be executed. Performance (in terms of operations per second per unit area) is reduced over the dedicated pipelines of FIRST, but this wins considerable flexibility. Data-dependent conditional execution is possible, as is mixing several differing processes on the same chip. This architecture seems well suited to low bandwidth applications (speech and low-bandwidth modems for example), and several impressive examples have been generated.

Although the Lager work has been predominantly in nMOS technology, workers at Leuven's IMEC are mapping similar architectures into CMOS [8]. Neither the FIRST nor Lager approaches are perfectly general, though each can achieve results which are close to optimal in specific cases. We shall return below to the theme of architectural generality.

4.2. Function Compilers

By concentrating on a single process form it is possible to develop very high level compilers for restricted applications. For example the CATHEDRAL system from Leuven [11] synthesises (bit-serial) digital filters from functional specifications.

Such a system is like a slice taken from a full compiler. The input may be purely functional; in this sense meeting our goal of true compilation. However, the compiler understands only one type of function.

4.3. Behavioural Compilers

All of the approaches discussed above may be classed as structural-input (in the last case functional). With a structural compiler the user directs the formation and connection of tangible hardware blocks. Our ideal compiler must transcend this level and interpret hardware structure from user-specified behaviour. This dissociation of behaviour from structure is the distinction of true compilation.

This is a hard step to take. Experienced users are sceptical of releasing structural control, and with good reason; there is scope in this translation for serious structural inefficiency. An extended runtime is the only penalty for this inefficiency in a software compiler. In silicon however, structural inefficiency can waste the primary (and exponentially expensive) resource of circuit area.

Apart from this key issue of efficiency, we must also await developments in parallel compiler technology, for silicon solutions are concurrent processors in the general form. This is likely to block the immediate development of true general purpose system compilers.

Meanwhile, limited developments will be possible in areas where parallelism is not hard to reveal. One such area with significant market pull is that of digital signal processing. Potential parallelism is relatively easily exposed in many cases because DSP algorithms tend to contain repetitive, independent structures.

Independence and repetition are easy to exploit architecturally; independent processes may be executed concurrently on independent processors, whilst repetition encourages the use of dedicated hardware forms over less efficient programmable structures. These features are attractive because efficiency and concurrency are prerequisites of competitive silicon solutions.

Real-time silicon architectures comprising linked independent processors already form the basis of Lager [10]. Although the processors are limited to the datapath form, an impressive range of DSP applications has already been demonstrated at speech and music bandwidths.

5. Road to the Holy Grail

Although we have now identified early forms of behavioural compilation, the ideal system compiler has yet to be unveiled. However, we are in a position to describe its general form, and offer a template for its future development. In this template, outlined in Figure 1, architecture will play the central role, to be supported above by language compilation, and below by silicon assembly.

5.1. Architecture

Compilers based on a single architectural form, such as Lager, CATHEDRAL and FIRST, can never offer a complete solution. We must recognise that architecture is the key factor governing the silicon area required to execute a given task (in real-time), and that area is an exponentially expensive resource. The product cost can be optimised only through architectural freedom. For example, a high-bandwidth FFT may be optimally executed around a dedicated hardware butterfly and state memory controller, whilst a low-bandwidth requirement may be better accommodated on a general-purpose datapath.

Future system compilers should therefore support a corpus of processor architectures. From the literature and experiences described above, we may identify at least the following candidate forms:

datapath (e.g. 2901)
accelerated datapath (e.g. plus multiplier)
algorithmic processors (e.g. serial FFT butterfly)
random logic / FSM processors (e.g. EXOR and count)

Perhaps the most fruitful research of this period will come from the development of a canonic architectural set.

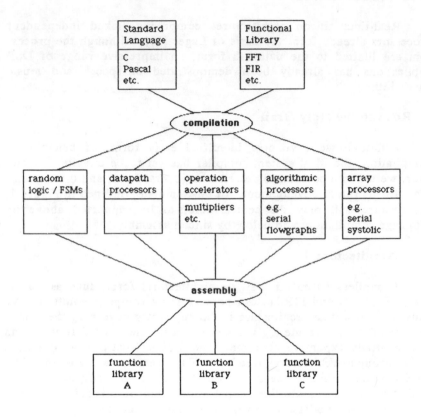

Figure 1

A Template for System Compilation:
High Level Specification -> Architecture -> Silicon

Each form must be optimally compiled for the execution of a given process; for example by varying the register and bus structures to suit a given datapath realisation. Excellent work on datapath generation has already been reported [13], [14].

5.2. Specification

There are two plausible alternatives to system specification; language (or behaviour) and function. Where language is used it will become increasingly attractive to adopt a standard form (C, Pascal, etc.). This is important for three reasons. The first is acceptability; the requirement to learn a new language is a significant deterrent to the user community. The second is transportability; or "compiler independence". Ideally a behaviour specified in a standard language should be capable of implementation through a variety of different system compilers. The third and strongest reason for this migration to a standard form, is the capability of algorithm development through standard software compilation on existing computer hardware. Of course the standard languages are not completely suited to the task of hardware compilation; they need to be embellished to express desired execution times (achieved by varying the degree of hardware concurrency) and fixed wordlength effects. Nevertheless the attractions outlined above are very strong.

An alternative approach to specification is to avoid programming *per se*, and allow the user instead to compose systems from a library of functions (FFT's, filters, etc.). Functional interfaces can be very attractive to the user, and may be linked to architectures by rule rather than by formal compilation.

The use of such techniques is exemplified by Silage [15], an applicative language for DSP, and Archer [16], a companion ARCHitecture interpretER, both of which have been used in the Lager system. In CATHEDRAL [11], filter specification is supported by FALCON, whilst knowledge-based architectural synthesis is automated by AMAI.

5.3. Assembly

As earlier parts of this paper have shown, the art of silicon assembly is already well advanced. It is practical to implement a custom datapath (for example) from standard microarchitectural functions (ALU, shifters, registers).

It is further desirable that the architectural corpus should be capable of realisation through a range of silicon assemblers and function libraries. To this end, the use of standard forms becomes attractive. However, we should be careful to preserve the flexibility of the silicon medium, for example to be able to add new functions (e.g. log approximation hardware) on demand. The automatic synthesis of these functions from boolean or other specifications is a further topic for current research. Again, some early results are encouraging [17], [18].

5.4. Conclusion

The principle of full system compilation from high-level specification appears to be feasible, especially in the area of DSP and real-time applications. Existing tools address the issue of silicon assembly, but this fulfills only the lower third of the full requirement for system compilation. future developments are required in silicon architectures and language(or function)-to-architecture compilation.

The proper solution of these problems will make possible a direct link from algorithm modeling and development, through to deliverable silicon solutions.

References

1. Southard, J.R., "MacPitts: An Approach to Silicon Compilation", Computer, Vol. 16, pp. 74-82 (December 1983).
2. Buric, M.R., Christensen, C., and Mathson, T.G., "The PLEX Project: VLSI Layouts of Microcomputers Generated by a Computer Program", IEEE Intl. Conf. on Computer-Aided Design, (Santa Clara, CA, September 1983).
3. Buric, M.R. and Matheson, T.G., "Silicon Compilation Environments", Proc. IEEE CICC, pp. 208-212 (Portland, OR, May 1985).
4. Johannsen, D.L., "Bristle Blocks: A Silicon Compiler", Proc. 16th DA Conf., (San Deigo, 1979).
5. Johnson, S.C., "VLSI Circuit Design Reaches the Level of Architectural Description", Electronics, pp. 121-128 (3rd May 1984).
6. Denyer, P.B. and Renshaw, D., "VLSI Signal Processing - A Bit-Serial Approach", Addison-Wesley (1985).
7. Ruetz, P.A., Pope, S.P. and Brodersen, R.W., "Computer Generation of Digital Filter Banks", IEEE Trans. on CAD, Vol. 5, No. 2, pp. 256-265, (1986).
8. De Vos, L., Jain, R., De Man, H., Ulbrich, W., "A Fast Adder-Based Multiplication Unit for Customised Digital Signal Processors, Proc. ICASSP 86, pp. 2163-2166, (Tokyo 1986).
9. Van Ginderdeuren, J., De Man, H., De Loore, B., Van Den Andenaerde, G., "Application Specific Integrated Filters for HIFI Digital", Proc. ICASSP 86, pp. 1537-1540.
10. Rabaey, J.M., Brodersen, R.W., "Experiences with Automatic Generation of Audio Band Digital Signal Processing Circuits", Proc. ICASSP 86, pp. 1541-1544, (Tokyo 1986).

11. Jain, R. et. al. "Custom design of a VLSI PCM-FDM transmultiplexer from system specifications to circuit layout using a computer-aided design system", IEEE JSSC, Vol. SC-21, (Feb., 1986).

12. Kahrs, M., "Silicon Compilation of a Very High Level Signal Processing Specification Language", pp. 228-238 in VLSI Signal Processing, ed. P.R. Cappello et. al., IEEE Press (1984).

13. Kowalski, T.J., "The VLSI Design Automation Assistant: Prototype System", 20th Design Automation Conference, paper 31.2, IEEE, (1983).

14. Jamier, R. and Jerraya, A.A., "APPOLON, A Data-Path Silicon Compiler", IEEE Circuits and Devices magazine, Vol 1 No. 13, pp 6-14, (May, 1985).

15. Hilfinger, P., "Silage: A high-level language and silicon compiler for digital signal processing", in Proc. CICC, Portland, OR, (May, 1985).

16. Rabaey, J.M., Pope, S.P. and Brodersen, R.W., "An Integrated Automated Layout Generation System for DSP Circuits", IEEE Trans. CAD, Vol CAD-4, No. 3, pp 285-296, (1985).

17. Kollaritsch, P.W. and Weste, N.H.E., "TOPOLOGIZER: An Expert System Translator of Transistor Connectivity to Symbolic Cell Layout", IEEE J. Solid State Circuits, Vol SC-20, No. 3, pp 799-804, (1985).

18. Asada, K. and Mavor, J. "MOSSYN: An MOS Circuit Synthesis Program Employing 3-Way Decomposition and Reduction Based on 7-Valued Logic", submitted to IEEE trans. on CAD of Integrated Circuits & Systems.

CHAPTER 2

A TOP DOWN APPROACH TO DSP LSI DESIGN USING DYNAMIC ALGORITHM TRACING AND FINITE WORD LENGTH SIMULATION.

Tsutomu KOBAYASHI, Toshiaki TANAKA and Osamu KARATSU
NTT Electrical Communications Laboratories
3-1, Morinosato Wakamiya, Atsugi-shi, Kanagawa 243-01, Japan

Abstract: A new top down approach to designing a digital signal processor LSI is proposed, based on newly developed algorithm evaluation tools: a finite word length simulator (FWLS) and a parallelism evaluator (PE). The FWLS evaluates the required accuracy and counts the total number of operations used in signal processing algorithms using real data such as speech signals or image data. The PE analyzes the parallel computability of the input programs which specify the target algorithm in high level languages, and estimates the total elapsed time of the specified algorithms in a given number of arithmetic facilities. A digital signal processor design method is proposed utilizing these tools and the automatic VLSI synthesizer [1] [2].

1. Introduction

Recent progress in semiconductor technology has brought an increase in real time applications of digital signal processing. A number of CAD tools have been developed and widely used in designing special-purpose and general-purpose DSP chips. However, several large gaps still exist between formulation of algorithms and implementation of VLSI logic design [3]. The purpose of this study is to bridge the gap between the algorithm level model and the RTL level model, and to realize a higher level design environment.

Starting with the algorithm level description, the conventional algorithmic program written in FORTRAN, a Finite Word Length Simulator (FWLS) is used to evaluate the speed and word length of the arithmetic operations. Next, a Parallelism Evaluator (PE) is used to analyze the parallelism and determine the minimum number of arithmetic units that satisfy the real time constraints. Rough images of the specific chip design can be obtained using these results. Here, RTL level specifications are described through a unified language HSL-FX [4].

Using design tools (e.g., logic synthesizer, logic simulator, circuits simulator, gate expander, chip floor planner, and layouter, [4]-[6]), an error free LSI can be designed. This paper focuses on early stage design methodology of the specific algorithm based DSP design.

2. Finite Word Length Simulator

In DSP-LSI design, the word length of facilities such as, execution units and registers, and the number of operations used in target algorithms, must be evaluated or estimated. Presently, these evaluations or estimations have been carried out manually, or by using specifically written individual simulation programs. The FWLS has been developed to eliminate such tedious and time-consuming work. The basic structure of the FWLS is shown in Fig. 1.

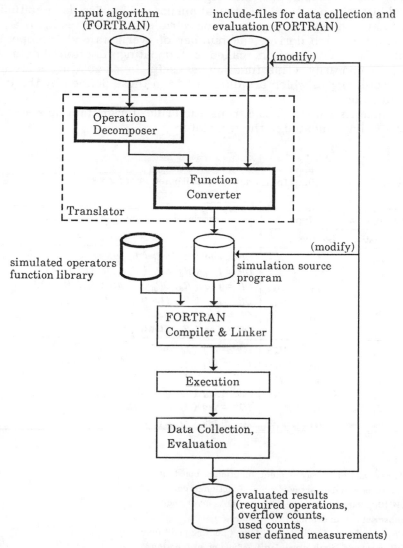

Fig. 1 Structure of finite word length simulator.
(Bold frames : original software-tools)

The arithmetic assignment statements including fundamental operations, such as **add, subtract, multiply,** and **divide (+,-,*,/)** in the input algorithm, are automatically compiled into one or two-term statements with a single operator; these operators are then converted to simulated finite word length operators, which are expressed as built-in subroutine packages. Simulated operators are FORTRAN callable functions specifically designed to simulate corresponding operations at specified word lengths.

The package has 36 simulated operators, including fixed point and floating point calculations, and logical operations. Three subroutines are additionally prepared for special purposes. The first is an initializing function which defines the simulation word length. The second is a probing function which derives the number of overflows and the operation counts. The last function, called a truncating function, limits word length temporarily. This function is useful in calculating statements using word lengths which are different from those defined by the initializing function.

Table 1 shows the basic translation rules for arithmetic statements which are implemented in the translator.

Table 1. Translation rules for fundamental arithmetic operators.

Input variable type			Translator produces the following output
Destination	Source		
R_d	R_{s1}	R_{s2}	$OP = feq\ (\ R_d,\ OP_f\ (\ R_{s1},\ R_{s2}\ )\ )$
R_d	R_{s1}	I_{s2}	$OP = feq\ (\ R_d,\ OP_f\ (\ R_{s1},\ floati\ (\ I_{s2}\ )\ )\ )$
R_d	I_{s1}	R_{s2}	$OP = feq\ (\ R_d,\ OP_f\ (\ floati\ (\ I_{s1}\ ),\ R_{s2}\ )\ )$
R_d	I_{s1}	I_{s2}	$OP = feq\ (\ R_d,\ floati\ (\ OP_f(\ I_{s1},\ I_{s2}\ )\ )\ )$
R_d	R_{s1}	-	$OP = feq\ (\ R_d,\ R_{s1}\ )$
R_d	I_{s1}	-	$OP = feq\ (\ R_d,\ floati\ (\ I_{s1}\ )\ )$
I_d	R_{s1}	R_{s2}	$IOP = ieq\ (\ I_d,\ intf\ (\ OP_f\ (\ R_{s1},\ R_{s2}\ )\ )\ )$
I_d	R_{s1}	I_{s2}	$IOP = ieq\ (\ I_d,\ intf\ (\ OP_f\ (\ R_{s1},\ floati\ (\ I_{s2}\ )\ )\ )\ )$
I_d	I_{s1}	R_{s2}	$IOP = ieq\ (\ I_d,\ intf\ (\ OP_f\ (\ floati\ (\ (\ I_{s1}\ ),\ R_{s2}\ )\ )\ )\ )$
I_d	I_{s1}	I_{s2}	$IOP = ieq\ (\ I_d,\ OP_i\ (\ I_{s1},\ I_{s2}\ )\ )$
I_d	R_{s1}	-	$IOP = ieq\ (\ I_d,\ intf\ (\ R_{s1}\ )\ )$
I_d	I_{s1}	-	$IOP = ieq\ (\ I_d,\ I_{s1}\ )$

notes

R ; Real variable, array element, function or constant

I ; Integer variable, array element, function or constant

OP, IOP ; Dummy variable, return value is error code

Subscripts (d: destination, s: source)

$OP_f :: = fadd\ |\ fsub\ |\ fmul\ |\ fdiv$ (floating point function)

$OP_i :: = iadd\ |\ isub\ |\ imul\ |\ idiv$ (fixed point function)

The translator produces output source statements according to these rules and the types of input variables. In addition, initializing and probing functions as well as user defined criteria measuring functions are inserted into the points designated by the user.

An example of the input algorithmic program and translated output called a Simulation Source Program is shown in Figs. 2(a), and 2(b). This example shows a modified lattice algorithm for PARCOR speech analysis [7].

```
 1   C.*********************************************************
 2   C.        subroutine mlatt(x,ip,ld,ref,alf,aaa)          *
 3   C.        description                                    *
 4   C.        Modified Lattice Method PARCOR analysis        *
 5   C.        args:                                          *
 6   C.        input   x: speech data                         *
 7   C.                ip: number of pole                      *
 8   C.                ld: data length of one frame            *
 9   C.        output  x: residual wave x(1),x(2),..x(ld-ip).  *
10   C.                ref: PARCOR (reflection) coefficients   *
11   C.                alf: Linear pridictive coefficients     *
12   C.                aaa: final residual power(sample-average) *
13   C.        note.... array x(*) is overwritten with residual*
14   C.                if you need not alf-- kill stepup routine*
15   C.--------------------------------------------------------*
16   C.*********************************************************
17   C.
18           subroutine mlatt(x,ip,ld,ref,alf,aaa)
19           dimension x(*),ref(*),alf(*),y(256)
20           call movrr(x,y,ld)
21           do 20 i=1,ip
22             w=dot(x,y(i+1),ld-i)
23             u=(dot(x,x,ld-i)+dot(y(i+1),y(i+1),ld-i))*0.5
24             ref(i)=w/u
25   C.        call stepup(ref(i),i,alf)
26             do 10 j=1,ld-i
27               xx=x(j)
28               x(j)=x(j)-ref(i)*y(i+j)
29               y(i+j)=y(i+j)-ref(i)*xx
30   10        continue
31   20      continue
32           call movrr(y(ip+1),x,ld-ip)
33           aaa=dot(x,x,ld-ip)/float(ld-ip)
34           return
35           end
```

Fig. 2 (a) An example of input algorithmic program.

```
 4   C.                              ..il residual*
15   C.-------------              --il stepup routine*
16   C.***********************.........***************************
17   C.
18           subroutine mlatt(x,ip,ld,ref,alf,aaa)
19           real*4 r4tmp0,r4tmp1,r4tmp2
20           dimension x(*),ref(*),alf(*),y(256)
21           call movrr(x,y,ld)
22           do 20 i=1,ip
23             OP=feq(w,dot(x,y(i+1),ld-i))
24             OP=feq(r4tmp0,fadd(dot(x,x,ld-i),dot(y(i+1),y(i+1),ld-i)))
25             OP=feq(u,fmul(r4tmp0,0.5))
26             OP=feq(ref(i),fdiv(w,u))
27   C.        call stepup(ref(i),i,alf)
28             do 10 j=1,ld-i
29               OP=feq(xx,x(j))
30               OP=feq(r4tmp1,fmul(ref(i),y(i+j)))
31               OP=feq(x(j),fsub(x(j),r4tmp1))
32               OP=feq(r4tmp2,fmul(ref(i),xx))
33               OP=feq(y(i+j),fsub(y(i+j),r4tmp2))
34   10        continue
35   20      continue
36           call movrr(y(ip+1),x,ld-ip)
37           OP=feq(aaa,fdiv(dot(x,x,ld-ip),float(ld-ip)))
38           return
39           end
```

Fig. 2 (b) Translator produced output source program.

Finite word length simulation is performed using the conventional 'COMPILE', 'LINK' and 'GO' with the simulation function library.

A frequency histogram at 18 bit floating point operations produced by FWLS for this example is shown in Fig. 3. The spectral distortion measurement using FWLS by fixed point operations is shown in Fig. 4.

It can be seen from the figure that 12 bits operations are sufficient to realize stable and low distortion hardware. Under these conditions, the distortion is less than 0.01 [dB^2] which is perceptually negligible.

As shown in the above example, a user can determine the optimum word length by evaluating the criteria, such as SNR or waveform distortion according to his specific needs.

Using this software tool, FWLS, finite word length simulation and estimation of the total amount of calculations can be performed without manually modifying the input program. The simulation speed of FWLS is almost 200 times faster than a conventional RTL simulator. The function translator used here has been generated by using 'lex' and 'yacc' utilities in the UNIX operating system.

```
<<< FLOATING POINT FORMAT SIMULATION RESULTS >>>
--- Conditions ---

NEXP=  6 [bit]   ; Exponent bit
MANT= 12 [bit]   ; Mantissa bit
IOVF= 1          ; Overflow  -> Max value
IUNF= 1          ; Underflow -> 0 Reset
IR  = 0          ; Truncate
```

FUNCTION	OVERFLOW	UNDERFLOW	USE
FEQ	0	0	31869
FNEG	0	0	0
FADD	0	0	8646
FSUB	0	0	5604
FMUL	0	0	14250
FDIV	0	0	13
FLOATI	0	0	0

```
TRUNCATION-FUNCTION( FTRUN ) USE COUNTS:

    ... FTRUN NOT USED
```

Fig.3 Frequency histogram obtained from finite word length floating point simulation.

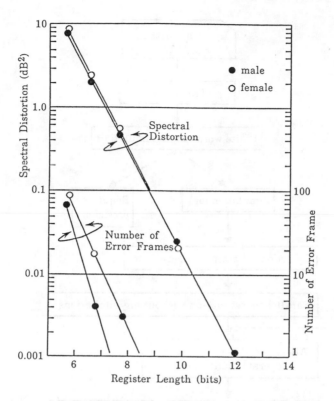

Fig. 4 Spectral distortion by fixcd point simulation.
The evaluated algorithm is a Modified Lattice PARCOR
Analysis. Input data is 8 [KHz], 14 [bits] sampled male and
female speech data, totally 7200 [frames]. Error frames
are defined by unstable reflection coefficients.

3. Parallelism Evaluator

The major purpose of the parallelism evaluator is to estimate the
required arithmetic capability of the target chip. The PE utilizes
address tracing and causality checking, which are simple but effective
techniques. The general flow chart of the PE is shown in Fig. 5.

(1) At the beginning, a dynamic trace file is created and logging
of all operations commences. Each operation is logged as a vector with
four elements: **operator,** two **operand source addresses,** and **operand
destination address.**

(2) After completing the dynamic trace for the specified part of
the source program, which is usually the largest loop consisting of all
concurrent processes in real time application, the relationships between
inputs and outputs are examined with respect to the following rules:

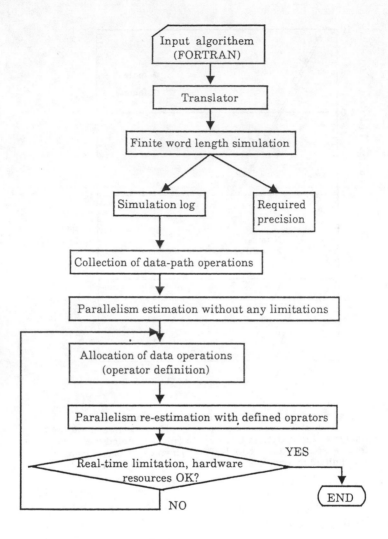

Fig. 5 General flow chart of parallelism evaluator.

(a) If step $n+1$'s inputs are independent of step n's outputs, step n and step $n+1$ can be executed in parallel; then vectors n and $n+1$ are entered into the parallel table as an Mth entry and vector $n+1$ is deleted as a logging entry.

(b) If step $n+1$'s input are one of step n's outputs, the step must be carried out sequentially, (this is the most likely case for a pipeline process); then $n+1$'s vector is entered as the $M+1$th entry of the table and the subsequent evaluation is continued.

(c) The above process (b) is continued until the end of the logging file.

After completion of tracing, a full parallel execution table and corresponding dynamic steps are derived.

*(d) Limitations to the number of arithmetic units and their functions are made, e.g., ALU1 [+ -], ALU2 [* +]; then the parallel table is retraced and the table is rewritten using the constraints.*

(e) The total number of dynamic steps is examined as to whether or not it meets real time constraints. If the results are acceptable, this process is terminated. Otherwise, Steps (d) to (e) are repeated.

(3) Accordingly, the number of arithmetic units and their functions can be determined. The block diagram design and corresponding control logics design are then carried out.

There are two approaches for the application of PE. A straightforward approach is to apply the PE to the whole concurrent program and determine the required number of arithmetic units and their functionalities. This approach is suitable for the initial design of a program controlled DSP or multi processor system. Another approach is to apply the PE to a specific part of the target program, this is quite suitable for interactive work in designing wired logics.

As an example of the latter case, the algorithm depicted in Fig. 2(a) is evaluated.

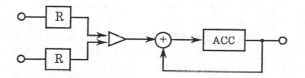

Fig. 6 (a)   Block diagram of macro-function **dot**

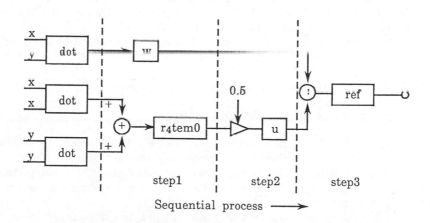

Fig. 6 (b)   Corresponding sketch of correlator block.

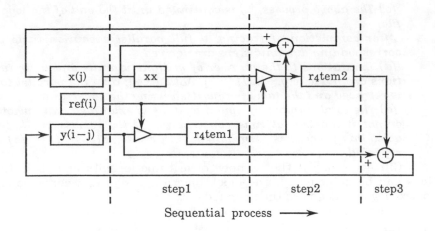

Sequential process ⟶

Fig.6(c) Corresponding sketch of lattice filter block.

The PE results suggest that all of the **dots** in Fig. 2(b) can be executed in parallel, and the rest of **fadd, fmul** and **fdiv** should be executed
sequentially as step 1 to step 3. Here, the function **dot** is a user-defined
macro-function which is evaluated independently. A block diagram of
the macro-function **dot**, is shown in Fig. 6(a). A diagram of this block,
referred to as the correlator block in the original algorithm is shown in
Fig. 6(b). A corresponding sketch of the lattice filter block, related to
the statements No. 28 to 34 in Fig. 2(b) is shown in Fig. 6(c). Note that
temporary latch variables **w, r4tmp0,** and **u** in Fig. 6(b) and **xx, r4tmp1,**
and **r4tmp2** in Fig. 6(c) can be included in the multiplier and the divider
respectively. The final block diagram is obtained by merging the duplicated variables and adding the required I/O interface.

A final block diagram for special purpose hardware of the modified
lattice PARCOR speech analyzer, (algorithm depicted in Fig. 4) is shown
in Fig. 7. Few limitations have been introduced in the number of arithmetic units in order to realize maximum parallelism.

4. Register Transfer Level Design Capture

Based on the above-mentioned initial design, register transfer
level blocks are described more precisely using HSL-FX. Then, more
detailed simulation is continued. An example of the RTL description list
corresponding to Fig. 7 is shown in Fig. 8. Here, the behavioral description for the function blocks is used.

It is not the purpose of this paper to provide details for lower level
design tools. References [4]-[6] will be helpful for understanding the
total DA system from RTL to layout.

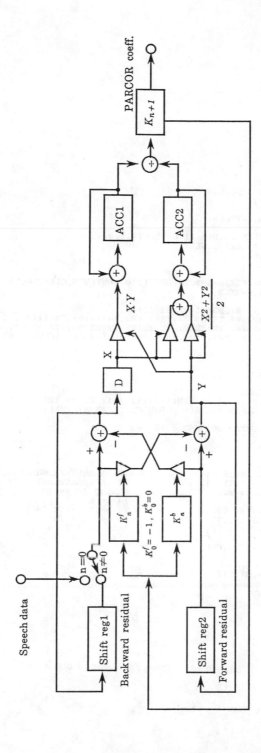

Fig. 7 Final sketch of block diagram. (Special purpose hardware for Modified Lattice PARCOR Analysis)

```
 1   IDENT: PARCOR;
 2   VERSION: NO2.00;
 3   DATE: 86/03/10;
 4   AUTHOR: TANAKA ;
 5   NAME: PARCOR;
 6   PURPOSE     :FSIM,GATEGEN;
 7   LEVEL       :TOP ;
 8      EXT:   CLK,SEL,READ,DIN<11:0>,DOUT<11:0>;
 9      INPUTS:    .DIN,.READ,.SEL;
10      OUTPUTS:   .DOUT;
11      CLOCK:     .CLK ;
12   TYPES:
13      SHREG(IN<11:0>,CLEAR#INPUTS,OUT<11:0>#OUTPUTS,CLK#CLOCK);
14      SHREG: SHREG1,SHREG2;
15   END-TYPES;
16      TERMINAL: S1<11:0>,S2<11:0>,S3<11:0>,
17               T1<23:0>,T2<23:0>,T3<11:0>,T4<11:0>,T5<11:0>,
18               T6<23:0>,T7<23:0>,T8<23:0>,T9<23:0>,T10<23:0>;
19      REGISTER: K<11:0>,KF<11:0>,KB<11:0>,ACC1<23:0>,ACC2<23:0>,D<11:0>;
20   BEHAVIOR-SECTION;
21      BEGIN
22        IF .READ THEN
23          RESET(D,ACC1,ACC2);
24        ELSE
25          IF .SEL THEN
26            S1 := .DIN; S2 := 12B111111111111; S3 := 12B0;
27            SHREG1.CLEAR := .SEL; SHREG2.CLEAR := .SEL;
28          ELSE
29            S1 := SHREG1.OUT; S2 := K; S3 := K;
30          END-IF;
31        END-IF;
32        AT .CLK DO
33          KF := S2;                     KB := S3;
34          T1 := MULT(KF,S1);            T2 := MULT(KB,SHREG2.OUT);
35          T3 := S1 - T2<23:12>;         T5 := SHREG2.OUT - T1<23:12>;
36          SHREG1.IN := T3;              SHREG2.IN := T5;
37          D := T3;                      T4 := D;
38          T6 := MULT(T4,T5);            T7 := MULT(T4,T4);
39          T8 := MULT(T5,T5);            T9 := T7 + T8;
40          ACC1 := ACC1 + T6;            ACC2 := ACC2 + T9;
41          T10 := DIV(ACC1,ACC2);        SHREG1.CLK := .CLK;
42          SHREG2.CLK := .CLK;
43        END-DO;
44        AT .READ DO
45          K := T10<23:12>;
46        END-DO;
47        .DOUT := K;
48      END;
49   END-SECTION;
50   END;
```

Fig. 8   RTL description from Fig.7 using HSL-FX.

24

5. Conclusion.

Newly developed DSP design tools, the Finite Word Length Simulator and the Parallelism Evaluator, have been described.

The FWLS makes it possible to determine the required precision and speed using the target algorithmic program without manual modification. The speed of FWLS is sufficiently fast, therefore real data base simulation is possible.

The PE enables estimating a minimum number of arithmetic facilities with respect to user defined constraints. As an extension of the PE, a pipeline evaluator is currently under investigation.

Using these tools, initial steps in DSP design can be accomplished without time-consuming manual work.

Further discussion is needed to develop automatic tools used in architecture level design. A combination of the tools discussed in this paper and knowledge-based synthesis tools such as those discussed in Ref. [8], provide promising approaches for further work.

References

[1] O.Karatsu et al. "An Automatic VLSI Synthesizer", Proc. of 1985 ISCAS, June 1985 Kyoto, pp. 403-406.

[2] O.Karatsu et al. "An Integrated Design Automation System for VLSI Circuits", IEEE Design & Test of Computers, vol. 1.2, No.5 Oct 1985, pp. 17-26.

[3] D.D.Gajski and R.H.Kuhn, "New VLSI Tools", Computer Dec 1983, pp. 11-14.

[4] T.Hoshino, O.Karatsu and T.Nakashima, "HSL-FX: A Unified Language for VLSI Design," Proc. 7th Symp. CHDL, Tokyo, Aug. 1985, pp. 321-406.

[5] K.Ueda, H.Kitazawa and I.Harada, "CHAMP: Chip Floor Plan for Hierarchical VLSI Layout Design," IEEE Trans. Computer Aided Design CAD-4, Jan. 1985, pp. 12-22.

[6] T.Adachi et al.,"Hierarchical Top-Down Layout Design Method for VLSI Chip" Proc. 19th DAC, Las Vegas, Nev. June 1982, pp.785-791.

[7] T.Kobayashi and H.Yamamoto, "Modified Lattice Type PARCOR Analysis Synthesis System", Acous. Soc. Japan vol. S77-06, May 1977. (in Japanese)

[8] K.Shirai,Y.Nagai and T.Takezawa., "Functional Logic Design System for Digital Signal Processors", Proc. VLSI'85, Kobe Aug. 1985, pp. 203-212.

CIRCUIT SYNTHESIS FOR SIGNAL PROCESSING
INCORPORATING ARCHITECTURAL CHOICES

Paul J. Ainslie, Thomas E. Fuhrman, Mira A. Majewski, and Fred N. Krull
General Motors Research Laboratories, Warren, Michigan 48090

Abstract

The AutoCircuit system for integrated circuit design allows the designer to capture and simulate a circuit design at a functional block diagram level. This paper describes functional blocks, called module generators, used in the AutoCircuit system. Each module generator is a software program which provides a graphical icon for design, simulation models, and a gate level netlist. AutoCircuit also incorporates some new approaches in processing architectures not previously available in general purpose circuit synthesis systems. Digital filter module generators in AutoCircuit synthesize signal processing architectures by automatically interconnecting lower-level modules using knowledge of filter architectures in their module generators. One architecture maximizes throughput at the expense of chip area, and another minimizes area at the possible expense of throughput. Examples of a digital filter generator and a filter circuit are presented. The designer manipulates the functional blocks in a graphical design environment which enables an evaluation of design alternatives while the designer is in the early stages of design tradeoffs.

The AutoCircuit system for integrated circuit design, developed at General Motors Research Laboratories (GMR), incorporates a methodology where the designer captures and simulates a circuit design at a functional block diagram level. The designer manipulates the blocks in a graphical design environment which enables an evaluation of design alternatives while the designer is in the early stages of design. This paper describes functional blocks, called module generators, used in the AutoCircuit system. Each module generator is a software program which provides a graphical icon for design, simulation models, and a gate level netlist. These modules can be used hierarchically to build more complex designs such as digital filters which incorporate an architecture synthesized by software.

Other Circuit Synthesis Systems

AT&T Bell Labs has developed a synthesis tool called the Functional Design System which is based on standard cells and generates macrocells such as counters and registers [1]. Reutz *et al.* at UC Berkeley have developed a digital filter layout generator which builds a filter from user inputs and a library of filter sections [2]. The ADAS system developed at Research Triangle Institute by Frank *et al.* supports the design and evaluation of architectures for signal processing and the supporting software using a directed graph and languaged-based approach [3]. Jasica *et al.* at the GE Corporate Research and Development Center have also developed a bit-serial module compiler for digital filtering [4].

A number of commercial firms have developed circuit synthesis systems, and most of these synthesize layout using low-level primitives such as transistors and interconnections. No commercially available systems include signal processing architectures specifically. VLSI Technology Inc. [5], Seattle Silicon Inc. [6] and Silicon Design Labs Inc. [7] all use a language-based description of low-level primitives, whereas Silicon Compilers Inc. [8] uses a set of geometric leaf cell primitives for synthesis. All of these approaches synthesize correct circuits, and some benchmark evaluations have been conducted by GMR [9]. Only the Genesil system from Silicon Compilers Inc. is targetted towards architectural development by providing rapid performance estimators [10].

Overview of the AutoCircuit System

AutoCircuit provides tools for design at a higher level than those of typical standard cell design systems such as GMR's AUTOCHIP system [11] or the Mentor Idea engineering workstation [12]. Figure 1 shows the elements of the AutoCircuit system. A new concept in AutoCircuit is the integration of digital signal processing design aids along with a general circuit synthesis system. Digital filter design aids distributed by the IEEE [13] have been front-ended with a menu system and are available in the AutoCircuit environment. A digital signal processing tool set called SIG from Lawrence Livermore National Laboratory has been installed and customized graphics drivers have been developed [14]. SIG provides analysis of signals using digital signal processing techniques such as FFT and convolution, and graphical display of both time series and frequency domain data. The digital filter design programs interface directly to the generators for the digital filter and ROM modules. A graphical editor and menu system provide the user interface to the database and the module generators. A commercial simulator CADAT provides the mixed level simulation in AutoCircuit [15]. Another tool, called AutoDraft, automatically synthesizes the gate level schematic for the circuit, which can be input to a standard cell design system [16]. A system for Automatic Test Generation, under development in another GMR project, works with AutoCircuit to provide full test coverage of possible fabrication defects in the circuit.

Module Generators in AutoCircuit

The module generators are computer programs which capture circuit design expertise into a series of program steps which synthesize the actual circuit. The designer must specify certain design parameters such as the number of bits in a datapath or the polarity of external signals. With this user input and the synthesis rules embedded in programs and tables, the module generators prepare five outputs:

1. a graphical icon for design
2. a functional simulation model
3. a netlist of the circuit using standard cell elements
4. a gate level model for simulation of the circuit
5. test vectors for testing the function of the circuit.

Currently 20 generators (listed in Table I) have been developed to synthesize different blocks. Many of these blocks use lower level generators to provide a hierarchical design. For example, the multiplexer generator uses a decoder and one or more selectors. The lowest level primitives used in the module generators are logic gates (AND, NAND, OR, NOR, XOR, XNOR, and INV) and flip-flops. In addition, layout generators for PLAs, RAMs and ROMs are assumed to be available.

Module	Description
adder	n-bit adder
addsub	n bit add/subtract
buffer	variable size buffer
compare	n-bit comparator
counter	n-bit counter
decoder	decoder for n outputs
digfilt	$2*n$ order digital filter
filtcore	2nd order filter section
gate	logic gate with n inputs
integrate	n-bit sign integrator
mux	multiplexer for n data paths
pla	PLA state machine
pssr	n-bit parallel-to-serial shifter
regfile	bank of n-bit registers
register	n-bit register
rom	n-bit by m-word ROM
selector	one-of-n data selector
ssr	n-bit serial shift register
switch	1-to-n switch
twosgen	n-bit twos-complement generator

Table I. Module generators in AutoCircuit.

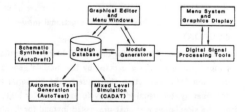

Figure 1. Block diagram of the AutoCircuit system.

AutoCircuit has two module generators which incorporate signal processing architectures in their synthesis routines. These module generators are for digital filtering and both implement general infinite impulse response (IIR) digital filters. The filter architectures were adapted from the distributed arithmetic technique of Peled and Liu [17] and both were implemented using lower level module generators from Table I. Other architectures which use distributed arithmetic for digital filtering have been reported by Chen for FIR filters [18] and by Sicuranza for nonlinear filters [19]. Figure 2a shows the architecture for a digital filter which maximizes throughput (at the expense of chip area) by duplicating the add/subtract and multiplexer circuitry for each section. This architecture is synthesized by the module generator FILTCORE. Another architecture, shown in Figure 2b, minimizes area by reusing the add/subtract circuitry for each second order section, at the possible expense of throughput. This architecture is synthesized by the module generator DIGFILT. The designer has the option to choose which architecture is used for the digital filter, simulate the function, and change the architecture completely by simply selecting the other module and supplying the same parameters. This concept gives the designer a great amount of flexibility, allowing the evaluation of alternative architectures without extensive and time-consuming redesign.

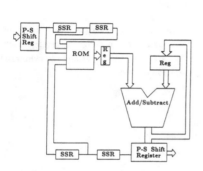

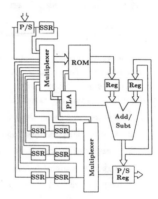

Figure 2a. Digital filter architecture used in the FILT-CORE module, adapted from [17]. A second order section is shown.

Figure 2b. Digital filter architecture used in DIGFILT module, adapted from [17]. A sixth order filter is shown.

A Filter Example

The filter design parameters shown in Table II were used for the example filter. The filter design and theoretical analysis is done by a program called EQIIR distributed by IEEE [13]. The filter coefficients generated by EQIIR for two second order sections of an elliptic IIR filter are listed in Table III. Figure 3 shows the symbol generated for graphical design in the AutoCircuit block editor. The DIGFILT module generator reads the filter coefficients from a data file prepared by EQIIR. In addition, the module generator needs to know the datapath width, the overall gain of the filter and the number of second order sections, information which is also available from EQIIR. Given these inputs, a functional model for the filter can be generated in the C language for use with CADAT.

The functional model for the digital filter was simulated using CADAT, with the impulse stimulus waveform for input automatically generated by the module generator. The simulation of the impulse response requires more than 1800 simulation steps due to the bit serial architecture in this implementation of the filter circuit. Consequently, the usual logic waveform display for logic simulation is not useful for this simulation. The data from the simulation was post-processed instead for display in SIG. Figure 4 shows the overlay of two time series plots as displayed in SIG: theoretical response as calculated by EQIIR, and simulated response from CADAT. The steady oscillation of the tail-end of the simulated response is due to bit noise. Figure 5 shows the overlay of the two magnitude plots from the FFT for the theoretical and simulated impulse responses. At this point

Filter type	lowpass	
Approximation	elliptic	
Sampling freq. (KHz)	4.0	
Cutoff freq (KHz)	0.500000	1.500000
Norm. cutoff freq.	0.785398	2.356194
Cutoff freq. S-dom.	0.414214	2.414213
Passband ripple (S)	0.020000	0.1755 dB
Stopband ripple (S)	0.001000	60.000 dB

Table II. Filter design parameters.

wordlength: 7

constant gain factor: 0.1000E+01

L	B2(L)	B1(L)	B0(L)	C1(L)	C0(L)
1	0.125000	0.2343750	0.125000	-0.6484375	0.156250
2	0.171875	0.2578125	0.171875	-0.6953125	0.578125

Table III. Filter wordlength, gain, and coefficients calculated by EQIIR and used by the DIGFILT module generator.

in the design, the designer may proceed with the gate level model, or iterate the filter design parameters and the filter realization in order to achieve a better match to the functional specification.

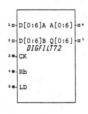

Figure 3. Digital filter graphical symbol generated automatically for the DIGFILT module.

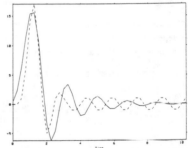

Figure 4. Impulse responses for the theoretical response as calculated by EQIIR (solid line) and the simulated response from CADAT (broken line).

A Circuit Example

Figure 6 shows a more extensive example using the digital filter synthesized above in the AutoCircuit design environment. This is a threshold circuit which filters an input waveform and compares the filter output with a threshold value latched in a register. In this example the input sampling rate is assumed to be 4 kHz with the significant data contained in the signal below 500 Hz. Consequently the filter cutoff is set at 500 Hz. In AutoCircuit, the circuit is captured in the design database by specifying the modules with relevant parameters and interconnecting the symbols graphically. The netlist for the simulation is prepared automatically. The input waveform for simulation was synthesized using SIG, adding together weighted components of 200 Hz, 1500 Hz and random noise. The threshold value latched in the register was chosen to provide a peak detection output at the GT output of the comparator.

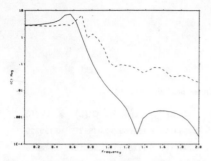

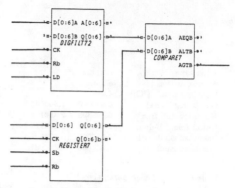

Figure 5. Magnitude plots from the FFT of the data in Fig. 4 for the theoretical response (solid line) and the simulated response (broken line). Data were windowed before transforms.

Figure 6. Example threshold detection circuit captured in AutoCircuit, using the digital filter from Fig. 3.

After simulation by CADAT the results were post-processed for display in SIG. Figure 7 shows the input signal, filtered signal and threshold-detection output for the circuit. The filter introduces a phase shift in the filter output as expected, but the essential peak information is present at the comparator output. The FFT spectra magnitude plots are shown in Figure 8 for the input and filtered output waveforms. The filter has effectively removed the 1500 Hz noise and supressed the high frequency random noise present in the input signal.

At this point in the design process the designer may either choose to continue with the design or redesign the filter depending on the circuit specification. The actual design and simulation time is sufficiently short (a few days at most) such that an iterative improvement of the circuit design is a reasonable choice.

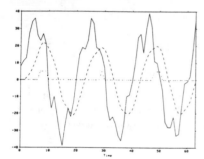

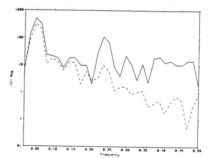

Figure 7. Simulation of the threshold circuit, showing the input (solid line), filter output (broken line) and threshold detected (dotted line).

Figure 8. Magnitude plots of the circuit input data (solid line) and filter output data (broken line) in Fig. 7., Data were windowed before transforms.

Discussion

The AutoCircuit system has integrated a number of high level tools for IC design, including block-level graphical design, mixed level simulation, automatic schematic synthesis, and digital signal processing tools. The module generators used beneath the building blocks in AutoCircuit were developed to synthesize circuits based on user parameters and software-coded design approaches. By using module generators hierarchically, complex modules such as digital filters can be synthesized.

AutoCircuit also incorporates some new approaches in processing architectures not previously available in general purpose circuit synthesis systems. The digital filter generators in AutoCircuit synthesize signal processing architectures by automatically interconnecting lower-level modules. Two digital filter generators incorporate knowledge of filter architectures in their respective module generators. Although each generator calls upon the same lower-level modules, the synthesized filter architectures are quite different. A designer can evaluate the two architectures and select the better one based on the application.

AutoCircuit uses standard cells as the lowest primitive, which may result in silicon area inefficiencies. In addition, capturing only two filter architectures in module generators restricts the designer's choices. However, the AutoCircuit approach of hierarchical design and simulation can easily be layered on more efficient cell generators such as those offered by silicon compilers. Once these low-level design roadblocks are removed from the design process, the designer is free to pursue alternative architectures. Similar development of module generators for other application areas may eventually lead to a extensive library of building blocks which a system designer could use for design.

References

[1] J. Dussault, C-C. Liaw, and M. M. Tong, "A high level synthesis tool for MOS chip design," Proc. 21st Design Automation Conference, June 1984, pp 308-314.

[2] P. A. Reutz, S. P. Pope, and R. W. Broderson, "Computer generation of digital filter banks," IEEE Transactions on Computer-Aided Design of Integrated Circuits and Systems, CAD-5, No. 2, April 1986, pp 256-265.

[3] G. A. Frank, D. L. Franke, and W. F. Ingogly, "An architecture design and assessment system," VLSI Design, Vol. VI, No. 8, August 1985, pp 30-43.

[4] J. R. Jasica, S. Noujaim, R. Hartley, and M. J. Hartman, "A bit- serial silicon compiler," Proc. ICCAD-85, Santa Clara, CA, November 1985, pp 91-93.

[5] VLSI Technology Inc., San Jose, CA.

[6] Seattle Silicon Technology Inc., Bellevue, WA.

[7] Silicon Design Labs Inc., Liberty Corner, NJ.

[8] Silicon Compilers Inc., San Jose, CA

[9] T. E. Fuhrman and F. J. Schauerte, "Design comparison: Hand-packed standard cells to silicon compilation," Proc. Electro-86, May 1986.

[10] G. F. DePalma, "Architecture experimentation," VLSI Design, Vol. VI, No. 11, November 1985, pp 80-86.

[11] P. J. Ainslie, G. D. Knight, and F. J. Schauerte, "Automating the design of integrated circuits for vehicle applications," SAE paper #850444, SAE International Congress and Exposition, Detroit, MI, February 1985.

[12] The Idea design system is a product of Mentor Graphics Corp., Beverton, OR, 97005-7191.

[13] IEEE Acoustics, Speech, and Signal Processing Society, Programs for Digital Signal Processing, New York, IEEE Press, 1979.

[14] D. Lager and S. Azevedo, "User's manual, SIG, a general-purpose signal processing program," UCID #19912, Lawrence Livermore National Laboratory, 1983, revised May 1985.

[15] CADAT is a product of HHB Systems, Inc., Mahwah, NJ.

[16 M. A. Majewski, T. E. Fuhrman, P. J. Ainslie, and F. N. Krull, "AutoDraft: automatic synthesis of circuit schematics," Proc. ICCAD-86, Santa Clara, CA, November 1986.

[17] A. Peled and B. Liu, Digital Signal Processing: Theory, Design, and Implementation, New York, Wiley and Sons, 1976.

[18] C-F. Chen, "Implementing FIR filters with distributed arithmetic," IEEE Trans. on Acoustics, Speech and Signal Processing, Vol. ASSP-33, No. 4, October 1985, pp 1318-1321.

[19] G. L. Sicuranza, "Nonlinear digital filter realization by distributed arithmetic," IEEE Trans. on Acoustics, Speech and Signal Processing, Vol. ASSP-33, No. 4, August 1985, pp 939-945.

CHAPTER 4

FIRST RESULTS AND DESIGN EXPERIENCE WITH THE SILICON COMPILER SYSTEM ALGIC

J.SCHUCK, M.GLESNER, M.LACKNER

Technical University Darmstadt
Institut fuer Halbleitertechnik
Schlossgartenstr.8, D-61 Darmstadt
FR Germany

ABSTRACT

ALGIC is a silicon compiler system for the full custom realization of digital signal processing circuits. Essential features of the **ALGIC** system are a flexible technology-independent design language, a strict separation of the silicon compiler software from the technology dependent parameters, a powerful cell generation concept based on a flexible cell library, a powerful floorplanning system and timing verification. Components necessary to build such a general system are described. Special emphasis is given to the description of an **ALGIC** design example.

1. INTRODUCTION

The event of VLSI technology offers many advantages for the integration of digital signal processing circuits. Key elements are temperature stability, long time stability, definite signal to noise ratio, insensivity to process irregulation, etc.

The implementation of real-time signal processing with its continuous data flow and its complex algorithms poses extreme computational demands which cannot always be met by general purpose processors. To meet the demands of DSP tasks the circuit has to be designed as a full custom realization. A custom VLSI design, however, is error prone, time consuming and very expensive.

Silicon compilers will in the future be capable to automate the design process thereby reducing the costs and time of a design dramatically. Actually some interesting approaches for an automatic layout generation have been presented (Ref. 1,2). The power of a layout generation system depends primarily on the complexity of the input language processor, an 'intelligent' control module, the layout generation modules and the floorplanning concept. Usually the silicon compiler systems are either restricted to special fields of applications and thereby realizing efficient layouts, or they are trying to cover a more general scope of application resulting in poor performance chips.

In the ALGIC (Automated Layout Generation for Integrated Circuits) system (Ref. 3) we have tried to overcome with these restrictions on the one hand by using a general PASCAL-like design language as an input of the system and on the other hand by using sophisticated tools like an optimizing PLA generator or parameterizable cells (arithmetic units, multipliers, adders, etc.) to produce dense high performance layout.

This paper presents the final realization of the **ALGIC** system and a design example of a digital Butterworth filter, realized as a three week design.

2. THE **ALGIC** SYSTEM

The flexible silicon compiler system **ALGIC** has been developed as a powerful design environment for an efficient layout generation of custom VLSI circuits. The short design cycles, attainable with the **ALGIC** system, are highly suitable for fast architectural studies and to find an efficient VLSI realization of DSP circuits.

The **ALGIC** system has been developed for two different approaches :

a) A general approach for high throughput systems which are described in a behavioural manner by an universal block structured input language. Thereby an efficient architecture is derived by converting the input description into an intermediate form and by evaluating a hierarchical structured control path which is linked with corresponding data path cells (Ref. 4).

b) A special approach, presented here, for a full custom realization of VLSI digital signal processing circuits (Ref. 5,6,7).

An overall survey of the compiler system is shown in figure 1. The design system processes a flexible technology independent language which is highly suitable for a structural and functional description of DSP circuits. The language is based on Pascal-like constructs including block-oriented concepts to support a hierarchical partitioning of the DSP circuit into smaller sub-systems. A hierarchical structured cell design methodology supports an efficient generation of the data path. Based on this design technique the cell layout of DSP functions can automatically be generated according to user specified circuit implementation (e.g. pipelining) and to the wordlength of signals. To increase the circuit performance powerful cell generators for the control path are included within the compiler.

2.1 THE DESIGN LANGUAGE

Starting from the algorithm which specifies a certain DSP function important constraints are involved for the choice and definition of the architecture for the customized circuit. In fact, a lot of constraints and important influences have to be observed like flexibility, optimability, data structure, data processing, architectural units, etc.. Because of these constraints an automatic translation from any signal processing algorithm to the corresponding architecture is restricted. Therefore the designer has to specify the architecture with regard to these constraints and this requires his full creativity and experience.

The translation of the architecture into the corresponding DSP circuit layout is the task of the **ALGIC** system. For that task the architecture and the DSP circuit have to be described on a hardware description level as an input to the silicon compiler system. Therefore the **ALGIC** language is intended as a flexible technology independent language.

In the language the following functions can be located:

- Description of the system frame

- Actions on the data
  (corresponds to the structural description of the data path)

- Control of these actions
  (corresponds to the functional description of control path)

Among the interesting characteristics of the language are PASCAL-like constructs to facilitate the learning of the language and block strukturing concepts to allow a hierarchical partitioning of the circuit to influence the floorplan results.

A survey of the language constructs is shown in figure 2.

2.2 THE CELL GENERATION

The cell generation of data path cells bases on a parameterisation concept. The cell layout generation is done by abbutting simple predesigned sub-cells, stored in the cell library including layout information and all essential attributes (electrical parameters, timing, I/O connections, size, etc.). The consequent application of this concept leads to a high flexibility and to a strong seperation of technology dependent parameters from the compiler software. In contrast to manual design automated layout generation based on a parameterized methodology reduces substantially costs and diminishes the probability of errors.

Among the available data path cells which can be generated for any precision and any function according to the user specified wordlength and circuit implementation are

a) different parameterized multipliers (Ref. 8,9)
   (speed, precision, area, pipelining, accumulation, rounding)
   - bit-parallel array multiplier
   - bit serial fully pipelined multiplier
   - fixed-coefficient multiplier '
   - pipelined GUILD-array

b) different adders
   - ripple carry adder
   - carry look ahead adder
   - carry select adder
   - carry save adder

c) different registers (dynamic, static, reset, shift)

d) counter, barrel shifter, RAM, simple gates, etc.

Specially for an efficient control path evaluation component layout generators have been implemented to increase the performance of the system:

- PLA/FSM generator using advanced folding techniques

- ROM generator
  (special features are bootstrapping circuitry and precharge
  concepts to increase speed and performance)

- Multiplexer/Demultiplexer generator with various output drivers

- Weinberger array generator

2.3 FLOORPLANNING IN **ALGIC**

Due to the different demands to a floorplanner in a Silicon-Compiler
environment a very flexible approach is used (Ref. 10). First of all different placement methods in a hierarchical structure (cluster generator)
are provided by the system:

- A mincut method (top-down) which uses a modified Fiduc-
cia/Mattheyses algorithm for the partitioning.

- A bottom-up method which places `good` module by `good` module successively on a `good` place, where the criterion `good` is controlled
by a flexible parameter set.

- A placement method based on simulated annealing with different
cost-functions is provided by the system.

These very different approaches allow a flexible placement and routing
process. In contrast to other floorplanners our system doesn't require a
fixed set of module dimensions. Modules can be specified with a set of
possible rectangular dimensions giving more flexibility to the module
generators. During the placement process relationships among area, connections, module dimensions and orientations are considered.

The fundamental key to this approach is a binary tree, in which every
node represents a cut in the layout where the leafes of the tree represent
modules. This data scheme leads to a slicing structure of the overall
layout. A sequence of different traversals are performed on the tree during the placement and routing process in order to determine the possible
module dimensions, positions and routing spaces. By using weights, certain
nets can be specified to be critical indicating that this length should be
taken preferable into account during the floorplanning process, i.e. for
electrical reasons. Furthermore it is possible to define relations between
modules (left, right, top, bottom) which are also considered. The advantages of the binary tree are significant: layouts based on this structure
can easily be routed inclusive VDD/GND routing since cyclic conflicts are
eliminated. Routing channels correspond to inner nodes of the tree, thus
channel widths can be accurately calculated via global routing which is
based on a heuristic Steiner tree search due to the NP-complexity of the
problem.

2.4 TIMING VERIFICATION IN **ALGIC**

The timing verifier estimates the necessary minimum clock period of the circuit for the worst and nominal case and identifies those pathes determining the clock period (Ref. 11,12).

The elementary cells of the circuit are hereby considered as simple delaytime elements neglecting their logical properties. For this reason, these elementary cells are represented as delaytime nodes in the internal structure. For the estimation of the nominal case clockperiod, the delaytime of an internal node is considered as a randomly distributed variable and for the worst case estimation it is considered as a single value. The timing verifier uses two different statistical models for distribution of the delaytime of an internal node, one is a modified beta-distribution, the other is a two-point distribution. The signal wires connecting the circuit cells are playing a double role in the internal graph. Because of their delaying property, they must also be represented by delaytime nodes in the internal structure, but on the other hand, the signal lines determines the connection between the elementary cells according to the signal flow on them. Therefore bidirectional signal lines must be represented by two nodes in the internal graph, one for each direction.

The timing verifier creates the internal data structure from the complete list of the elementary cells of the circuit and the connections among them which are derived by the **ALGIC** input language. The only necessary information about the delaytime of a single cell is its minimum, maximum and typical value, which must be determined by its designer. If the meantime and the standard deviation of this delay is not given, which is optional, the timing verifier estimates them from the values above by two equations modelling the cell delay as an appropriate beta-distribution. The timing values of the elementary cells as well as the electrical behaviour of the cell inputs/outputs are stored in the cell library.

The delay of a signal interconnect must be estimated indirectely from its characteristics. With given specific resistance and capacitance of the different interconnect layers, the time constant of the wire can be calculated from its length. After the general placement, the wire length must be estimated from the coordinates of its terminal points, but after the routing, the real wire can be extracted. The delay of the wire can now also be considered as a randomly distributed variable, due to the dependency of the output resistance of its loading gate from the signal transition. Approximating the nonlinear resistance of the signal outputs by simple linear pull-up and pull-down resistances, the interconnect can be modelled as a RC-tree (Ref. 13).

The feature making this timing verifier best suitable for large scale integrated circuits, is that all procedures above have a running time, which is strictly proportional for the forward and backward algorithm or at least almost proportional to the number of internal delaytime nodes for the case of the estimation of the clockperiod.

3. DESIGN CYCLE

The **ALGIC** design of a DSP circuit bases on an iterative process. The user starts with the specification of the DSP task (computation, precision, scan frequency, speed, etc.) to be addressed. The next step is to translate the processing algorithm to a corresponding architecture. The architecture will now be described with the **ALGIC** design language. On this step special emphasis must be given to the hierarchical partitioning of the circuit and to the selection of data path cells.

Before starting the **ALGIC** system the options for cell generation (e.g. PLA/FSM folding, input/output decoding of ROM), data and control path evaluation (e.g. specification of netweights to influence the place-ment, preplacement of 'critical' cells) and the parameters of the floor-planning subsystem (e.g. placement constraints) have to be specified.

Running the **ALGIC** system compile time and run time errors as well as messages and results of the cell generators, floorplanning tools and timing verification are listed in a protocol file. For the verification of the generated circuit the following information are available :

- cell size and cell I/O
- circuit area
- floorplan (including routing)
- minimum clock frequency, critical path, timing slacks
- power consumption, etc.

If the DSP task is matched the **ALGIC** system is finally started to generate the circuit layout. Otherwise it may be necessary to change the placement parameters, run time options, hierarchy of the input description, selection of data path cells or, if case of need, to optimize the architec-ture of the DSP circuit to match the DSP task. Thus the user can interac-tively study the effects of architectural and other modifications on layout area, power consumption and timing behaviour and can optimize the design without needing to generate an actual layout. A typical design cycle is shown in figure 3.

4. DESIGN OF DIGITAL BUTTERWORTH FILTER

To demonstrate the capability of the **ALGIC** system a layout design of a digital Butterworth filter was performed by one designer within three weeks. The realized Butterworth filter has a low-pass characteristic with the following characteristical datas:

- Cutoff frequency   : 10 kHz
- Stop frequency   : 13 kHz
- Insertion loss   : -0.7 dB
- Stop band attenuation : -20 dB
- Scan frequency   : 40 kHz
- Filter order   : 6

The Butterworth filter has been realized by a series arrangement of three Biquad filters. The flowgraph of a single Biquad filter is given in figure 4. Because a one to one implementation of such Biquad filters would be too area and power consumptive, an implementation of a word multiplexed architecture (shown in fig. 5) was choosen (Ref. 14). This architecture requires only one multiplier. The multiplications are performed consecutively. The registers RREG1 and RREG2 are arranged to store the intermediate results of the accumulation. The registers REG1 to REG6 (one register for each filter order) have to store the 'past' signal values. The 15 coefficients (5 for each Biquad filter) are stored in the ROM. To avoid a direct feedback coupling, the dynamical register MULREG have been inserted between the multiplier and the full adder. The whole DSP circuit is controlled by a finite state machine. An overflow control circuitry has been created to detect arithmetic overflow during accumulation.

This architecture has been converted into the corresponding **ALGIC** input description. An extract of the input description, treating the accumulation part of the circuit, is shown in fig. 6. The description is hierarchical structured. The accumulation functions are concentrated in the procedure **accu** . The procedure head includes the formal parameters which specifies the ports to the outside of the subsystem. In the declaration part, the internal terminals and local procedures and functions are specified. Functions which directly refer to a cell in the library are marked by the keyword LIB. In the block statement part (enclosed by BEGIN and END) the structural description of the subsystem **accu** is given.

The final floorplan, optained after several passes through the design cycle, is shown in fig. 7. The circuit layout (fig. 8) is currently fabricated on the 5 um NMOS-technology line of the Technical University Darmstadt.

The characteristical datas of the VLSI realization of the filter are listed below:

architecture   : word multiplexed trippel Biquad filter
clock frequency   : 600 kHz
data word length   : 8 bit
complexity   : ca. 8000 transistors
chip area   : 5 mm * 4 mm
power consumption   : ca. 300 mW

5. CONCLUSION

The basic concepts and ideas of the **ALGIC** compiler have extensively been discussed. Starting from the **ALGIC** input language we have shown the different modules which are necessary to transform the source text down to the layout level. A short design example was given to demonstrate the features and first results of the silicon compiler.

The **ALGIC** system is already running in an experimental version and is able to generate efficient chip layouts.

REFERENCES

/1/ J.M.Rabaey, S.P.Pope, R.W.Broderson:
"An Integrated Automated Layout Generation System for DSP circuits",
IEEE transactions on Computer Aided Design,
Vol.CAD-4.,NO.3,1985, pp.285-296

/2/ P.B.Denyer, A.F.Murray, D.Renshaw:
"FIRST: Prospect and Retrospect",
IEEE Workshop on VLSI Signal Processing,
Los Angelos, Nov. 1984, pp. 252-263

/3/ M.Glesner, H.Joepen, J.Schuck, N.Wehn:
"Silicon Compilation from HDL and Similiar sources",
published in "Hardware description languages" by North Holland,
Volume 7, Summer 1986

/4/ H.Joepen, M.Glesner:
"Architecture Construction for a General Silicon Compiler System",
Proc. of ICCD 85-Conf., Port Chester, N.Y., Oct.85, pp.312-316

/5/ M.Glesner, J.Schuck, H.Joepen:
"A Flexible Silicon Compiler for Digital Signal Processing Circuits",
Proc. of ICCD 84-Conf., Port Chester, N.Y., Oct.84, pp.845-850

/6/ J.Schuck, M.Glesner, H.Joepen:
"ALGIC - A Flexible Silicon Compiler System for DSP Circuits",
Los Angelos, Nov. 1984, pp.216-227

/7/ J.Schuck, M.Glesner, H.Joepen, N.Wehn:
"ALGIC - A Silicon Compiler for Digital Signal Processing Circuits:
Implementation and Design Experience", Integrated Circuit Technology
Conference, Limerick, Sept. 1986

/8/ J.Schuck, M.Glesner:
"Layout Generation for Multipliers in VLSI-Digital Signal Processing",
ESSCIRC, Sept.84, Edinburgh, pp.165-170

/9/ S.Meier, H.Schlappner:
"Design of multipliers for Digital Signal Processing",
Studienarbeit TH Darmstadt, Institut fuer Halbleitertechnik, Dec. 83

/10/ N.Wehn:
"HOPPLA - A Powerful Floorplanner for VLSI Circuits",
Internal Report, TH Darmstadt, Institut fuer Halbleitertechnik, Feb.86

/11/ M.Glesner, J.Schuck, R.B.Steck:
"SCAT - A New Timing Verifier in a Silicon Compiler System",
Proc. 23th Design Automation Conf., Las Vegas, June 86,pp.220-226

/12/ R.B.Hitchcock:
"Timing Verification and the Timing Analysis Program",
Proc. 19th Design Automation conf. 1982, Las Vegas, pp.594-604

/13/ Lin t., C.A.Mead:
"Signal Delay in General Networks with Application to Timing
Simulation of Digital Integrated Circuits",
Proc. Conf. Adv. Res. in VLSI, MIT (1984), pp.93-99

/14/ S.Meier:
"Architectural Studies for the Realization of Integrated DSP circuits",
Diplomarbeit, TH Darmstadt, Institut fuer Halbleitertechnik, Jan. 86

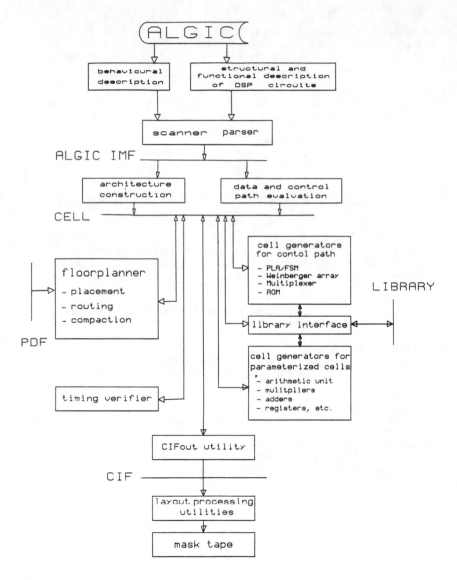

Fig. 1: Overall Survey of ALGIC Silicon Compiler

```
SYSTEM
DEF      ... ENDDEF
CONST    ... ENDCONST
VAR      ... ENDVAR
PORT     ... ENDPORT
TERMINAL ... ENDTERMINAL
REG      ... ENDREG
BEGIN    ... END
PARBEGIN ... PAREND
IF       ... THEN          ... ELSE
SET      ... TO/DOWNTO     ... DO
ON       ... DO WITH CLOCK ... ELSE
CASE     ... DO
FOR      ... TO/DOWNTO     ... DO
WHILE    ... DO

PROCEDURE
FUNCTION
LIB
IN,OUT,INOUT,TRISTATE
AND,NAND,OR,NOR,EXOR,NEXOR,NOT
<,>,<=,>=,<>,=
ABS,ROUND,COMPL,TRUNCATE
```

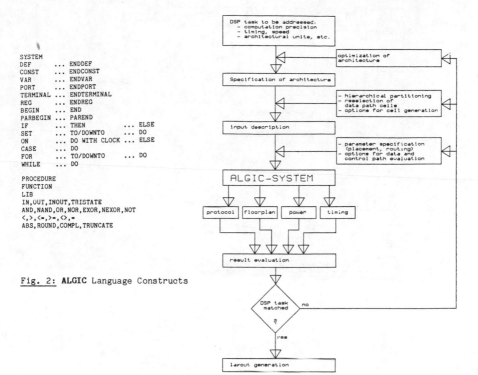

Fig. 2: **ALGIC** Language Constructs

Fig. 3: Design Cycle

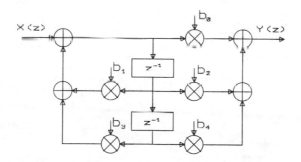

Fig. 4: Flowgraph of Biquad Filter

41

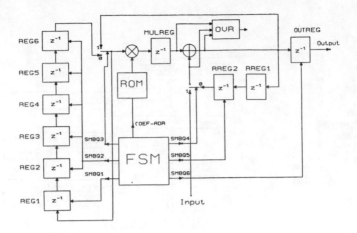

Fig. 5: Data Path of Butterworth Filter

```
PROCEDURE accu (PORT IN accu_in[0..7], sig_in[0..wo_le], smak[0..1];
                     OUT accu_out[0..wo_le], overflow ENDPORT);

  DEF

  TERMINAL
     add_in1[0..wo_le], add_in2[0..wo_le], rreg1_out[0..wo_le],
     rreg2_out[0..wo_le], cin, add_out[0..wo_le], nsum[0..wo_le],
     cout, ncout
  ENDTERMINAL;

(*****************************************************************)

  PROCEDURE ovr_control (PORT IN a, b, c, nc; OUT ovr ENDPORT);

    DEF

    TERMINAL
       na, nb, ovr1, ovr2
    ENDTERMINAL;

    FUNCTION nand3 (PORT IN in1, in2, in3 ENDPORT) : 1..1;
       LIB;

    FUNCTION nand2 (PORT IN in1, in2 ENDPORT) : 1..1 ;
       LIB;

    FUNCTION inv (PORT IN ein ENDPORT) : 1..1 ;
       LIB;

    ENDDEF;

    BEGIN

       na := inv (a);
       nb := inv (b);

       ovr1 := nand3 (a, b, nc);
       ovr2 := nand3 (na, nb, c);
       ovr  := nand2 (ovr1, ovr2)

    END;

(*****************************************************************)

  ENDDEF;

  BEGIN

    reg_dyn (accu_in, add_in2);

    IF smak [0]
       THEN
          add_in1 := sig_in
       ELSE
          add_in1 := rreg2_out;

    cin := 0;

    full_add (add_in1, add_in2, cin, add_out, nsum, cout, ncout);

    SET i := 0 TO wo_le DO  sbd2 (add_out[i], accu_out[i]);

    ovr_control(add_in1[0],add_in2[0],accu_out[0],nsum[0],overflow);

    reg_dyn (accu_out,rreg1_out);                 (* RREG1 *)
    reg_reset (rreg1_out, smak[1] ,rreg2_out)     (* RREG2 *)

  END;
```

Fig. 6: ALGIC Description of Accumulator

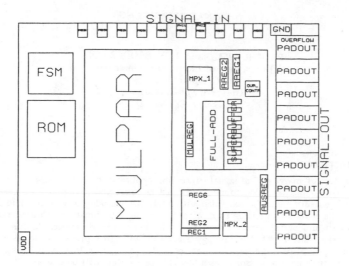

Fig. 7: Floorplan of Butterworth Filter

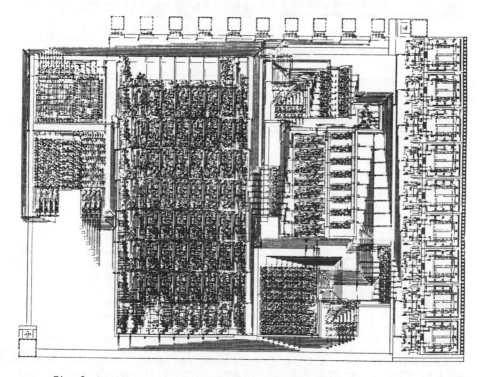

Fig. 8: Layout of Butterworth Filter

RECENT DEVELOPMENTS IN A/D INTERFACES FOR

SIGNAL PROCESSING

PAUL R. GRAY and R. W. BRODERSEN
Department of Electrical Engineering
and Computer Sciences
University of California
Berkeley, California

1. INTRODUCTION

A key aspect of many digital signal processing applications is the necessity of handling signals that originate in the analog domain and must be converted to digital form prior to processing, and of generating signals that must ultimately drive analog transducers and actuators and must thus be coverted to analog form after processing. The impact of these analog-digital interfaces on cost and performance in a given application can be dramatic. As the continuing reduction in feature size in VLSI technology allows digital signal processing to address lower and lower cost applications, the implementation of the analog interface at low cost, either on the same chip as the signal processor or as a low cost peripheral, will be crucial.

Functions that must be performed in such interfaces may include fixed gain or programmable gain signal preamplification, analog anti-alias filtering as well as other signal preprocessing, analog multiplexing, sampling, quantization, subsequent digital decimation, anti-aliasing and post-processing, and provision of a precision voltage reference. In the D/A interface function, interpolation, D/A conversion, analog reconstruction filtering and analog output drive functions are also often required. A generic block diagram of a typical A/D interface is shown in Fig. 1.

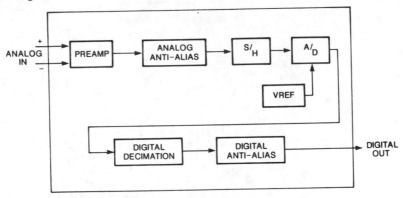

1. Typical A/D Interface Block Diagram

A typical example of an interface of the type shown in Fig. 1 is the PCM telephony single-chip codec-filter that is currently produced in very large quantity by a number of manufacturers. (1,2) These devices perform most of the functions above at a linearity equivalent to about 7 bits, peak SNR of 40dB, resolution of about 13 bits, and sample rate of 8kHz. Many signal processing applications require much higher linearity, peak SNR, and sample rate than those of which these devices are capable. Also, in low-cost applications it may be desirable to include the A/D interface directly on the signal processing chip itself, requiring that the interface be compatible with standard digital technology that often does not include a precision capacitor.

Within the past several years, a number of new circuit approaches have emerged that promise to dramatically reduce the cost of high performance of A/D interfaces. One important development is the increasing use of oversampling feedback coders to implement voiceband A/D interfaces that do not depend on component matching for linearity and that can be implemented in standard digital technologies lacking precision capacitors. A second important development is the self-calibration technique for eliminating the traditional dependence of integral linearity on component matching in successive approximate A/D converters. A third is the use of pipelined feedforward techniques for very high speed high resolution interfaces. Also, the performance of monolithic CMOS bandgap voltage references has improved dramatically through the use of parasitic lateral bipolar transistors(3). Switched capacitor antialiasing and signal preconditioning circuitry has improved both in terms of maximum clock rate capability(4) and in terms of silicon area through the use of circuit techniques that allow these filters to take better advantage of the scaling of technological feature size.(5)

The impact of this progress will be to allow the implementation of high-performance (ie 12-14 bit) A/D interfaces ranging from voiceband to video frequencies either on the same chip with the signal processing function or as a low-cost peripheral component.

In this paper, we attempt to summarize the most important recent developments in the A/D converter field that impact the architecture and design of A/D interfaces like the one shown in Fig. 1. In section 2, circuit techniques for the self-calibration of successive approximation A/D converters are described. In section 3, the concepts of oversampling and noise shaping are described. In section 4, pipelined architectures are described. In section 5 some likely future trends are summarized.

2. SELF-CALIBRATION OF MONOLITHIC SUCCESSIVE APPROXIMATION A/D CONVERTERS

Mismatches in precision components have long limited the achievable linearity in monolithic successive approximation A/D converters applicable to voiceband signal processing. Using on-chip intelligence these errors can now be removed through calibration, eliminating the need for laser trimming or other expensive adjustments in high-resolution DACs and ADCs.

2.1 Role of Component Matching in High Speed DAC Linearity

A successive approximation A/D converter usually consists of a high-speed DAC, voltage comparator, and successvie approximation logic. The logic causes the DAC together with the comparator to carry out a binary search for the value of the input through n successive comparisons where n is the number of bits in the conversion. The linearity of the result is directly determined by the linearity of the DAC. Traditionally, the DAC has been implemented with an array of precision passive components, either capacitors or resistors, and implemented in such a way that the linearity of the transfer characteristic is directly determined by the precision of the matching or ratioing of the passive components. This in turn required that for high-precision monolithic DACs some sort of trimming of the values of the capacitors or resistors was required to achieve the linearities of 1/2 lsb above 10 bits because of the limited as-fabricated matching achievable in monolithic components.

An important class of high speed DAC that partially circumvents this problem is the so-called "inherently monotonic" DAC.(6,7) In this case the structure is modified so that the DAC achieves an inherently high degree of differential nonlinearity(ie is relatively uniform step size). The resistor string DAC is a good example of an inherently monotonic structure. Another is the segmented structure of weighted current sources. In many applications, the differential linearity is much more critical than integral linearity, and in those cases the inherently monotonic approache is very useful. However, the integral linearity of such DACs still is limited by the component matching achievable in the technology, and in critical applications such as the front end of echo cancelled modems, better linearity is often required.

2.2 Self-Calibrating Charge-Redistribution A/D Converter

The basic concept of self-calibration involves the use of on-chip intelligence to measure the actual ratio errors in the critical passive components and then introduce appropriate analog or digital corrections during the A/D conversion process. This concept could be applied to any type of DAC, including the resistor-based types. For the sake of illustration we use the charge-redistribution converter, shown in Fig. 2. The DAC consists simply of a binary array of capacitors, a set of switches, and a reference potential V_{ref}. The DAC output voltage is just

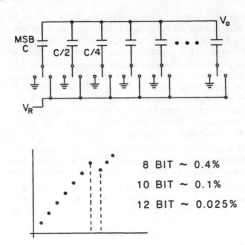

8 BIT ∼ 0.4%

10 BIT ∼ 0.1%

12 BIT ∼ 0.025%

DIFF. NONLIN. ∼ $\dfrac{\Delta C}{2C}$

AT MAJOR CARRY

2. Mismatch effects in a binary weighted capacitor array DAC

the top plate voltage of the array. The design of this type of DAC has been extensively described in the literature. Nominally, the largest capacitor should be exactly equal to the sum of the remaining capacitors. If this is not true, then nonlinearity through the major carry of the transfer charactersitic results. In order to achieve 1/2 lsb of integral nonlinearity at 12 bits, the matching accuracy must be approximately 0.025%. This is well beyond he capability of monolithic processes unless the components are trimmed.

Fortunately, the value of the mismatch can be quantized with a relatively simple switching sequence proposed by Lee(8). Consider the circuit of Fig. 3, in which two approximately

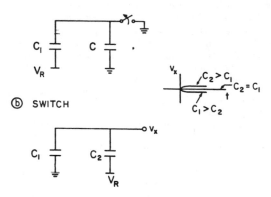

3. Conceptual measurement of the ratio error between two nominally identical capacitors

equal capacitors are switched between ground and V_{ref}. Initially, C_1 is connected to V_{ref} and C_2 is connected to ground. The top plate is initially grounded. The top plate switch is then opened and the bottom plates reversed. If the capacitors were identical there would be no change in the top plate voltage. If they are different, there will be a change in the top plate voltage that is a direct measure of the mismatch. This switching sequence is carried out on the capacitor array of Fig. 2 by making C_1 be the largest capacitor and C_2 be the parallel connection of the rest of the array. The residual top plate voltage can then be quantized by simply carrying out the remaining part of the normal A/D conversion sequence, giving a quantitative measure of the ratio error. This process can be repeated on the remaining capacitors. The table of numbers resulting can then be used in a correction process in that the error in each capacitor is inserted by means of a sub-DAC during conversion. An experimental prototype of such a converter, investigated by Lee[8], is shown in Fig. 4.

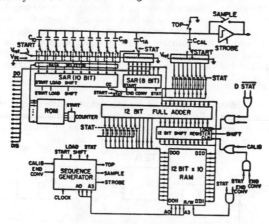

4. Block diagram of an experimental self-correcting A/D converter

The example given above illustrates the self-calibration concept, which in fact can take many forms and be applied to a variety of different DAC and ADC configurations. For example, it particularly effective when applied to the algorithmic converter, in which only one parameter, the loop gain, need be calibrated. (9,10)

Self-calibration as a technique for implementing practical interfaces for signal processing is very attractive in the sense that it lessens the component matching problem, but also has some disadvantages. The necessity for the system to initiate periodic recalibrations is a burden not present in conventional ADCs. Also, the internal accuracy required during the calibration is normally about 2 bits higher than during normal conversions, which increases the requirements on the system noise floor. Finally, it is critical that enough diagnostics be included in the interface so that it is possible to verify that a valid calibration has in fact taken place without disruption by supply noise or other spurious signals.

3. OVERSAMPLED FEEDBACK CODERS

Self-calibration is one approach to the realization of ratio-independent, high-resolution, medium-speed A/D converters. Another is to use a low-resolution quantizer sampling at a rate much higher than the desired output sampling rate, and digitally average the quantization noise over many input samples to achieve the desired resolution. This approach is most attractive when the resolution of the quantizer can be reduced all the way to 1 bit, essentially eliminating the necessity of matching components to achieve DAC linearity. These techniques have been under study for many years,(11,12,13) but have recently become competitive with other approaches because the digital averaging circuitry can be implemented in an economical amount of silicon area using scaled technologies. A second factor that has increased the suitability of these coders for voiceband application is that in newer scaled MOS technologies the achievable clocking rate for analog sampled data circuitry is sufficiently fast that a one-bit feedback coder can in fact achieve 13-14 bit resolution in a voiceband A/D interface.

Oversampled feedback coders depend on two comcepts: oversampling and noise shaping. These concepts will be discussed separately.

3.1 Oversampling Concept

The quantization noise introduced in the A/D conversion process can in many circumstances be usefully approximated as being white and uniformly spread from DC to half the sampling rate, assuming the signal amplitude is large compared to the step size. Since the total energy contained in the quantization noise is a constant determined by the step size, that portion of the noise lying in a given spectrum can be reduced by increasing the sample rate, at the rate of 3dB per octave of sample rate increase. Viewed another way, if the number of samples taken per unit time is doubled and each pair of samples produced is averaged to produce new samples at the original sample rate, the signal component of the samples will add linearly while the quantization noise component in the samples will add as uncorrelated random variables. Thus the signal-to-noise ratio will improve by 3dB.

In principle, one could utilize the approach of simply increasing the sampling rate to directly improve the SNR in, for example, speech encoders. There are two drawbacks to this, however. The first is that the rate of improvement is relatively slow; in order to get the equivalent of 1 additional bit of resolution the sample rate must be increased by a factor of 4. A second reason is that the underlying assumption that the signal amplitude is much larger than the step size will be violated for small signals if this process is carried very far.

3.2 Noise Shaping

Both of these drawbacks can be solved by incorporating the quantizer in a feedback loop and having it operate on the difference between the signal and a recent estimate of the signal (differential coding). A general block diagram of this type of coder is shown in Fig. 5. By

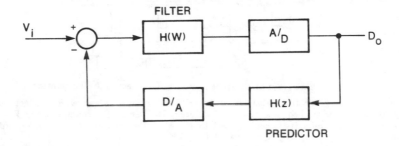

5. Block diagram of differential feedback coder

incorporating an integrator or more complex transfer function in the forward portion of this feedback loop, it is possible to make the noise transfer function from the quantizer to the output different from that of the signal and effectively introduce one or more zeroes at the origin in the noise transfer function.

Commercial application of this type of coder to date has focussed principally on two variations, interpolative coders(14) and sigma-delta coders. The principal difference between them is that the interpolative coder incorporates a companding DAC in the feedback loop that allows a lower sampling rate than in sigma-delta coders. Sigma-delta coders usually utilize a 1-bit DAC as the feedback element and thus must operate at much higher sampling rates.

3.3 Sigma-Delta Coders

The sigma-delta coder is most often implemented with a one-bit DAC in the feedback loop in place of the high-resolution DAC in the interpolative case. The resulting DAC output waveform when a sinusoid is encoded is shown in Fig. 6. Note that this waveform is a pulse density modulated square wave of amplitude equal to the full scale reference voltage, and as a result the total energy in the waveform is quite large. When the signal is small, the quantization noise energy is larger than the signal by a factor that is much larger than in the interpo-

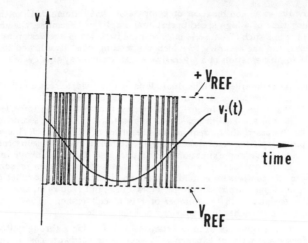

6. Typical DAC output waveform resulting from quantization of a sine wave.

lative case. The distribution of noise energy in this waveform is illustrated in Fig. 7. For a second order implementation(15), the noise distribution has a double zero at the origin, which leads to the ability to achieve large dynamic range with relatively low oversampling ratios. The sigma-delta coder has the fundamental advantage that the overall linearity of the coder can be made independent of component matching, and that the quantization noise level is approximately independent of signal level. The primariy disadvantages are the necessity of sampling at relatively high rates and the complexity of the decimation filter.

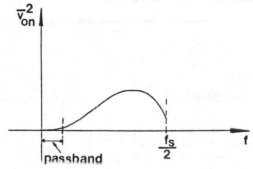

7. Typical output noise spectrum from a sigma-delta coder.

Most current implementations of the sigma-delta coder use a second-order filter in the forward path giving two zeroes at the origin in the noise transfer function from the quantizer input to the output(16,17). This configuration coupled with an appropriate decimation and bandlimiting filter can achieve a dynamic range of approximately 90dB with an oversampling ratio of 128 to one. Because of the enhancement of the out-of-band noise in the output of the coder, more stopband rejection is required in the decimation filter than in the first-order case. A typical implementation of the decimation filter would use a transversal filter with a $SINC^3$ frequency response giving third-order transmission zeroes at each harmonic of the resampling frequency. This type of FIR filter can be realized with a simple dedicated combination of registers and accumulators with no need for a multiplier. The final band shaping and decimation is often done in an IIR filter with a decimation ratio of two or four.

4. PIPELINED FEEDFORWARD A/D CONVERTERS

One convenient approach to the classification of A/D conversion techniques is by means of the number of clock cycles required to complete one output sample, where a clock cycle is understood to involve some combination of comparator delay time, DAC settling time, and perhaps the settling time of a high speed operational amplifier. Of course, the amount of time actually required for one such clock cycle is a complex function of the design of the converter, the technology used, and the accuracy to which the settling must be allowed to reach, but for a monolithic MOS implementation in a given line width technology this cycle time is a reasonable benchmark.

Serial conversion techniques such as dual slope or quad slope, require on the order of 2^n cycles where n is the number of bits in the conversion. These techniques have very low throughput rates but can be made insensitive to many types of analog imperfections such as gain and offset errors, and are used in low-speed applications. A one-bit second-order feedback noise shaping coder discussed above, requires on the order of $(2^{0.4n+1})$, clock cycles, or 100 to 200 per final output sample for resolution in the 13 bit range, and represents the next stage of speed capability. Successive approximation techniques, also discussed above, require n clock intervals, where n is the number of bits, assuming that self-calibration or some other technique can eliminate the component matching problem. Finally, fully parallel converters can achieve a throughout of one output sample per clock cycle, but require an amount of hardware that increases exponentially with the number of bits of conversion. The hardware cost has generally held parallel approaches to resolutions of 8 bits or less.

One architecture that appears to have promise for achieving a throughput rate at or near one sample per clock cycle without exponentially increasing hardware cost is the feedforward or pipelined approach, illustrated in Fig. 8. The converter is made up of elemental stages con-

PIPELINED ADC ELEMENT:

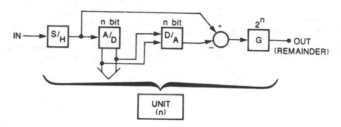

PIPELINED A/D: **(FEED FORWARD)**

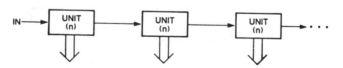

8. Block diagram of generic feedforward pipelined concerter.

siting of a low- resolution ADC and DAC, together with interstage sample/hold and amplifier elements. Each stage can have relatively low resolution, making the associated hardware complexity fairly low. In each stage, a coarse A/D conversion is performed, and the result is fed to a DAC. For maximum throughput, the ADC element would typcially be a flash converter. The difference between the DAC output and the input is amplified by an appropriate factor and fed to the next stage input sample/hold. A 12-bit A/D converter, for example, could be made up of three four-bit elements.

The pipelined architecture has been in use for many years in high-perfomance hybrid and discrete component converters. Translating these to monolithic implementation reqruies extensive use of error-correction techniques and self-calibration of the linearity of the DACs and the interstage gain factors. It also requires the development of extremely fast interstage analog amplifiers. There has been significant progress in the last few years in this area. (18)

4. Summary and projections for the future

Recent circuit innovations in the A/D conversion area will result in monolithic interfaces spanning the entire range of sampling rates that are lower in cost than those available today. For voiceband applications, it appears likely that oversampling techniques with digital decimation will find wide acceptance, particularly in applications where the interface must be implemented on the same chip with the processor. Self-calibrated successive approximation and pipelined converters are likely to find application at higher sampling rates.

A key question to be resolved in the implementation of the A/D interface for a given application is the location of the A/D interface in the signal path, and the amount of analog signal pre-processing to be implemented prior to A/D conversion. Moving the A/D interface closer to the analog domain reduces the amount of analog processing required, but often dramatically increases the sampling rate and resolution required in the A/D converter function. Thus this choice is seldom simply an issue of the relative cost per pole of analog filters vs digital filters. A good example is the case of a full duplex voiceband data modem, in which the presence of an echo signal that can be 40dB or more larger than the desired signal implies that if the band separation filtering is done digitally the resolution of the A/D converter must be in the 13 bit range. However, if the bandsplit filtering is done in the analog domain the echo does not pass through the A/D converter and the A/D needs a resolution of only 6-8 bits. The optimum choice for a given application is often a complex function of the nature of the signal statistics, the technology being used, and the circuit technqiues available to address the problem.

This work was supported by the National Science Foundation under Grant ECS-8023872 under Grant ECS-8100012.

REFERENCES

1. B. K. Ahuja, M. R. Dwarkanath, T. E. Seidel, and D. G. Marsh, "A Single-Ship CMOS Codec with Filters, "Digest of Technical Papers, 1981 International Solid-State Circuits Conference, New York, N. Y., February, 1981.

2. B. K. Ahuja, W. M. Baxter, and P. R. Gray, "A Programmable CMOS Dual-Channel Interface Processor"., IEEE Journal of Solid State Circuits, Vol SC-19, no 6, December, 1984

3. M. G. R. Degrauwe, E. Vittoz, H. Oguey, O. Leuthold, " A Family of CMOS Compatible Bandgap References", Digest of Technical Papers, 1985 International Solid State Circuits Conference, New York, N. Y., February, 1985

4. K. Matsui, et al, "CMOS Video Filters Using Switched Capacitor 14Mhz Circuits", Digest of Technical Papers, 1985 International Solid-State Circuits Conference, New York, N. Y. February, 1985

5. C. K. Wang, R. P. Castello, and P. R. Gray, "A Scalable High- performance Switched Capacitor Filter", IEEE Journal of Solid-State Circuits, vol SC-21, no 1, February, 1986

6. B. Fotouhi and D. A. Hodges, " An MOS 12b Monotonic 25us A/D Converter." Digest of Technical Papers, 1979 International Solid- State Circuits Conference, February, 1979

7. D. Hester, K. S. Tan, and C. R. Hewes, " A Monolithic Data Acquisition Channel" , Digest of Technical Papers, 1983 International Solid State Circuits Conference, February, 1983

8. H-S Lee, D. A. Hodges, and P. R. Gray, " A Self-Calibrating 12b, 12us CMOS ADC" Digest of Technical Papers, 1984 International Solid-State Circuits Conference, February, 1984

9. P. W. Li, M. Chin, P. R. Gray, and R. Castello, " A Ratio-Independent Algorithmic Analog-Digital Conversion Technique, Digest of Technical Papers, 1984 International Solid-State Circuits Conference, February, 1984

10. C. Shih and P. R. Gray, "Reference-Refreshing Cyclic A/D and D/A Converters", IEEE Journal of Solid-State Circuits, vol SC-21, no 4, August, 1986

11. James C. Candy, "A Use of Limit Cycle Oscillations to Obtain Robust Analog-to-Digital Converters", IEEE Transactions on Communciations, vol COM-22, no 3, March 1974

12. H. Inose, Y. Yasuda, and J. Murakami, "A Telkemetering System by Code Modulation-Delta-Sigma Modulation", IRE Transactions on Space Electronics and Telemetry, vol SET-8, September, 1962

13. J. D. Everhard, " A Single-Channel PCM Codec," IEEE Transactions on Communications, COM-27, February, 1979

14. B. A. Wooley and J. C. Candy, " An Integrated Per-channel Encoder Based on Interpolation" , IEEE Journal of Solid-State Circuits, vol SC-14, February, 1979

15. James. C. Candy, "A Use of Double Integration in Sigma Delta Modulation," IEEE Transactions on Communcications", vol COMM-33, no. 3, March 1985

16. P. Defraye,et al, "A 3 mu CMOS Digital Codec with Programmable Echo Cancellation and Gain Setting", IEEE Journal of Solid-State Circuits, vol SC-20, no. 3, June 1985

17. R. Koch and B. Heise, "A 120khz Sigma-Delta A/D Converter", Digest of Technical Papers, 1986 International Solid State Circuits Conference, Anaheim, California, February, 1986

18. L. C. Yiu and P. R. Gray, " A Pipelined 13-bit 250ksample/sec A/D Converter", to be published

CHAPTER 6

High Performance Data Conversion with VLSI Technology

Jeff Teza
Director of Product Marketing
Brooktree Corp.
San Diego Ca.

Abstract

Segmented architectures in digital-to-analog (DAC) and analog-to-digital (ADC) converters are taking advantage of the tremendous integration of VLSI technology. These architectures take advantage of CMOS and BIPOLAR's scaling feature sizes, CAD tools, and wafer economies. High performance mixed analog and digital circuits are being merged into LSI and VLSI systems on a chip. The same technology building 16 and 32 bit Microprocessors, DSP engines, and memories is capable of producing video speed DACs and ADCs. "Mixed-signal" VLSI chips are now possible in a variety of applications.

Architectures

Minimizing the number of components was the design wisdom when active elements were expensive. VLSI technology suggests a more appropriate approach is to use many circuit elements, taking advantage of density and relative device matching. Indeed, laser trimming or other techniques for achieving precise elements is more costly than using hundreds or even thousands of transistors in an IC design.

Segmented data converter architectures are conceptually a very natural way to do conversion. This approach to converting signals dates back to the 1950s. More elegant architectures implementable in VLSI have been cited in the last few years. These architectures are in contrast to earlier binary weighted schemes such as R-2R ladders and binary weighted current or voltage sources. The binary weighted approach requires circuit element ratios to track each other over time and temperature. This is difficult to achieve without trimming the designs in the manufacturing process.

In an N bit highly segmented data converter, 2^N elements are used to create the analog or digital output. For a "flash" ADC (FADC) these elements are voltage comparators. In a DAC these elements are current or voltage sources. Very high speeds are attainable with this architecture. Using this number of elements along with the associated switches, decoders, resistor dividers etc. requires a large amount of active area in an IC design. It is this chip area which constrains this approach.

In an N bit highly segmented D/A converter 2^N-1 circuit elements (typically current sources) are selected by a "thermometer" type of decoder to produce the output. This thermometer decoder adds one more circuit element to the output as the input increments by one (Figure 1). In this scheme relative element matching of only 50% yields a design with +- 1/2 LSB differential linearity (DL). The large area consumed by decoding, switches etc. makes it difficult to construct converters over 10 bits in a manufacturable die size.

Fewer than 2^N-1 circuit elements can be used to build an N bit DAC. For example in a 6:2 segmented 8 bit DAC design, 2^N or 64 current sources would be selected by decoding the 6 MSBs of the input. Another $2^N -1$ or 3 current sources would be selected by the decoded 3 LSBs of the input, this is a 4x savings in elements over the highly segmented approach. The MSB current sources would be 4x larger than the 3 LSB sources. This would require the MSB current sources to have relative matching as close as the LSB step size, or 12% for a +- 1/2 LSB design. If the current sources do not match this well, the DAC may become non-monotonic as a large source is substituted for the sum of the three smaller ones. Integral linearity (IL) is another important consideration in picking a segmentation scheme. IL varies inversely with the square root of area. Thus IL forms another important consideration when reducing the number of active elements. By improving the relative device matching by 4x, circuit area can be decreased by 16x. This square relation between active element area and element matching follows from fabrication technology dimensional control.

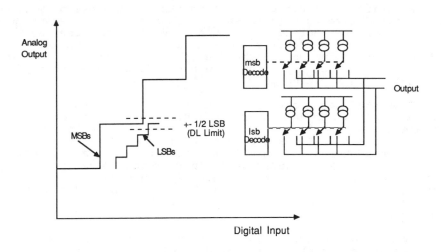

FIGURE 1: Highly segmented DAC

MOS transistor relative matching of less than .5% and capacitors in the .1% range is practical with MOS 2 um semiconductor technology. Bipolar transistor matching of .2% and resistors in the .1% range is achievable with BIPOLAR 2um semiconductor technology.

The general concept of a tradeoff between area and linearity is dramatically shown in Figure 2. Here an 8 bit 75 Mhz CMOS DAC with a die size of 3.78 x 4.52 mm using a 6:2 segmentation yields a +- 1/4 LSB Differential Linearity (DL) and +- 1/2 LSB Integral Linearity (IL) design. The second die is a triple 50 Mhz CMOS DAC with a 4.42 x 4.93 mm die size using 4:4 segmentation but achieves only +- 1 LSB IL and DL. The second die contains 2 more DACs with only a 28% increase in silicon area, paying for the added density with poorer linearities.

8 bit 75 Mhz CMOS DAC

Triple 8 bit 50 Mhz CMOS DAC
FIGURE 2: Tradeoff between linearity and die size

Sharing Semiconductor Resources

Highly segmented DAC and FADC architectures take advantage of digital CMOS and BIPOLAR fabrication technology trends (switching speed and density). To understand how these architectures will evolve, it's important to understanding where this fabrication technology is leading them.

Figure 3 shows semiconductor feature size trends over time. Scaling limits are beginning to slow down the historical trend lines. Feature sizes are now being reduced by 1/2 every 6 years. Element matching is proportional to these feature sizes. Thus we should be able to match elements 2x more accurately in the near future. Because of cleaner manufacturing environments manufacturable die areas are going up on the order of 2x every 8 years.

Power supply voltages for high density logic and memory circuits may be lowered from todays 5 volts down into the 3 volt range. This is to avoid certain physics problems in sub 1um transistors. This is undesirable from both the FADC and DAC points of view. Segmented DACs using current sources need enough voltage to provide adequate biasing of the current sources and still maintain some output voltage compliance. Single step FADCs have a limit of the practical quantization "Q steps" set by transistor matching. This is in the 4 mV range. Thus 10 bit FADCs require 1024 x 4 mV or 4 volts of supply to build reliable comparators. Multiple step bipolar FADCs and current output DACs are the most practical approaches for low voltage processes.

Isolating digital circuit noise from coupling into a high speed data converter built on the same IC is a major problem. Advanced CMOS IC processes use a twin-well technology which can help to isolate circuit elements from noise occuring in a common substrate. Careful layout is still required, and much of this know how is empirically derived. CAD simulation tools are unable to predict these problems effectively. This adds substantial risk to mixed-signal VLSI system designs.

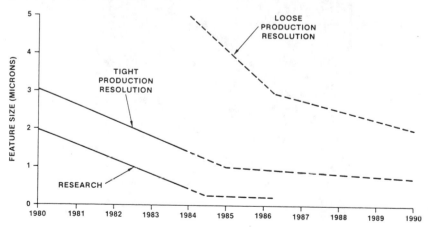

FIGURE 3: Semiconductor scaling trends (Courtesy of ICE Corp.)

VLSI Implications

The near future holds a promise for the use of standard cell design tools to construct high speed D/A and A/D converters. DAC "cells" that are optimized for the application and then placed and interconnected by software tools in an automated design process appear most promising. Experimental work at constructing 10 - 15 Mhz 8 bit DACs using various silicon compilers as well as cell based design techniques used in the converters shown, promise video speed analog "building blocks" in the next few years.

Figure 4 shows a RAMDAC which incorporates 6K bits of dual-ported RAM with three 8 bit D/A converters all operating at 70 Mhz. Integrating memory and logic with the data converters is easily achieved on an LSI chip using the segmented architecture approach to the D/A converters.

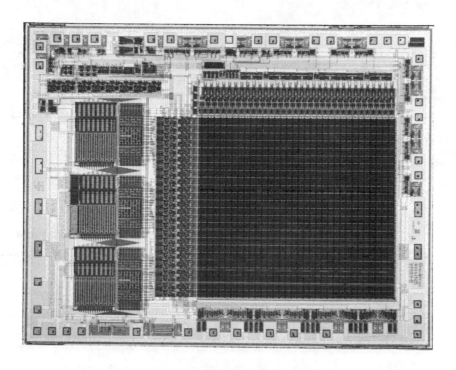

FIGURE 4: Integrated Triple DAC'S and RAM

In video applications differential phase (DP) and differential gain (DG) of 2% and 1° is a typical target for high quality video. This prevents the chrominance and luminance from mixing with each other. Video sampling rates of 2-4x the chrominance subcarrier frequency (7-14 Mhz) is also required to avoid aliasing. These numbers are easily achievable in both CMOS and Bipolar technology, as well as much higher speeds for computer generated images.

FADCs for digitizing broadcast quality video must also meet these DP and DG requirements as well as maintaining bit error rates on the order of 1 in 10^6. Thus video designers often opt for the best A/D affordable and spec much higher resolution DACs, limiting the cumulative error sources as much as possible. Monolithic video CODECS which achieve these sorts of requirements are currently being fabricated. Digital TV and other high volume applications are also researching fully integrated 8 bit FADCs and 8 or more bit DACs on a general purpose DSP chip.

Applications such as radar and multiple frequency band communication systems require converters meeting signal to noise (SNR) and noise power ratio (NPR) constraints. FADCs exhibit a theoretical SNR of approximately 6N+2 dB where N=# of bits. Thus an ideal 8 bit FADC would have 50 dB and a 12 bit DAC 74 dB SNR's. These are hard to achieve in practice by 1 or 2 bits creating an "effective bits" view of FADCs by some of these users. It is unlikely applications requiring >50 dB of SNR will be able to integrate FADCs onto a single VLSI chip. FADC complexities and noise sensitivities do not make it practical in the near future. 12 bit DACs are however candidates for integration in these applications.

Conclusions

A number of factors have been described that influence the architecture of a data converter. These include; number of elements, relative device matching, differential linearity and integral linearity. Also less predictable factors such as; power supplies and isolation must be thoroughly understood. Several near optimum DAC implementations were described.

Dynamic DACs using capacitors as current sources and autozeroing comparators in FADCs promise advances in integration not unlike the contribution of the DRAM in memory technology.

[1] Fujio T., "High-Definition Television Systems," Proceedings of the IEEE, pp 646-655, April 1985.
[2] Packard Dennis, Brooktree Corp. Personal interview.
[3] Van de Grift, Van de Plassche, "A Monolithic 8-Bit Video A/D Converter," IEEE Journal of Solid-State Circuits, Vol SC-19, pp 374-378, June 1984.

CHAPTER 7

An Integrated Analog Optical Motion Sensor

John Tanner and Carver Mead

California Institute of Technology
Computer Science Department, 256-80
Pasadena, California

1 Introduction

Future machines that interact flexibly with their environment must process raw sensory data and extract meaningful information from them. Vision is a valuable means of gathering a variety of information about the external environment. The extraction of motion in the visual field, although only a small part of vision processing, provides signals useful in tracking moving objects and gives clues about an object's extent and distance away.

This paper describes the theory and implementation of an integrated system that reports the uniform motion of a visual scene. We have built a VLSI circuit that reports the motion of an image focused directly on it. The chip contains an integrated photosensor array to sense the image and has closely coupled custom circuits to perform computation and data extraction.

The integrated optical motion detector was designed to use local analog image intensity information as much as possible to extract image motion. We combined a new high performance photosensor with analog computation elements and used a novel approach to extracting velocity information from a uniformly moving image. The new motion detector has a number of features that overcome the shortcomings of previous designs [1,2]:

- The continuous, non-clocked analog photosensor has been demonstrated to operate over more than four orders of magnitude of light intensity [3]; this is a much greater range than video camera based vision systems.

- The system analyzes information in analog light intensity variations in the image. Sharp edges can be utilized but are not required. The image contrast requirements are small.

- Local image gradients rather than global patterns are used; global notions such as object boundaries are not needed.

- The analog circuitry prevents the information loss inherent in thresholding or digitization.

- Temporal aliasing is avoided by continuous computation; there is no clocking or temporal sampling (there is spatial sampling).

2   A Two-Dimensional Analog Motion Detector

An algorithm for two-dimensional velocity detection must address the problem that arises from an inherent ambiguity between motions along the two axes. This ambiguity occurs when the field of view is limited, as is the view through an aperture. The *aperture problem* is well known for binary-valued images.

Figure 1 shows the view through a square aperture. A black-and-white image containing a single straight edge is moving at some velocity; the position of the edge at a later time is shown by the dashed line. The velocity cannot be uniquely determined from these two "snapshots". There is an infinite family of possible velocities, as illustrated by the arrows. We can view the image velocity components v_x and v_y as the x and y coordinates in a *velocity plane*. In this plane, the actual velocity of the image defines a point. The family of possible image velocities defines a *constraint line* in velocity space that has the same orientation as does the edge in physical space. To be consistent with the visual information from the local aperture, the actual velocity point is constrained to lie on the line in velocity space.

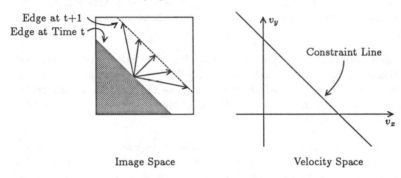

Image Space   Velocity Space

Figure 1: The aperture problem: local information is not sufficient to determine two-dimensional velocity uniquely. The family of possible velocities, denoted by the arrows in image space, define the constraint line in velocity space. The constraint line has the same orientation as does the edge in physical space.

Note that Figure 1 illustrates the ambiguity problem but does *not* depict the operation of our system. The velocity detector described in this paper (1) represents intensity values continuously—the images we consider are *not* just black and white; and represents time continuously—there is *no* notion of snapshots of the image or of clocking in this system.

Using analog values for intensities and gradients does not eliminate the ambiguity problem. Figure 2 shows the intensity plot of an image that contains gray-scale information and varies smoothly in intensity throughout. Brightness is plotted as height above the image plane for each point in the x-y image plane and is a function of the objects in view. As the image moves, the shape of the intensity surface will stay the same but will translate in the image plane. A stationary sensor will detect changes in intensity due to the movement of the image. The change in intensity with time, $\frac{\partial I}{\partial t}$, is a change in height of the intensity surface at a fixed point. The local sensor cannot tell whether its upward or downward movement is due to motion of the intensity surface along only the x axis, only the y axis, or to motion along both axes. Two of these possibilities are shown as arrows in Figure 2. The inherent ambiguity cannot be resolved by strictly local information.

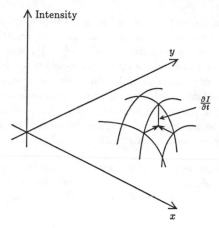

Figure 2: The intensity surface of a two-dimensional image. The change in intensity from a present high value to a future lower one, $\frac{\partial I}{\partial t}$, could be the result of many possible motions of the intensity surface. The arrows represent two possible motions.

Following Horn and Schunk [4], the expression that relates the intensity derivatives to the velocity is:

$$\frac{\partial I}{\partial t} = -\frac{\partial I}{\partial x}v_x - \frac{\partial I}{\partial y}v_y. \tag{1}$$

The intensity of Figure 2, a function of the two spatial variables x and y, has spatial derivatives $\frac{\partial I}{\partial x}$ and $\frac{\partial I}{\partial y}$. The x and y components of velocity are v_x and v_y, and $\frac{\partial I}{\partial t}$ is the change in intensity with respect to time at a stationary observation point.

This equation shows that the values of three local derivatives of the intensity do not allow velocity to be uniquely determined; there is an inherent ambiguity.

The local intensity derivatives do provide some useful information—they constrain the possible values of the x and y components of velocity, just as the black-and-white images in the aperture problem constrained the velocity. Writing Equation 1 in the form of the line equation $Ax + By + C = 0$, we get:

$$\frac{\partial I}{\partial x}v_x + \frac{\partial I}{\partial y}v_y + \frac{\partial I}{\partial t} = 0.$$

Each local set of three derivatives defines a line in the velocity plane along which the actual velocity must lie. The slope of this constraint line is $-\frac{\partial I}{\partial x}/\frac{\partial I}{\partial y}$. If we view the gray-scale image as having a fuzzy "edge" with orientation perpendicular to the intensity gradient, the constraint line has the same orientation in velocity space as the "edge" has in physical space. This is the same orientation as the constraint line of the black-and-white image.

The aperture problem for binary-valued images is just a special case of the general two-dimensional velocity ambiguity. Local images, gray-scale or black-and-white, can provide only a family of possible velocities. This set of velocities can be represented by the coefficients of the equation for the constraint line.

It is much easier to determine the constraint line if the analog information is retained. For gray scale images, the coefficients of the constraint line equation are just the three partial derivatives that can be measured locally—$\frac{\partial I}{\partial x}$, $\frac{\partial I}{\partial y}$, and $\frac{\partial I}{\partial t}$. An image with continuous intensity values can be made into a black-and-white image by thresholding. Determining the orientation of the edge of the binary-valued image (and so its constraint line) is a more global problem of determining the boundary between black regions and white regions and fitting a line to the boundary. To determine the velocity constraint line, it is much easier to measure the coefficients locally than to throw away the information and then try to reconstruct it with a global process.

The ambiguity of a single local set of measurements can be resolved by using another set of local values from a nearby location. These values define another line in the velocity plane. The intersection of these two lines uniquely determines the actual velocity (Figure 3).

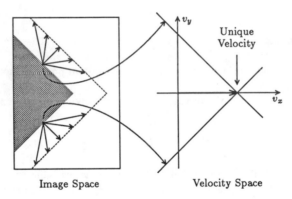

Figure 3: Uniquely determining velocity by the intersection of constraint lines.

2.1 Solution of Simultaneous Constraints

In practice, to find the actual velocity, we use the constraint contributions from all sites on the sensor array. Using only a small number of sites that are close together relative to an object size results in a few constraining lines in the velocity plane that are nearly parallel. A small error in any of the derivatives or in the constraint solver can then result in a large error in the computed velocity. Errors will be kept to a minimum when two lines in the velocity plane cross at right angles. This intersection occurs when there are contributions from two sites on edges that are perpendicular. An "edge" in this case is used loosely to mean a line perpendicular to the direction of greatest intensity change. Taking contributions from a large number of sites will then insure that we have pairs of orthogonal constraints for any reasonable image.

The barber-pole is a well-known example of an illusion that occurs because the orthogonality of constraint lines cannot be assured. In this illusion, the rotating cylinder produces a purely horizontal velocity. Our vision system erroneously reports "seeing" a vertical velocity. Images such as gratings and stripe patterns with intensity variations along only one axis cause this mistake. All constraint lines are coincident, so their intersection is not unique. It is not possible for man, beast, computer, or chip to disambiguate the motion of such a pattern.

In practice, there is no such thing as a perfect stripe pattern. The ability to correctly disambiguate velocity is thus a matter of degree. Our chip should reliably report the actual velocity unless the signals resulting from intensity variation along one axis lie below the noise level.

2.2 Constraint-Solving Circuits

Our constraint-solving circuit contains a set of global wires that distribute a best guess of velocity to all the individual constraint-generating sites (Figure 4). Each locale performs some computation to check whether the global velocity satisfies its constraint. If there is an error, circuitry within the local site then supplies a "force" that tends to move the global velocity toward satisfying the local constraint. The global velocity components are represented as analog voltages on the set of global wires. The correcting forces are currents that charge or discharge the global wires.

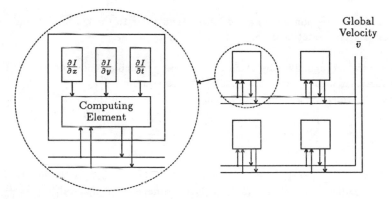

Figure 4: Block diagram of the constraint solver cell and array. The three locally derived derivatives, $\frac{\partial I}{\partial x}$, $\frac{\partial I}{\partial y}$, and $\frac{\partial I}{\partial t}$, along with the present value of the globally distributed velocity components are inputs to a circuit that continuously computes a correction to the global velocities.

Finding the intersection of many lines is an over-constrained problem. Any errors will result in a region of intersection in which the real desired point is most likely to lie. To compute a most-probable intersection point (velocity), we must know what types of errors to expect, to define "most probable" and select on the basis of that definition a forcing function that varies with detected error. In the absence of rigor, we can make reasonable guesses for the forcing function. It should be monotonic—the greater the error, the harder we should try to move it in the right direction. Because a forcing function linear in error distance is easiest to implement, we have selected it. The constraint solver minimizes the error by finding the least-squares fit of the velocity point to all the constraint lines.

The direction of the correction force should be perpendicular to the constraint line (Figure 5). Based on local information alone, the cell knows that the global velocity should lie on its locally defined constraint line, but it knows nothing about where along the line the global velocity lies. If a non-orthogonal component in the correction force were generated by the cell, an undue preference for the position of the global velocity on the constraint line would be expressed.

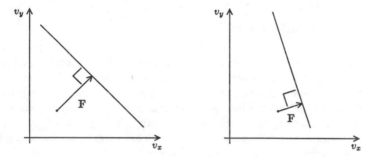

Figure 5: The correction force should be perpendicular to the constraint line in order to express only information obtained from the image.

We ensure orthogonality by using a direct perpendicular construction for the correction force. If we rearrange the constraint line equation $\frac{\partial I}{\partial x}v_x + \frac{\partial I}{\partial y}v_y + \frac{\partial I}{\partial t} = 0$ from the implicit form $Ax + By + C = 0$ to the slope-intercept form $y = \mathrm{m}x + \mathrm{b}$, we get:

$$v_y = -\frac{\frac{\partial I}{\partial x}}{\frac{\partial I}{\partial y}}\,v_x - \frac{\frac{\partial I}{\partial t}}{\frac{\partial I}{\partial y}}\,.$$

The slope of the constraint line, where defined, is $\mathrm{m} = -\frac{\partial I}{\partial x}/\frac{\partial I}{\partial y}$, so the slope of the desired perpendicular correction force is then the negative reciprocal, $-\frac{1}{\mathrm{m}} = \frac{\partial I}{\partial y}/\frac{\partial I}{\partial x}$. A unit vector in this direction is:

$$\widehat{\Delta\mathbf{v}} = \frac{\nabla\mathbf{I}}{|\nabla\mathbf{I}|} = \left\langle \frac{\frac{\partial I}{\partial x}}{\sqrt{\frac{\partial I}{\partial x}^2 + \frac{\partial I}{\partial y}^2}}, \frac{\frac{\partial I}{\partial y}}{\sqrt{\frac{\partial I}{\partial x}^2 + \frac{\partial I}{\partial y}^2}} \right\rangle.$$

The magnitude of the correction force should be greater if the present point in velocity space is farther away from the constraint line. The force should be zero for points lying on the constraint line. The direction of the force should always be perpendicular to the constraint line, and should have a sign such that the global velocity point will move toward the constraint line. A forcing function that is linear with error distance fulfills all these requirements and can be easily computed as:

$$D = \frac{\frac{\partial I}{\partial x}v_x + \frac{\partial I}{\partial y}v_y + \frac{\partial I}{\partial t}}{\sqrt{\frac{\partial I}{\partial x}^2 + \frac{\partial I}{\partial y}^2}}.$$

If we just substitute the values for the velocity components into the line equation and normalize by the quantity under the radical, we get D, a signed distance. The magnitude of D is the distance from the present velocity point $\langle v_x, v_y \rangle$ to the constraint line. The sign of D indicates on which side of the line the point is and therefore the direction of the correcting force. The vector, $\Delta\mathbf{v}$, from the current velocity to the point on the constraint line is then:

$$\Delta\mathbf{v} = D \cdot \widehat{\Delta\mathbf{v}} = \left\langle D\frac{\frac{\partial I}{\partial x}}{\sqrt{\frac{\partial I}{\partial x}^2 + \frac{\partial I}{\partial y}^2}}, D\frac{\frac{\partial I}{\partial y}}{\sqrt{\frac{\partial I}{\partial x}^2 + \frac{\partial I}{\partial y}^2}}\right\rangle$$

$$= \left\langle \frac{(\frac{\partial I}{\partial x}v_x + \frac{\partial I}{\partial y}v_y + \frac{\partial I}{\partial t})\frac{\partial I}{\partial x}}{\frac{\partial I}{\partial x}^2 + \frac{\partial I}{\partial y}^2}, \frac{(\frac{\partial I}{\partial x}v_x + \frac{\partial I}{\partial y}v_y + \frac{\partial I}{\partial t})\frac{\partial I}{\partial y}}{\frac{\partial I}{\partial x}^2 + \frac{\partial I}{\partial y}^2}\right\rangle.$$

Each cell should produce a force (electrical current), $\mathbf{F}$, that will tend to move the global velocity proportional to the detected error, $\Delta\mathbf{v}$. We would also like to scale this correcting force according to our confidence in the local data, C:

$$\mathbf{F} = C \cdot \Delta\mathbf{v}.$$

There is more information in a higher contrast edge, or at least there is a higher signal-to-noise ratio. A greater weight should be afforded to the correcting forces in those higher contrast areas. Our measure of contrast in the image is the intensity gradient, a vector quantity $\nabla\mathbf{I} = \langle \frac{\partial I}{\partial x}, \frac{\partial I}{\partial y} \rangle$. Confidence is related to the magnitude of the gradient, $|\nabla\mathbf{I}| = (\frac{\partial I}{\partial x}^2 + \frac{\partial I}{\partial y}^2)^{\frac{1}{2}}$. If we choose our confidence, C, to be the square of the magnitude of the intensity gradient, we have:

$$C = |\nabla\mathbf{I}|^2 = \frac{\partial I}{\partial x}^2 + \frac{\partial I}{\partial y}^2.$$

This choice greatly simplifies the correcting force calculation by canceling out the denominator. Our force equation becomes:

$$\mathbf{F} = \left\langle (\frac{\partial I}{\partial x}v_x + \frac{\partial I}{\partial y}v_y + \frac{\partial I}{\partial t})\frac{\partial I}{\partial x}, (\frac{\partial I}{\partial x}v_x + \frac{\partial I}{\partial y}v_y + \frac{\partial I}{\partial t})\frac{\partial I}{\partial y}\right\rangle.$$

Writing the two components of this vector equation separately we have:

$$
\begin{aligned}
F_x &= (\frac{\partial I}{\partial x}v_x + \frac{\partial I}{\partial y}v_y + \frac{\partial I}{\partial t})\frac{\partial I}{\partial x} \\
F_y &= (\frac{\partial I}{\partial x}v_x + \frac{\partial I}{\partial y}v_y + \frac{\partial I}{\partial t})\frac{\partial I}{\partial y}.
\end{aligned}
\tag{2}
$$

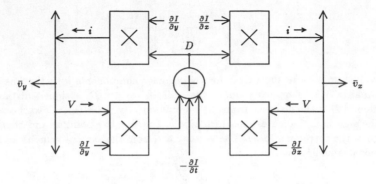

Figure 6: Block diagram for each cell's motion detection circuitry.

Analog computational elements within each cell of a two-dimensional array implement the calculations described by the above pair of equations. The block diagram of an implementation of the orthogonal two-dimensional formulation is shown in Figure 6.

The two-dimensional constraint solver can be viewed in terms of feedback. Each cell computes an error, the signed scalar quantity D, the distance in velocity space from the global average velocity to the locally known constraint line. This distance error is used as feedback to correct the system. It is multiplied by the appropriate two-dimensional vector perpendicular to the constraint line, to generate a correction force in the same direction to correct the global velocity vector.

The constraint solving system for two-dimensional motion detection is collective in nature. Local information, weighted by confidence, is aggregated to compute a global result. Each cell performs a simple calculation based on moving the global velocity state into closer agreement with its locally measured information. The collective behavior that emerges is the tracking of the intersection of constraint lines to solve the two-dimensional ambiguity, when possible, and report accurately the two-dimensional analog velocity of the image.

3   Results

The 8×8 array chip has been tested extensively by electronically simulating motion and by projecting actual moving images. The first set of experiments used an electronically controlled light source to apply an intensity field to the chip that varied spatially across the chip and varied with time. The space and time derivatives of intensity were controlled to simulate a moving intensity pattern while the velocity outputs from the chip were monitored. A second set of experiments focused actual images onto the chips and measured the chip's response. The constraint line behavior was verified and the correction forces were mapped for different images.

3.1 Characterization of the Motion Output

The motion simulation test setup is shown in Figure 7. Any changes in the light intensity were assumed to be due only to motion of the image, not to changes in the illumination level. Rapidly changing the illumination level under experimental control allowed the motion of a spatial intensity gradient to be simulated.

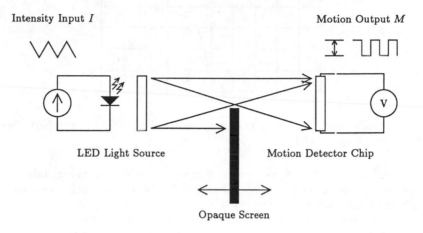

Figure 7: The test setup to simulate motion electronically. An LED cast light directly on the motion detector chip. Varying the LED current produced a controlled $\frac{\partial I}{\partial t}$. An opaque screen made a shadow edge on the chip. The distance from the chip to the screen controlled the sharpness of the edge, $\frac{\partial I}{\partial x}$.

A time derivative was generated by changing the current through the LED light source. A triangle wave intensity, used in these experiments, made a $\frac{\partial I}{\partial t}$ that is a square wave. The magnitude of the $\frac{\partial I}{\partial t}$ square wave is the slope of the triangle wave, which is dependent on the amplitude and frequency of the triangle wave. Frequency was used to vary $\frac{\partial I}{\partial t}$.

An opaque screen between the LED and the chip partially occluded the light and caused a spatial derivative of intensity (edge) to fall on the chip. Moving the screen closer to the chip made a greater $\frac{\partial I}{\partial x}$; that is, a sharper edge. The position of the screen was adjusted until the measured spatial gradient was the desired value.

When a spatial intensity gradient that varied in time as a triangle wave was applied to the motion detector chip, the differential voltage on the chip's velocity outputs was a square wave. Figure 8 shows an oscilloscope trace of the LED input current and the velocity output of the chip. For these experiments, the screen producing the spatial gradient was aligned with the y axis of the chip. As the intensities varied with time, the y component of velocity reported by the chip was very nearly zero.

With experimental control of $\frac{\partial I}{\partial x}$ and $\frac{\partial I}{\partial t}$, the velocity output of the chip was tested to verify that the reciprocal relationship for velocity, $v_x = -\frac{\partial I}{\partial t}/\frac{\partial I}{\partial x}$, held. This relationship

CH1 DC 500mV 10ms AVG; CH2 DC 50mV 10ms AVG;

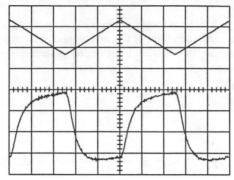

Figure 8: Oscilloscope traces of LED current (top) and reported velocity from the chip (bottom) for frequency of 20Hz.

comes from Equation 1 with v_y set to zero. Two sets of measurements were taken. First, reported velocity as a function of $\frac{\partial I}{\partial t}$ was plotted for fixed values of $\frac{\partial I}{\partial x}$; we expected a straight line graph. Second, the triangle wave frequency generating $\frac{\partial I}{\partial t}$ was held constant and the spatial gradient was varied; the expected curve was a hyperbola. For all plots, the reported velocity was the amplitude of the square wave of the x component output from the chip.

Figure 9 plots reported velocity versus $\frac{\partial I}{\partial t}$ (frequency) for three fixed values of $\frac{\partial I}{\partial x}$. The straight lines represent the theoretical proportional behavior, as shown on the log-log plot. The experimental results matched the predicted ones in the range from 1Hz to 40Hz. Beyond that frequency, the amplitude of the output rolled off.

Figure 10 is a plot of reported velocity on the vertical axis versus applied $\frac{\partial I}{\partial x}$ on the horizontal axis for three fixed values of $\frac{\partial I}{\partial t}$. The curves approximate a hyperbola over most of their range. The behavior to the left of the peaks deviates significantly from a hyperbola.

As $\frac{\partial I}{\partial x}$ decreases, the reported velocity should increase; the resistance of the circuitry will increase, however, causing it to have less effect on the reported velocity. When $\frac{\partial I}{\partial x}$ equals zero, the reported velocity could take on any value because it is not affected at all by the local cells. The current source loads on the v_x and v_y lines in our implementation have large but finite resistance so, in the absence of any information from the visual field, the reported velocity will tend toward zero. The schematic representing this effect is shown in Figure 11. For simplicity, we consider only the one-dimensional case, which can be derived from the two-dimensional case by setting $\frac{\partial I}{\partial y} = 0$. For large $\frac{\partial I}{\partial x}$ (contrast ratios), the resistance of the local circuits is much smaller than that of the loads, so the load's effect on reported velocity is negligible. As the contrast ratio is reduced, the load resistance must be taken into account.

So far we have referred to the output of the chip as the velocity v. We have now

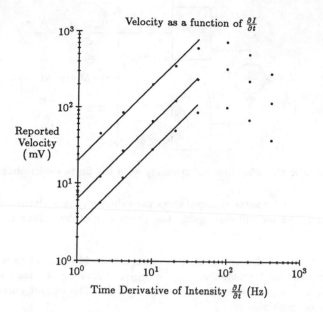

Figure 9: Measured motion response of the chip as a function of $\frac{\partial I}{\partial t}$ for fixed values of $\frac{\partial I}{\partial x}$. The straight lines represent an ideal linear response.

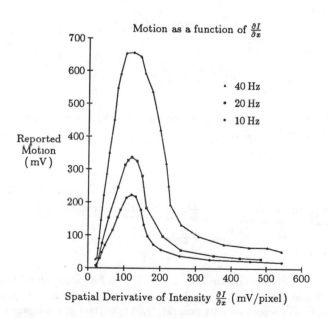

Figure 10: Measured motion response of the chip as a function of $\frac{\partial I}{\partial x}$ for fixed values of $\frac{\partial I}{\partial t}$. The response approximates a hyperbola over most of its range, as expected for velocity. Near zero, the response is linear with $\frac{\partial I}{\partial x}$.

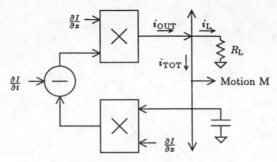

Figure 11: Schematic of the motion cell circuitry with the finite load resistance, R_L.

shown, however, that under some circumstances the values on these global output lines may not be velocity, so we will distinguish the chip's output by calling it "reported motion," M.

Because $M = \int i_{TOT}\, dt$, the condition for the system to be in steady state is $i_{TOT} = 0$. Before considering the load resistance, $i_{TOT} = i_{OUT}$. Including the load resistance, $i_{TOT} = i_{OUT} - i_L$ so for steady state $i_{OUT} = i_L = M\frac{1}{R_L}$. The output current, i_{OUT}, produced by the top multiplier is:

$$i_{OUT} = (M\tfrac{\partial I}{\partial x} - \tfrac{\partial I}{\partial t})\tfrac{\partial I}{\partial x}.$$

Steady state becomes:

$$M\frac{1}{R_L} = (M\tfrac{\partial I}{\partial x} - \tfrac{\partial I}{\partial t})\tfrac{\partial I}{\partial x}.$$

Solving for M we get:

$$M = \frac{\partial I}{\partial t}\frac{\frac{\partial I}{\partial x}}{\left(\frac{\partial I}{\partial x}\right)^2 + \frac{1}{R_L}}. \tag{3}$$

A plot of this mathematical function is shown in Figure 12. For sufficiently large $\frac{\partial I}{\partial x}$, $\left(\frac{\partial I}{\partial x}\right)^2 \gg \frac{1}{R_L}$, so Equation 3 reduces to:

$$M = -\frac{\frac{\partial I}{\partial t}}{\frac{\partial I}{\partial x}} = v. \tag{4}$$

As $\frac{\partial I}{\partial x}$ approaches zero, so that $\left(\frac{\partial I}{\partial x}\right)^2 \ll \frac{1}{R_L}$, the equation becomes:

$$M = \tfrac{\partial I}{\partial t}\tfrac{\partial I}{\partial x}R_L. \tag{5}$$

This analysis allows us to determine the chip's behavior for images that have different contrast ratios. When there is sufficient difference between light and dark areas, the motion detector chip will report velocity accurately. As contrast in the image is reduced, the motion output M will change smoothly to become a function proportional to both $\frac{\partial I}{\partial t}$ and $\frac{\partial I}{\partial x}$. It seems a particularly graceful way for a system to fail as the contrast ratio in its field of view is reduced to the point at which velocity can no longer be extracted. For a strict velocity detector, the zero-contrast case is undefined, so a device

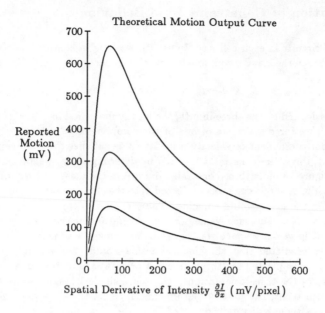

Spatial Derivative of Intensity $\frac{\partial I}{\partial x}$ (mV/pixel)

Figure 12: Plot of the theoretical motion response curve as a function of $\frac{\partial I}{\partial x}$ according to Equation 3. This curve approximates a hyperbolic response to the right and a linear response near zero.

that reported "true velocity" could take on any value in the absence of information. Our motion detector behaves better—it reports zero motion when it can detect no spatial intensity variation.

Note that, if the multiplicative definition of motion is desired over the entire operating range, the motion detector chip can be easily made to do this calculation. Setting the control current to zero on the feedback multiplier makes $i_{\text{OUT}} = \frac{\partial I}{\partial t}\frac{\partial I}{\partial x}$. This current can be turned into a voltage by the load resistor or a higher performance current sensing arrangement off chip.

Over the complete range of $\frac{\partial I}{\partial x}$'s, and in particular in both the hyperbolic and linear regimes of motion, Equations 4 and 5 show that the magnitude of the motion response should be proportional to $\frac{\partial I}{\partial t}$. The three curves of Figure 10, taken at frequencies of 10Hz, 20Hz, and 40Hz are scaled versions of the same curve and so bear out this proportionality over the range of $\frac{\partial I}{\partial x}$. Figure 9 shows a plot of reported motion as a function of $\frac{\partial I}{\partial t}$. The three curves are for fixed $\frac{\partial I}{\partial x}$'s, one chosen for each of the regimes of operation and one for midway in the transition region between them.

3.2 Verification of Constraint Line Behavior

The collection of circuits in each cell (see Figure 6), working in concert, tries to satisfy the constraint between the x and y components of velocity according to the line equation:

$$\frac{\partial I}{\partial x}v_x + \frac{\partial I}{\partial y}v_y + \frac{\partial I}{\partial t} = 0.$$

This constraint is defined by the three inputs, the locally measured intensity derivatives, $\frac{\partial I}{\partial x}$, $\frac{\partial I}{\partial y}$, and $\frac{\partial I}{\partial t}$. If we force the value of one of the components of velocity, the circuit will drive the other component of velocity until its value satisfies the constraint. For a given image input, the entire constraint line can be determined by sweeping the forced velocity value. Figure 13 plots three constraint lines from the measured response of the motion detector chip. A single edge was projected onto the chip so that all the constraint lines of each cell in the array would coincide. The x component of velocity was driven to a sequence of values. For each value, the chip determined the y component and the resulting point in velocity space was plotted. The image was not moving relative to the chip, so we predicted that the constraint line would pass through the origin. The constraint line was plotted for three different orientations of the edge. To ensure the relative angles of the three orientations, we used a single triangle as the image for each trial. Between trials, the part of the image falling on the chip was adjusted by translations only. Although the data deviated from the ideal slightly, this experiment clearly demonstrated the constraint line behavior of the motion detector chip.

3.3 Velocity Space Maps

To demonstrate the two-dimensional collective operation of the motion detector chip, we focused an image of a single high contrast edge onto the chip. Because the image was stationary, the chip should report zero motion. The global output lines were driven externally to take on a sequence of values chosen to scan the velocity space in a regular grid. For each x,y pair of voltages driven onto the chip, the chip responded with a current intended to move the global point in velocity space into agreement with the velocity of its image input; namely, zero velocity. These resulting x,y pairs of currents were measured for each point and displayed as a small vector originating at the forced point in velocity space. The resulting maps of these vectors (Figures 14 through 16) show the direction and magnitude of the force exerted by the chip on the global velocity lines. The point of stability—the attractor point—is near zero, as it should be. The amount by which the chip pulls as the global line moves farther away from the attractor depends on the structure of the applied image. A one-dimensional image, such as the single edge used in this experiment, provides information about the velocity only perpendicular to the edge. Thus, the chip should pull harder when the velocity lines are forced away from the real velocity in a direction perpendicular to the one-dimensional image stimulus. The image contains less information about the velocity parallel to the applied edge, so forced displacements of velocity away from zero in that direction result in much smaller restoring forces.

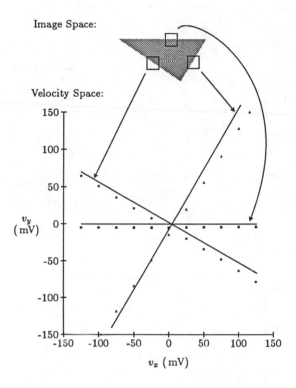

Figure 13: Demonstration of the constraint line behavior of the motion detector chip. The x component of velocity was swept while v_x versus v_y was plotted. The three trials were for the edge in the image oriented at 0°, 60°, and −30°. The lines are the ideal constraint lines.

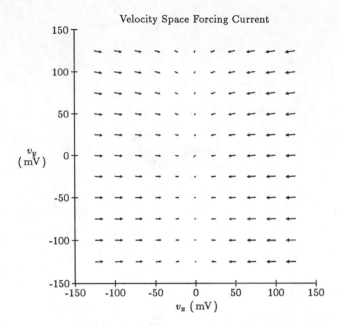

Figure 14: Velocity space map of the restoring forces generated by the motion detector chip in response to an edge at 90°.

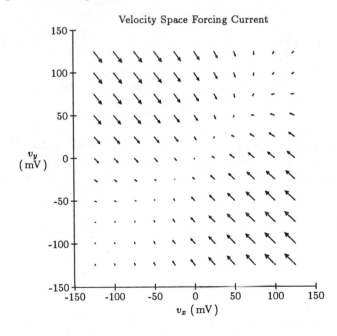

Figure 15: Velocity space map of the restoring forces generated by the motion detector chip in response to an edge at 45°.

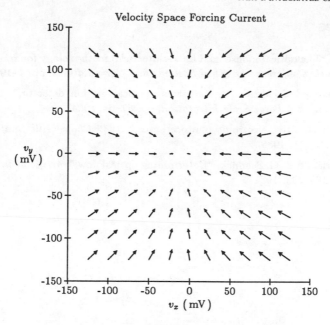

Figure 16: Velocity space map of restoring force for a bright circle on a dark background. Edges of all orientations are represented.

4   Conclusions

We have designed and built the first integrated motion detector using analog circuit elements and collective computation. We formulated the algorithm and designed the architecture to exhibit collective behavior through the aggregation of locally derived quantities. We used closely coupled analog photosensors and small analog computational elements making feasible the parallel operation of large arrays of sensors and computing elements. The system makes extensive use of local intensity information and therefore operates in the presence of global intensity gradients. This local computing also eliminates the need for any prior higher-level processing like edge detection or object recognition.

The motion detector architecture, along with local analog computational elements, provides a dense, reliable means of processing high bandwidth visual data. The motion detector chip demonstrates the suitability of analog VLSI circuits for processing of sensory data.

The motion detector is one of the first of a growing class of systems that employ collective computation. As we gain experience with this type of system, we should expand the range of application beyond that of processing sensory data. The motion detector proved to be a good first example. Although it has a crisp, solid mathematical foundation, the motion detector exhibits the collective behavior that is so necessary for processing sensory data from the fuzzy real world.

References

[1] R. F. Lyon, "The optical mouse, and an architectural methodology for smart digital sensors," *CMU Conference on VLSI Systems and Computations*, pp. 1–19, 1981.

[2] J. E. Tanner and C. Mead, "A correlating optical motion detector," *1984 MIT Conference on Very Large Scale Integration*, pp. 57–64, 1984.

[3] C. Mead, "A sensitive electronic photoreceptor," *1985 Chapel Hill Conference on VLSI*, pp. 463–471, 1985.

[4] B. K. P. Horn and B. G. Schunck, "Determining optical flow," *Artificial Intelligence*, vol. 17, pp. 185–203, 1981.

Part II

VLSI ALGORITHMS AND ARCHITECTURES

CHAPTER 8

ON THE ANALYSIS OF
SYNCHRONOUS COMPUTING ARRAYS[1]

J.M. Jover[2], T. Kailath, H. Lev-Ari and S.K. Rao[2]

Information Systems Laboratory, Stanford University
Stanford, California 94305

ABSTRACT: This paper is concerned with the analysis of synchronous, special purpose, multiple-processor systems (including, e.g., systolic arrays), i.e., the problem of determining the algorithm executed by the system. There has been some prior work in this area, especially by Melhem and Rheinboldt (1984), who were the first to obtain a general solution to the analysis problem. Our approach is different, combining ideas well known in system theory with certain graph-theoretical concepts. The key to our analysis method is the notion of *equivalence* between iterative algorithms. We show that the analysis problem amounts to finding a suitable equivalence transformation, and we formulate an efficient procedure for carrying out this tranformation.

1   Introduction

In this paper we are concerned with the *analysis problem* of determining the algorithm executed by a given synchronous, special-purpose, multiple-processor array. The problem arises because such arrays (or architectures) are often designed heuristically. Several formulations have been suggested in the computer science literature to solve a simpler related problem often called *verification*, in which one wants to check that a *given* algorithm is indeed implemented by the architecture. In the analysis problem we are given the topology of the network, the function performed by each processor (including timing information), and the input data streams, and wish to determine the algorithm performed by the array.

Previous work on verification is due to H.T. Kung and C.E. Leiserson (1979); M. Chen and C. Mead (1982); H. Lev-Ari (1983); H.T. Kung and W.T. Lin (1983); C.J. Kuo, B. Lévy, and B. Musicus (1984); and, E. Tidén (1984). The most general results, encompassing both verification and analysis, appear in a paper of R. Melhem and W.C. Rheinboldt (1984). The methods known so far tend to be somewhat involved and of limited generality. The main contribution of this paper is a new approach and a simple and general solution to the analysis problem, based on ideas from system theory.

The key to our approach is that we view the analysis problem as part of a cycle: starting from an algorithm, we design (or synthesize) a physical circuit; then we can complete the cycle (*i.e.*, solve the analysis problem) by properly retracing our steps to recover the original algorithm (or rather one that it is equivalent to it in an appropriate sense). The first step in this cycle is to represent a given iterative algorithm, i.e., a set of relations between

---

[1]This work was supported in part by the U.S. Army Research Office, under Contract DAAG29-83-K-0028, by the Air Force Office of Scientific Research, Air Force Systems Command under Contract AF-83-0228, by the Department of the Navy (NAVELEX) under Contract N00039-84-C-0211, by NASA Headquarters, Center for Aeronautics and Space Information Sciences (CASIS) under Grant NAGW-419, and by the Department of the Navy, Office of Naval Research, under Contract N00014-85-K-0612 (CORE).

[2]J.M. Jover and S.K. Rao are currently with AT&T Bell Laboratories, Holmdel, NJ 07733.

sequences of data, by a so-called *signal flow graph* (SFG), which shows interconnections between blocks that perform ideal mathematical operations (*i.e.*, take no time to compute). The next step in the cycle is to modify the chosen SFG to obtain a *logical circuit* (*i.e.*, a hardware implementation with physical modules that compute the same functions as the blocks in the SFG but in nonzero time and with some explicit delays). For the *analysis problem* we have to reverse the above path by modifying the logical circuit to obtain a SFG, and thereby an associated algorithm, equivalent to the one we started with.

Therefore to solve the analysis problem it is helpful to understand the design phase, which is our first object of attention in this paper. The whole cycle will be explored in some detail using a simple example from linear system theory; in fact this example was the one that helped us to understand the analysis problem in a context more familiar to us, since for linear systems the design and analysis problems are well understood and there are well established techniques, such as z-transforms and block-diagram-manipulations, to solve them (see, e.g., Kailath 1980).

2 Algorithms, SFGs, and Logical Circuits

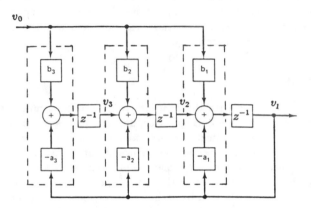

Figure 1: Observer canonical form (modified from Kailath (1980), p. 43). We define nodes $\{v_0, v_1, v_2, v_3\}$ as shown: one at the input point, and the others at the outputs of the z^{-1} blocks.

Consider the iterative expression

$$y(k) = b_1 u(k-1) - a_1 y(k-1) + b_2 u(k-2) - a_2 y(k-2) + b_3 u(k-3) - a_3 y(k-3) \quad (1)$$

which describes the relation between two sequences, $u(\bullet)$ and $y(\bullet)$, that constitute a so-called linear filter. This filter produces a sequence of output values $\{y(k)\}$, given, at each k, certain past values of $y(\bullet)$ and of an *input sequence* $u(\bullet)$. Representations of this algorithm using simple building blocks—adders, multipliers, and separators (or index-shifting blocks)—can be set up in *many ways* (see, e.g, Kailath, 1980, Ch. 2). One of these, the so-called observer form, is shown as a *signal flow graph* (SFG) in Figure 1, where we have used a convention (arising from the use of what are called z-transforms) common in system theory of labeling the separator blocks by the symbol z^{-1}.

Any signal-flow-graph is a *network of connected blocks*. The interconnecting wires propagate *sequences* of data elements, which we shall call *variables*. The points at which variables appear will be called *nodes*. Thus, for instance, the variable $x_1(k)$ denotes the sequence of data elements that appears (for $k = 0, 1, \ldots$) at the output (i.e., at the node v_1) of the linear filter in Figure 1. The *processors* (=blocks) of a SFG transform one or several input variables into a single output variable. In general, this transformation need not be linear. The set of all variables and all the transformations determined by the processors constitutes the *iterative algorithm* performed by the SFG.

Now, with the important convention that arithmetic operations are *instantaneous*, i.e., that the input and output quantities have the same indices, while the separators (or z^{-1} blocks) shift the indices by unity, we can write the following (so-called "state" [3]) equations

$$\begin{cases} x_1(k) & = b_1 u(k-1) - a_1 x_1(k-1) + x_2(k-1) \\ x_2(k) & = b_2 u(k-1) - a_2 x_1(k-1) + x_3(k-1) \\ x_3(k) & = b_3 u(k-1) - a_3 x_1(k-1) \\ y(k) & = x_1(k) \end{cases} \tag{2}$$

Notice that these state equations actually represent an *aggregated* SFG corresponding to the modules described by broken lines in Figure 1. Conversely, it is easy to see how to draw this aggregated SFG from the equations (2).

To summarize the above discussion, we can say that generally the first step in obtaining a physical implementation is to start with an input-output description and then to convert it, perhaps via the intermediate step of constructing some (aggregated) SFG representation, into an *iterative algorithm*, which is a set of equations of the form

$$\begin{aligned} x_1(k) & = f_1 \{x_1(k-s_{11}), x_2(k-s_{21}), \ldots, x_n(k-s_{n1})\} \\ x_2(k) & = f_2 \{x_1(k-s_{12}), x_2(k-s_{22}), \ldots, x_n(k-s_{n2})\} \\ & \vdots \\ x_n(k) & = f_n \{x_1(k-s_{1n}), x_2(k-s_{2n}), \ldots, x_n(k-s_{nn})\} \end{aligned} \tag{3}$$

where k is the index of iteration, and s_{ij} are known as the *index displacements*. We emphasize that this conversion procedure is highly nonunique: there are many algorithms that can implement a given input-output map. However, there is a one-to-one correspondence between equations (3) and the corresponding (aggregated) signal-flow graph.

SFGs are not truly "physical" implementations of mathematical algorithms such as (2) or (3), because in any physical hardware implementation, the arithmetic operations will *not* be instantaneous. One way to accommodate these physical constraints (and to interpret the SFG as a physical system) is by taking the iteration interval (i.e., the physical time separation between sequence elements) to be very large, so that the arithmetic operations in each computing module will all be completed before the next iteration begins, i.e., before the next data sample is entered into the system. A more efficient procedure, likely to result in smaller iteration intervals, is to determine a "schedule" of the times at which each operation should be performed, as explained next.

We shall confine ourselves to digital implementations, in which we have an underlying clock, whose period will be taken as the basic time unit. Then the time required for additions and multiplications (or other arithmetic operations) will be measured as integral multiples of clock cycles. We shall not concern ourselves with the details of what happens *within* any particular clock cycle.

---

[3]The values $\{x_1(k), x_2(k), x_3(k)\}$ describe the "state" of the system at time k, in the sense that knowing them and $\{u(l), l \geq k\}$ we can compute $\{y(l), l \geq k\}$ irrespective of the prior values of the $x_i(\bullet)$, *i.e.*, of $\{x_1(j), x_2(j), x_3(k), j < k\}$.

The main goal of the scheduling procedure is to determine an appropriate *iteration interval*, i.e., the physical time (measured in clock cycles) between two consecutive data at any point in the system (this will be the same at all points in a synchronous system), and any additional delays required, called *shimming delays*, that may have to be added to the processing and transmission delays of the system to ensure that the proper elements in the various sequences are interacting correctly.

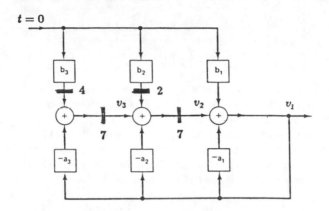

Figure 2: Logical circuit for the observer canonical form.

Several algorithms for scheduling have been developed. Here we use the ideas of Jagadish *et al.*, 1986 (see also Jagadish, 1985, and Rao, 1985) to determine a scheduling for the observer canonical form. Suppose that multiplication (and data transfer) takes 7 clock cycles, addition (and transfer) take 3 clock cycles, and a pure transfer of data along an interconnecting wire takes 1 clock cycle. Applying the scheduling procedure of Jagadish *et al.* results in a physical implementation of the observer canonical form as shown in Figure 2, in which the blocks represent hardware components with a computational delay as assumed above, and with additional (shimming) delays whose value (in multiples of clock cycles) is indicated next to them. Digital designers usually call such a figure a logical (circuit) diagram.[4] It should be noted that the scheduling procedure is highly nonunique and, therefore, that several different logical circuit diagrams can be associated with a given SFG. We shall elaborate on this nonuniqueness in the following section.

3   Logical graphs and algorithm graphs

For many purposes, especially timing analysis, it is convenient to redraw the logical circuit diagram as what we shall call a logical graph G, see Figure 3. To draw G, we put down a vertex of the graph for each node of the logical circuit, and connect the vertices with edges that represent the directed paths between the nodes of the logical circuit. With an edge from v_i to v_j we associate a positive weight d_{ij} corresponding to the total computational

---

[4]The adjective 'logical' arises from the fact that the hardware is based on so-called 'logical' components obeying the rules of Boolean logic (algebra).

and propagation delay for that path. For example, from v_2 to v_1 we have an edge with weight $d_{21} = 3$, a self loop from v_1 to itself with weight $d_{11} = 1 + 7 + 3 = 11$, a path from v_1 to v_2 with weight $d_{12} = 1 + 1 + 7 + 3 + 7 = 19$, and so on. With each vertex v_i, we associate a sequence $\{x_i(k)\}$. For example, for the observer form we can write the function performed at each vertex as follows

$$
\begin{cases}
x_1() & = b_1 u() - a_1 x_1() + x_2() \\
x_2() & = b_2 u() - a_2 x_1() + x_1() \\
x_3() & = b_3 u() - a_3 x_1() \\
y() & = x_1().
\end{cases}
\tag{4}
$$

Note that we cannot write directly from the logical graph G the actual index dependences as we did in eq. (2) from the SFG. Finding these index dependencies (i.e., the index displacements s_{ij} in (3)) is the heart of the analysis problem.

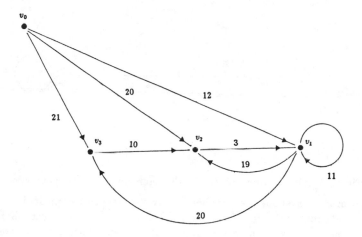

Figure 3: Logical graph G for the aggregated observer canonical form.

3.1 Algorithm graphs

To facilitate the reconstruction of an iterative algorithm from a logical graph G of some physical implementation, we have to obtain a representation of iterative algorithms that is similar in form to the logical graph. For this purpose we introduce a so called *algorithm graph* G^*: this, like G, has a vertex for each variable in the algorithm, and its edges represent the index dependencies, i.e., the weight of the directed edge connecting v_i (the vertex representing $x_i()$) to v_j equals the non-negative index displacement s_{ij}. Thus, *the algorithm graph is a precise image of the iterative algorithm* (3), i.e., there exists a one-to-one correspondence between iterative algorithms and their graphs. The difference between the algorithm graph G^* and the SFG is that in G^* we focus only on the index dependencies, ignoring the details of the actual functional computations. As stated before, it is the determination of a correct set of index dependencies from the timing information in the logical graph that is the key to the solution of the analysis problem.

3.2   Relating G and $G^\star$

The logical graph G and the algorithm graph $G^\star$ have the same topology; the only difference is that in $G^\star$ the edge weights represent the number of separators in the path while in G they represent physical delay. For instance, compare the logical graph for the observer canonical form (Figure 3) with the corresponding algorithm graph (Figure 4). Moreover, there is a simple relation between the index displacement s_{ij} (=number of separators) associated with an edge in $G^\star$ and the physical delay d_{ij} associated with an edge in G. To make this relation explicit we need to analyze the way synchronous systems work.

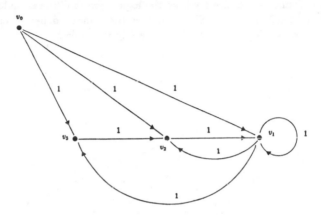

Figure 4: Algorithm graph $G^\star$ for the aggregated observer canonical form.

In synchronous systems the time between two consecutive elements in any sequence is constant (and equal to the iteration interval, δ). If in addition we assume that such systems are time-invariant, as we do in logical graphs, all the computations (at the vertices) involve data arriving at some multiple of the iteration interval. Consider, for example, the graph G for the observer form (Figure 3). Apply the first element, $u(1)$, of an input sequence $u(\bullet)$ at time $\lambda_0 = 0$; by definition, the rest of the elements will be generated every δ clock cycles. Let us denote by λ_i the time instant at which the vertex v_i generates the output $x_i(1)$.

Recall that d_{ij} (resp. s_{ij}) is the physical delay in the logical graph G (resp. the index displacement in the algorithm graph $G^\star$) along a directed path connecting the vertex v_i to the vertex v_j. Since v_i generates $x_i(1)$ at time λ_i, $x_i(1)$ arrives at vertex v_j at time $\lambda_i + d_{ij}$. However, we cannot equate λ_j to $\lambda_i + \delta_{ij}$, because λ_j will depend upon the number of separators in the path from v_i to v_j, which in turn will fix the actual iteration in which the input $x_i(1)$ is operated on at vertex v_j. In our case, the path from v_i to v_j has s_{ij} separators and therefore, the vertex v_j will associate the input $x_i(1)$ with the $(1 + s_{ij})$th iteration rather than with the first iteration. Consequently, v_j will generate $x_j(1)$ at time $\lambda_j \equiv \lambda_i + d_{ij} - s_{ij}\delta$ where δ denotes the iteration interval. It follows that, for *every path* from v_i to v_j,

$$d_{ij} = (\lambda_j - \lambda_i) + s_{ij}\delta \qquad (5)$$

This is the basic equation that must be satisfied by every logical graph G that implements a given algorithm graph $G^\star$.

An important consequence of the basic equation is that for every (directed) loop the total computational delay is a multiple of the iteration interval, i.e., $v_i = v_j$ implies that $d_{ii} = s_{ii}\delta$. Thus, if we define δ_{max} as the *greatest common divisor* (gcd) of the computational delays around all loops, we conclude that the actual iteration interval δ must be a divisor of δ_{max}. Moreover, since every loop in a graph is a combination of *fundamental loops* (i.e., loops determined by a spanning tree), δ_{max} can be computed as the gcd of the computational delays around fundamental loops alone.

Notice that the basic equation (5) also implies that an algorithm graph G^* has an implementation G only if it is *computable*, i.e., the sum of index displacements around each directed loop must be strictly positive.

After this discussion, the analysis problem can now be restated as follows: given a logic graph G and the λ_i associated with the input vertices, find an algorithm graph G^* and a set of $\{\lambda_i\}$ that satisfy (5).

4   An analysis procedure

We now present an analysis procedure that constructs a set of λ_i and an algorithm graph G^* with *nonnegative* s_{ij} for any given logical graph G. First we form a *rooted tree* consisting of the shortest paths from the input vertex v_0 to the remaining vertices of G. If there is more than one source (input) vertex we first extend the graph by adding a vertex v_0 with an edge from v_0 to each of the source vertices, and then construct a shortest path rooted tree. Next we set $s_{ij} = 0$ for the edges contained in the rooted tree. This determines λ_i for all vertices and we can use equation (5) to compute the index displacements s_{ij} for the edges that are not in the rooted tree.

Our procedure can be formally summarized as follows:

Procedure 1 *Given a logical graph G, an iteration interval delta and the time λ_i at which the first element of each input sequence $\{u_i\}$ is entered, determine an algorithm graph G^* such that G is an implementation of G^*. The procedure consists of the following steps:*

1. *Extend the graph G by adding a vertex v_0 with an edge from v_0 to each of the source vertices v_i, and with weight λ_i. Call this graph G_{ext}.*

2. *Form a spanning tree with the minimum-delay path from v_0 to every vertex in G_{ext}. For every vertex v_j let $\lambda_j = \lambda_i + d_{ij}$ where d_{ij} is the delay on the edge that connects to v_j its closest vertex, v_i.*

3. *For each link e_{ij} in the tree formed above compute*
$$d_{ij}^* = \lambda_i - \lambda_j + d_{ij}.$$
For each self-loop with weight d_{ii} assign
$$d_{ii}^* = d_{ii}.$$

4. *Compute $\delta_{max} = gcd\{d_{ij}^*\ \ \forall i,j\}$. If δ is not a divisor of δ_{max} then the given δ is not valid for the graph G.*

5. *Associate a weight $s_{ij} \geq 0$ with each link e_{ij} in the tree formed in step 2*
$$s_{ij} = d_{ij}^*/\delta_{max}.$$
Associate zero weight, $s_{ij} = 0$, to the edges in the tree.

There are several algorithms for implementing step 2 of the procedure (shortest-path tree), such as those in Dantzig (1975). For a comprehensive presentation of these algorithms see, *e.g.*, Even (1979).

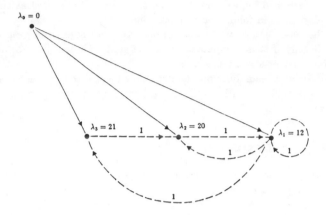

Figure 5: Shortest-path tree and the corresponding algorithm graph for the aggregated observer canonical form.

As a simple example, consider the observer form of Section 1. Notice that an extension (step 1 of our procedure) is not required, because $\lambda_0 = 0$. The corresponding shortest-path tree and the resulting index displacement are described in Figure 5. From this graph we can easily write the following state equations

$$\left\{ \begin{array}{ll} x_1(k) & = b_1 u(k) - a_1 x_1(k-1) + x_2(k-1) \\ x_2(k) & = b_2 u(k) - a_2 x_1(k-1) + x_1(k-1) \\ x_3(k) & = b_3 u(k) - a_3 x_1(k-1). \end{array} \right. \tag{6}$$

We now observe that the graphs G^* in Figures 4 and 5 are different, and consequently that the resulting algorithms (eqs. (2) and (6)) are *apparently* different. However, the difference in the algorithms in eqs. (2) and (6) amounts only to shifts in the indices (we can change the indices for the sequence $u(\bullet)$ in eq. (6) from k to $k-1$ and obtain eq. (2)) and therefore both algorithms are *equivalent* in the sense that they determine the same set of recursions.

This observation holds for every algorithm graph G^* and for arbitrary shifts in the indices of data sequences. In other words, if we redefine $\tilde{x}_i(k) := x_i(k - c_i)$ and replace $x_i(\cdot)$ by $\tilde{x}_i(\cdot)$ in the algorithm graph, the computations remain unchanged, even though the index displacements become modified, i.e., s_{ij} is replaced by

$$\tilde{s}_{ij} := s_{ij} + c_i - c_j \tag{7}$$

We shall say that two algorithm graphs are *equivalent* if they are isomorphic (in the graph theoretic sense) and if their index displacement can be related by a set of shifts c_i, as described in (7).

Our analysis procedure recovers an algorithm that is equivalent to the one used in the design of the given logical graph G^*. To prove this notice that the algorithm G^* recovered by our procedure satisfies the equation $\lambda_j = \lambda_i + d_{ij} - s_{ij}\delta$, while the original algorithm $\tilde{G}^*$, used to design the logical circuit G, satisfies the equation $\tilde{\lambda}_j = \tilde{\lambda}_i + d_{ij} - \tilde{s}_{ij}\delta$. Assuming all graphs in consideration have been extended, we can assume for the root vertex v_0 that $\lambda_0 = 0 = \tilde{\lambda}_0$. Consequently, $\tilde{\lambda}_i - \lambda_i = (s_{i0} - \tilde{s}_{i0})\delta$ so that we can replace $\tilde{\lambda}_i$ by $\lambda_i + c_i\delta$, where $c_i := s_{i0} - \tilde{s}_{i0}$, and we conclude that

$$\tilde{s}_{ij}\delta = \tilde{\lambda}_i - \tilde{\lambda}_j + d_{ij} = \lambda_i - \lambda_j + d_{ij} + (c_i - c_j)\delta = (s_{ij} + c_i - c_j)\delta$$

which establishes the equivalence between G^* and $\tilde{G}^*$

It is also easy to verify that the index displacements generated by Procedure 1 are positive and integer. To prove that $s_{ij} \geq 0$ note that s_{ij} will be nonnegative if, and only if, $d_{ij}^* \geq 0$. Since we have formed a shortest-path tree from v_0, it means that $\lambda_i + d_{ij} \geq \lambda_j$ and, therefore, that $d_{ij}^* = \lambda_i - \lambda_j + d_{ij} \geq 0$. Thus, $s_{ij} \geq 0$. Since δ_{max} is a divisor of every d_{ij}^*, it is obvious that s_{ij}, as computed in Procedure 1, will also be integer.

5   Concluding Remarks

We have given a simple procedure for recovering (within a natural equivalence) the iterative algorithm executed by a given special purpose synchronous computing array. The solution is based on reversing (modulo equivalence) the process by which one can translate an iterative algorithm into a logical circuit. A general theory for such conversion has recently been developed by S.K. Rao (1985) and H.V. Jagadish (1985) in their work on the analysis and synthesis of what they call *Regular Iterative Arrays* (RIAs). The synchronous circuits studied in this paper can be identified as a special class of RIAs (with a one-dimensional index space). Therefore, the algebraic techniques developed, in particular by Rao (1985) can be applied to generalize the results of this paper to other classes of RIAs (with multi-dimensional index spaces). In particular, regularity of spatial structure, as in systolic arrays, can be exploited to reduce the study of such systems to that of a single module (see, e.g., Jover (1985)).

Our analysis procedure is restricted to logical graphs in which every vertex is reachable from at least one source (input) vertex. It can be shown that logical graph not possessing this property can be modified by adding edges from the root v_0 to some of the unreachable vertices, so that subsequently our analysis procedure can be applied (see Jover (1985)).

Acknowledgements

We wish to thank Sandra M. Young for preparing the illustrations.

References

1. M.C. Chen and C.A. Mead, "Concurrent Algorithms as Space-Time Recursion Equations," *Proceedings of the USC Workshop on VLSI and Modern Signal Processing*, Los Angeles, November 1982. Also in *VLSI and Modern Signal Processing*, S.Y. Kung, H.J. Whitehouse, and T. Kailath, ed., Englewood Cliffs: Prentice Hall, 1985.

2. G.B. Dantzig, "On the shortest route through a network," in *Studies in Graph Theory*, D.R. Fulkerson, ed., MAA Studies in Mathematics, Volume 11. MAA, 1975.

3. S. Even, *Graph Algorithms*, Rockville, MD.: Computer Science Press, 1979.

4. D.E. Heller and I.C.F. Ipsen, "Systolic Networks for Orthogonal Equivalence Decompositions," *SIAM Journal on Statistical and Scientific Computing*, Vol. 4, No. 2, pp. 261–269, June 1983.

5. H.V. Jagadish, R.G. Mathews, T. Kailath, and J.A. Newkirk, "A Study of Pipelining in Computing Arrays," to appear in *IEEE Transactions on Computers*, 1986.

6. H.V. Jagadish, "Techniques for the Design of Parallel and Pipelined VLSI Systems for Numerical Computation," Ph.D. dissertation, Department of Electrical Engineering, Stanford University, December 1985.

7. J.M. Jover, "On the Modeling and Analysis of Systolic and Systolic-Type Arrays," Ph.D. dissertation, Department of Electrical Engineering, Stanford University, December 1985.

8. T. Kailath, *Linear Systems*, Englewood Cliffs: Prentice Hall, 1980.

9. D.E. Knuth, *The Art of Computer Programming*, Volume 3: Sorting and Searching. Reading, Mass.: Addison-Wesley, 1973.

10. H.T. Kung and C.E. Leiserson, "Systolic arrays (for VLSI)," *Sparse Matrix Proceedings 1978*, I.S. Duff and G.W. Stewart, ed., Society for Industrial and Applied Mathematics, pp. 256–282, 1979.

11. H.T. Kung and W.T. Lin, "An Algebra for VLSI Algorithm Design," *Proceedings of the Conference on Elliptic Problem Solvers*, Monterey, California, January 1983.

12. C.J. Kuo, B.C. Levy, B.R. Musicus, "The Specification and Verification of Systolic Wave Algorithms," *1984 USC Workshop on VLSI Signal Processing*, Los Angeles, California, November 1984.

13. H.W. Lang, M. Schimmler, H. Schmeck, and H. Schröder, "Systolic Sorting on a Mesh-Connected Network." *IEEE Transactions on Computers*, vol. C–14, No. 7, pp. 652–658, July 1985.

14. H. Lev-Ari, "Modular Computing Networks: A New Methodology for Analysis and Design of Parallel Algorithms/Architectures," Integrated Systems Inc., Report #29, Palo Alto, California, December 1983.

15. R. Melhem and W.C. Rheinboldt, "A Mathematical Model for the Verification of Systolic Arrays," *SIAM Journal on Computing*, vol. 13, No. 3, pp. 541–565, Aug. 1984.

16. S.K. Rao, "Regular Iterative Algorithms and Their Implementations on Processor Arrays," Ph.D. dissertation, Department of Electrical Engineering, Stanford University, October 1985.

17. R. Schreiber, "On Systolic Arrays Methods for Band Matrix Factorizations," Department of Numerical Analysis and Computing Science, The Royal Institute of Technology (Sweden), Technical Report TRITA–NA–8316, 1983.

18. E. Tidén, "Verification of Systolic Arrays—A Case Study," Department of Numerical Analysis and Computing Science, The Royal Institute of Technology (Sweden), Technical Report TRITA–NA–8403, 1984.

Appendix: EXAMPLES

Back Substitution

This array was developed by Kung and Leiserson (1979) and also verified by Melhem and Rheinboldt (1984). Figure 6 depicts the system, the processors' functions, the graph G, and a shortest-path tree. This time, we have associated a computational delay of 10 for each of the edges in the graph G; this choice corresponds to the usual one for systolic arrays: all the outputs to a cell are produced at the same time even if some of the outputs may take less time to compute.

The input sequences are as follows: zeros at the vertex v_6, the elements of a column vector, b, at vertex v_1, and the diagonals of a banded, lower-triangular matrix, A at vertices v_2–v_5 (the main diagonal at v_2, the first subdiagonal at v_3, and so on). The input sequences are as follows:

$$\left\{ \begin{array}{l} u_1 = \{b_1, b_2, b_3, \ldots\} \\ u_2 = \{a_{11}, a_{22}, a_{33}, \ldots\} \\ u_3 = \{a_{21}, a_{32}, a_{43}, \ldots\} \\ u_4 = \{a_{31}, a_{42}, a_{53}, \ldots\} \\ u_5 = \{a_{41}, a_{52}, a_{63}, \ldots\} \end{array} \right.$$

We are also given the times λ_i, at which the first input of each sequence is entered; they are as follows

$$\lambda_1 = \lambda_2 = 30, \quad \lambda_3 = 40, \quad \lambda_4 = 50, \quad \lambda_5 = 60, \quad \lambda_6 = 0.$$

These times correspond to the weights in the edges connecting the root (not shown) and the input vertices.

The output is at vertex v_{13}; we can write it as

$$y_{13} = \{y(1), y(2), y(3), \ldots\}$$

Figure 6(d) shows the states in addition to the shortest path. We computed them using $\delta_{max} = gcd\{20, 40, 60\} = 20$. From this figure we can write the following equations

$$\left\{ \begin{array}{l} x_7(k) = x_{12}(k-3) \cdot u_5(k-3) + u_6(k) \\ x_8(k) = x_{11}(k-2) \cdot u_4(k-2) + x_7(k) \\ x_9(k) = x_{10}(k-1) \cdot u_3(k-1) + x_8(k) \end{array} \right.$$

$$\left\{ \begin{array}{l} x_{10}(k) = (u_1(k) - x_9(k))/u_2(k) \\ x_{11}(k) = x_{10}(k) \\ x_{12}(k) = x_{11}(k) \\ y_{13}(k) = x_{12}(k) \end{array} \right.$$

Substituting these equations we get

$$y_{13}(k) = (u_1(k) - x_9(k))/u_2(k)$$

where

$$x_9(k) = y(k-1) \cdot u_3(k-1) + y(k-2) \cdot u_4(k-2) + y(k-3) \cdot u_5(k-3) + u_6(k)$$

Substituting now into the input and output sequences we obtain

$$y_{13}(k) = (b_k - x_9(k))/a_{kk}$$

where

$$x_9(k) = y(k-1) \cdot a_{k,k-1} + y(k-2) \cdot a_{k,k-2} + y(k-3) \cdot a_{k,k-3}$$

These equations are the well known method of solving a triangular system, called back substitution. Therefore, the array discussed performs back substitution.

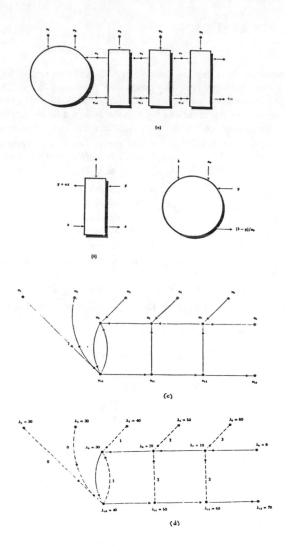

Figure 6: Back substitution array: (a) system, (b) processors' functions, (c) graph G, and (d) a shortest-path tree.

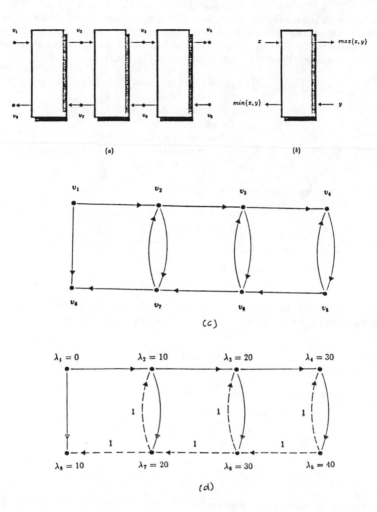

Figure 7: Sorting array: (a) system, (b) processors' functions, (c) graph G, and (d) the shortest-path tree.

Sorting

This array was developed by H.T. Kung and first reported and verified by Melhem and Rheinboldt (1984). The system sorts a sequence of n real numbers, $u_1 = \{u_1(1), u_1(2), \ldots, u_1(k)\}$ by using the linear array of $n-1$ processors depicted in Figure 7. The output sequence, $y_8 = \{y_1(1), y_1(2), \ldots, y_1(k)\}$ is sorted in ascending order. In this example, we assume all the computational delays to be equal and of value 10 since all the operations are well balanced and should take the same amount of time. The iteration interval given is 20. We have only one source, so we take it directly as the root with $\lambda_1 = 0$. Figure 7(c) and (d) show the graph and the shortest-path tree (in this case is unique). From it we can readily write the equations

$$\begin{cases} x_2(k) = max\{u_1(k), x_7(k-1)\} \\ x_3(k) = max\{x_2(k), x_6(k-1)\} \\ x_4(k) = max\{x_3(k), x_5(k-1)\} \\ x_5(k) = x_4(k) \\ x_6(k) = min\{x_3(k), x_5(k-1)\} \\ x_7(k) = min\{x_2(k), x_6(k-1)\} \\ y_8(k) = min\{u_1(k), x_7(k-1)\} \end{cases}$$

which can be rearranged as follows, after substituting $x_5(k) = x_4(k)$

$$\begin{cases} x_2(k) = max\{u_1(k), x_7(k-1)\} \\ y_8(k) = min\{u_1(k), x_7(k-1)\} \\ x_3(k) = max\{x_2(k), x_6(k-1)\} \\ x_7(k) = min\{x_2(k), x_6(k-1)\} \\ x_4(k) = max\{x_3(k), x_4(k-1)\} \\ x_6(k) = min\{x_3(k), x_4(k-1)\} \end{cases}$$

These equations correspond to the so-called bubble sort (see Knuth, 1973). Other types of sorting algorithms could be implemented, see, for instance Rao (1985) and Lang et al. (1985).

CHAPTER 9

Designing Systolic Arrays with DIASTOL(*)

Patrice FRISON, Pierrick GACHET, Patrice QUINTON

IRISA, Campus de Beaulieu,
35042 RENNES-Cedex
FRANCE

1. Introduction

These last few years, there has been a fastly growing interest for systolic arrays as a particular means of implementing special-purpose VLSI architectures. Domains that could most benefit from this kind of organization are signal processing and numerical analysis [1].

The DIASTOL system we present here is based on the formal method described by Quinton [2], [3]. DIASTOL is a system whose purpose is to allow systolic architectures to be designed quickly and reliably. The design process starts from the equations of the algorithm to be implemented. These equations are entered into the system and transformed until they become so-called *uniform recurrent equations* (URE). Then the synthesis program helps the designer building various systolic arrays by using the *dependence mapping* procedure. This results in an abstract (draft) description of the design, comprising the number of cells, the topology of the array, its connectivity, the timing of the data movements, and the cell fonctionnality. This abstract specification can be used as a starting point for the functional design of the cells. Depending on the type of the calculations performed by the elementary cells, DIASTOL offers several design styles based either on parallel pipelined, parallel skewed or bit-serial hardware operators. The hardware design process consists in associating consistently operators with the function symbols of the URE. The detailed timing of the design is then automatically found from the characteristics of these operators, by using a refinement of the dependence mapping procedure. A library of basic operators makes its possible to obtain quickly a functional description of a chip implementing the algorithm. The goal that is pursued is to attain designs that are specified at the bit level, as described for example by McCanny et al.[4].

This paper gives an informal overview of the operation of DIASTOL. A more formal treatment can be found in [2], [3] and [5]. In the second section, we present a simplified model of hardware operators that allows several common design styles to be represented. In section three, the principles of the dependence mapping procedure are briefly recalled and exemplified

(*) This work is partially supported by the french Research Co-ordinated Program C^3.

on the matrix multiplication algorithm. The extension of this method to the detailed design of the cells is also presented. In section four, exemples of designs for the matrix multiplication are developed.

2. A Simple Model of Hardware Operators

Our goal in this section is to give an abstract model of hardware operators, so that the timing characteristics of parallel, parallel skewed or bit-serial operators can be described within a single framework.

2.1 Presentation of the model

For the purpose of this paper, we modelize an hardware operator as a "black box" Op (figure 1) having several input ports I_i and one output O (multi-output operators are not considered here for the sake of simplicity). Each input (or output) port has $n(I_i)$ (respectively $n(O)$) bits. Associated with the operator is a function symbol f(Op) and several attributes that describe its timing behaviour. We assume that this timing is expressed relative to a reference time, in term of discrete time steps such that clock cycles or phases. Each operator is characterized by its type, i.e. *parallel* or *bit-serial*, and a set of timing paramaters. The *period* of the operator, denoted as $\psi(Op)$ is the time that must separate the input (resp. the output) of successive data entering the operator. The *skew* of the operator, denoted as $\sigma(Op)$ is the time that separates successive bits of the input or output data. When an operator is skewed, we assume that the data have the same skew, and that they enter the operator *least significant bit first*. Finally, to each input or output port P of the operator, is attached an *offset* denoted as $\omega(P)$, that defines at what time the first bit of the input is entered or delivered. In figure 2, several operators are schematized. Operator a) has skew 0 and period 1. The offset of its inputs are 0, and the offset of its output is 2. This represents a two stage pipeline parallel operator, with *latency* 2. Operator b) is a parallel skewed operator with the same input and output as in operator a). The bit skew is 1. Such an operator is useful in VLSI implementation in order to break delays due to the propagation of long signals (such as carries for example). Finally, operator c) is a bit-serial four bit input output operator. Its period is 4. The offset of the second input is 1, which indicates that the second input has to be delayed by one cycle relative to the first one. Notice that there may exist relationships between the parameters of an operator: for example, the period of a bit-serial operator must be a multiple of its skew.

2.2 Examples of operators

For the purpose of this paper, we shall consider a particular case of bit-serial operators. A well

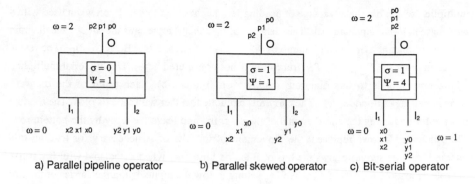

Figure 1 - Examples of operators

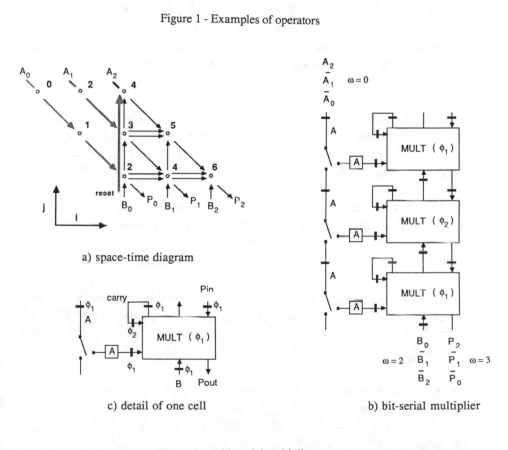

Figure 2 - A bit-serial multiplier

known bit-serial multiplier has been described by Lyon[6]. We describe here a similar multiplier, which has the advantage of needing less latches than that of Lyon when used with a two phase non-overlapping clock. The diagram of figure 2(a) represents the time-space diagram of the multiplication of two n bit numbers $A = A_{n-1} \cdots A_0$ and $B = B_{n-1} \cdots B_0$. The result is a 2n number $P = P_{2n-1} \cdots P_0$ (truncation is not considered here). The bit-serial multiplier which corresponds to this diagram is shown on figure 2(b). Each point of the diagram corresponds to a full-adder cell. The carry bits circulate together with the bits of A. The product bits are formed along the anti-diagonal of the diagram, and leave the array by the bottom row. The shadowed arrows represent the movement of the bits of A necessary for loading this operand. A reset signal follows the bits of B and is high only when associated with B_0. It has the role of loading the bits of A, as well as setting to low the inputs P and the carries of the full adders of the line i=0. The multiplier contains n full-adder cells that work on opposite clock phases. For example, cell 0 is clocked on ϕ_1 whereas cell 1 is clocked on ϕ_2. The partial product bits flow from the top to the bottom, as well as the bits of A. The bits of B and the reset bits flow from the bottom to the top. The exact timing of this operator is best described in figure 2(a). One can see that the operands have skew 2. The period of the operator is 2n. Finally, the offset of A, B and P are respectively 0, n-1 and n. The main difference between this design and that of Lyon is that the B bits and the reset bits do not need being latched during one full clock cycle between consecutive full adder cells, which saves four latches for each cell. On the other hand, it is fair to recognize that Lyon's design has a simpler timing, as A and B would have the same offset. Notice that our design can be extended easily to handle two's complement binary numbers by using Booth recoding algorithm.

A straightforward bit-serial adder for two 2n bit binary numbers is shown in figure 3. It has period 2n and skew 2. The offset of the inputs is 0, and the offset of the output is 1.

3. Detailed design of systolic arrays using dependence mapping

In this section, we describe the principles of the DIASTOL system, and we show how the basic method can be extended to handle the design of the cells.

3.1. Introduction

Consider the matrix multiplication of two NxN matrices C=AB. The coefficients of C are given by the following equation :

$$\forall\ i,j : 1 \leq i \leq N,\ 1 \leq j \leq N,\ c_{ij} = \sum_{k=1}^{N} a_{ik} b_{kj} \qquad (1)$$

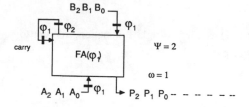

Figure 3 - A bit-serial adder

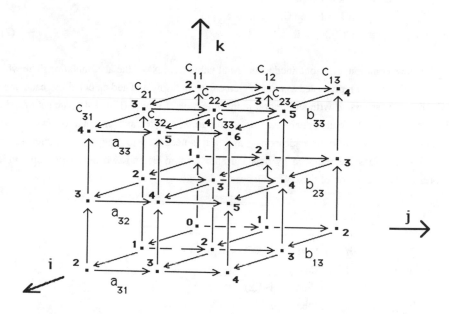

Figure 4 - Dependence graph for the matrix multiplication

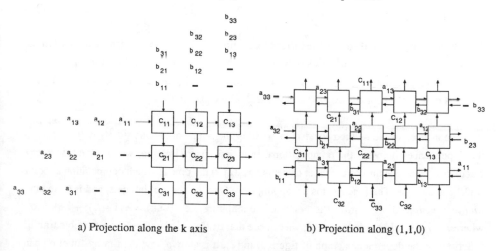

a) Projection along the k axis

b) Projection along (1,1,0)

Figure 5 - Two different systolic arrays for matrix multiplication

The fist step of the method consists in rewriting (1) as a left-to-right summation, in order to avoid the use of unbounded arity operators. This gives

$$\forall \ i,j,k : 1 \leq i \leq N, \ 1 \leq j \leq N, \ 1 \leq k \leq N,$$

$$C(i,j,k) = \ \text{if } k = 1 \text{ then } 0$$
$$\text{if } k \neq 1 \text{ then } C(i,j,k-1) + a_{ik}b_{kj}$$
$$c_{ij} \equiv C(i,j,N) \tag{2}$$

In this new equation, c_{ij} is obtained as the final value $C(i,j,N)$ of the accumulation. The whole computation can be represented as a set of elementary multiply and add calculations, each one of which being associated with an integral point of a cube of size N (figure 4). The set of points that are considered is called the *domain of computation,* denoted as D in what follows. In order to obtain a design without broadcasting a and b, it is necessary to pipeline the values a_{ik} and b_{kj}. These values are replaced by new variables A and B, and two new equations which represent the circulation of these variables are added, i.e. :

$$\forall \ i,j,k : 1 \leq i \leq N, \ 1 \leq j \leq N, \ 1 \leq k \leq N,$$

$$C(i,j,k) - \ \text{if } k = 1 \text{ then } 0$$
$$\text{if } k \neq 1 \text{ then } C(i,j,k-1) + A(i,j,k) \times B(i,j,k)$$
$$A(i,j,k) = \ \text{if } j = 1 \text{ then } a_{ik}$$
$$\text{if } j \neq 1 \text{ then } A(i,j-1,k)$$
$$B(i,j,k) = \ \text{if } i = 1 \text{ then } b_{kj}$$
$$\text{if } j \neq 1 \text{ then } B(i-1,j,k)$$
$$c_{ij} \equiv C(i,j,N) \tag{3}$$

Equation (3) is a special form of recurrence equations called *uniform recurrence equation* (URE) [7] as a computation depends only of neighbouring points, in a uniform manner. In our case, the computation at point (i,j,k) depends on points $(i,j,k-1)$, $(i,j-1,k)$, and $(i-1,j,k)$. Figure 4 depicts the *dependence graph* associated with equation (3).

The second step of the method consists in ordering the points of D consistently with respect to the dependences between the computations. In a first approximation, we assume that every calculation takes one unit of time. Let us denote as $t(i,j,k)$ the time at which computation at point (i,j,k) can be done. The function t is a mapping from D to the set Z of integers, and is called the *timing function*. If we restrict t to be an affine mapping, i.e. $t(i,j,k) = \lambda_1 i + \lambda_2 j + \lambda_3 k + \alpha$, where $\lambda_1, \lambda_2, \lambda_3$, and α belong to Z, we can see that $t(i,j,k) = i + j + k - 3$ is a possible timing function for the dependence graph of figure 4. In [2], it is shown how the timing-function can be computed automatically.

The final step of the method consists in mapping the dependence graph on a finite set of processing elements. We call *allocation function*, and denote as a(i,j,k) the mapping from Z^3 to itself that define the number of the processor where computation at point (i,j,k) is done. The simplest way is to use a projection of the space R^3. Depending of the direction of this projection, data will move or not in the final design. Consider first the projection of the cube of figure 4 along the k axis. The corresponding design is depicted in figure 5a). A given processor P_{ij} carries out the computations associated with points of segments parallel to the k axis. Therefore, c_{ij} is computed on P_{ij}. The coefficients of matrix A move from the left to the right, and those of B from the top to the bottom. Other well-known designs for the matrix multiplication can be obtained by using other directions of projection. Figure 5(b) shows a design obtained using the projection along vector (1,1,0). In this design, we can see that the c_{ij}'s move from bottom to top, whereas the a_{ik}'s and b_{kj}'s flow respectively from left to right and right to left. Notice that in this design, the cells are working only every other cycle. This is clearly seen on the dependence graph, as computations mapped on the same processor are separated by two cycles instead of one as in the design of figure 5(a).

3.3. Generalization of the method

The previous section described informally the dependence mapping procedure. This method can be use to produce quickly "drafts" of systolic arrays. In this section, we shall deal with the detailed design of the cells, taking into account the hardware technological aspects of the design of a chip. The approach taken in DIASTOL makes use of a set of predefined hardware operators, which are modelized using the parameters defined in section 2. Such operators are used to replace corresponding function symbols in the equations. The timing and allocation functions that have been defined in section 3.1. are generalized in order to allow the delays introduced by the operators as well as the timing of the data (at the bit level) to be computed automatically. Depending on the complexity of the functions in the URE, several styles of operators can be chosen. Here, we shall consider only fully parallel or bit-serial operators, as those described in section 2. Other examples can be found in [5].

Let us return to the URE of equation (3). In order to separate the multiply and plus functions defining C(i,j,k), we split the corresponding equation by adding a new variable P. This gives :

$\forall$ i,j,k : 1≤i≤N, 1≤j≤N, 1≤k≤N,

$$C(i,j,k) = \quad \text{if } k = 1 \text{ then } 0$$
$$\text{if } k \neq 1 \text{ then } C(i,j,k-1) + P(i,j,k)$$
$$P(i,j,k) = \quad A(i,j,k) \times B(i,j,k)$$

$$A(i,j,k) \;=\; \text{if } j = 1 \text{ then } a_{ik}$$
$$\text{if } j \neq 1 \text{ then } A(i,j-1,k)$$
$$B(i,j,k) \;=\; \text{if } i = 1 \text{ then } b_{kj}$$
$$\text{if } j \neq 1 \text{ then } B(i-1,j,k)$$
$$c_{ij} \;\equiv\; C(i,j,N) \tag{4}$$

We now attach one hardware operator to each equation (except of course to the $\equiv$ statement), in such a way that the timing parameters of all the operators are defined relative to the same unit, and that all have the same skew σ. To an identity equation, we attach a delay in order to avoid the possible broadcasting of a data. A delay has obviously a period 1, and the offset of its input and ouptut are respectively 0 and 1. Notice however that our method still works if no delay is attached to the identity. It may then result is what Kung[8] calls semi-systolic arrays. This intermediate step is called a *virtual mapping* (figure 6) of the computations. In order to obtain the *physical mapping* that we seek, we must schedule the computations by means of a timing function and map them on processing cells. By doing so, extra delays will be introduced automatically between the operators, as we shall see in the following.

In order to schedule the computations, we associate with each variable V of the equations an affine timing function $t(i,j,k) = \lambda_1 i + \lambda_2 j + \lambda_3 k + \alpha_V$, that gives the time (in term of the common unit of the operators) at which *the first bit* of $V(i,j,k)$ is computed. It must be pointed out that λ_1, λ_2, and λ_3 are kept independent of V. In other words, all the variables will be computed at the same rate, possibly with a slight constant time offset which depends on the relative values of the scalars α_V. Let us consider simultaneously the timing function and the allocation function. Consider the projection along the vector $u = (0, 0, 1)$, which gives the design of figure 5(a). The parameter to be determined are therefore $\lambda_1, \lambda_2, \lambda_3, \alpha_C, \alpha_A, \alpha_B$, and α_P. Consider the equation $C(i,j,k) = C(i,j,k-1) + P(i,j,k)$. Assume that we associate with this equation an adder with period ψ and skew σ, and let $\omega(O)$, $\omega(I_C)$, and $\omega(I_P)$ be the offset associated respectively to the output and the inputs C and P of the adder. Clearly $C(i,j,k)$ cannot be computed before the time at which $C(i,j,k-1)$ is available augmented by the time needed by the operator to produce $C(i,j,k-1)$, namely $\omega(O) - \omega(I_C)$. Therefore, we must have :

$$t_C(i,j,k) - t_C(i,j,k-1) \;=\; \lambda_3 \geq \omega(O) - \omega(I_C)$$

Applying a similar reasoning to $C(i,j,k)$ and $P(i,j,k)$ gives :

$$t_C(i,j,k) - t_P(i,j,k-1) \;=\; \alpha_C - \alpha_P \geq \omega(O) - \omega(I_P)$$

Such constraints can be obtained for each equation. They are called *latency constraints*. Another type of constraints involve the period of the operators and the direction of projection. Consider two computations that are mapped on the same processing cell. Assume that these computations are associated with points z_1 and z_2 of D. As a given equation is mapped on the same operator, the interval of time that separates the computation of this equation for successive points must be greater than or equal to the period of the operator. Consider again the case of C(i,j,k). As C(i,j,k) and C(i,j,k+1) are computed by the same operator (since u = (0,0,1)), we must have :

$$t_C(i,j,k+1) - t_C(i,j,k) \geq \psi$$

or equivalently, $\lambda^T u \geq \psi$ where $\lambda^T u$ denotes the dot product of λ and u. As the same condition holds for all the operators, we finally have the constraint :

$$\lambda^T u \geq \text{Max } \{\psi\}$$

which is called the *period constraint*. Since the timing-functions are linear, both period and latency constraints result in a system of linear inequalities. By solving the corresponding linear program, one can find admissible values of the parameters. Once the parameters are chosen, it is possible to compute the number of delays to be inserted between operators. As an example, consider the case of the delay to be inserted between the output P(i,j,k) of the multiplier and the corresponding input of the adder in the matrix multiplication cell. P(i,j,k) is available at time $t_P(i,j,k)$. On the other hand, as the result of the adder must be produced at time $t_C(i,j,k)$, its input P(i,j,k) must arrive at time $t_C(i,j,k) - (\omega(C) - \omega(P))$. Therefore, the number Δ of delays to be inserted is :

$$\Delta = t_C(i,j,k) - (\omega(C) - \omega(P)) - t_P(i,j,k) = \alpha_C - \alpha_P - (\omega(C) - \omega(P))$$

Moreover, one can notice that the number of delays (and consequently the total number of delays in a cell) is a linear function of the parameters. Therefore, in order to minimize this number, it suffices to solve the corresponding integer programming problem. It should be pointed out that this method is close to that proposed recently by Goossens et al.[9] for the delay management problem in bit-serial architectures. It can however be applied to other design styles, and embeds in a single framework the calculation of the systolic organization and the delay management problem.

4 Application to the matrix multiplication

In this section, we describe two implementations of the matrix multiplication systolic array, using either purely parallel operators, or bit serial operators.

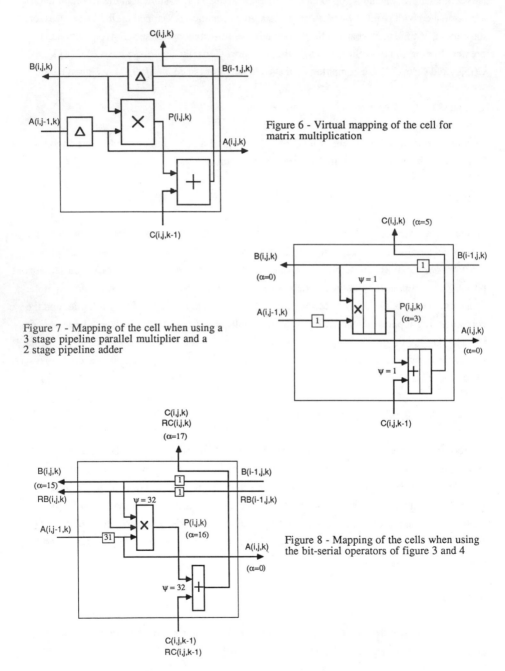

Figure 6 - Virtual mapping of the cell for matrix multiplication

Figure 7 - Mapping of the cell when using a 3 stage pipeline parallel multiplier and a 2 stage pipeline adder

Figure 8 - Mapping of the cells when using the bit-serial operators of figure 3 and 4

4.1. Parallel pipelined operators

Consider first that the multiply operator is a three stage pipelined operator (input offset = 0, output offset = 3, and period $\psi_x = 1$), and that the adder is a two stage pipelined operator (input offset = 0, output offset = 2, and period $\psi_x = 1$). The delays associated with the identity are assumed to be parallel registers. Consider the design resulting of a projection along vector u = (1,1,0), as sketched in figure 5(b). The latency constraints given by equation (3) are :

From C(i,j,k) : $\quad \lambda_3 \geq \omega(C) = 2$
$$\alpha_C - \alpha_P \geq \omega(C) = 2$$
From P(i,j,k) : $\quad \alpha_P - \alpha_A \geq \omega(P) = 3$
$$\alpha_P - \alpha_B \geq \omega(P) = 3$$
From A(i,j,k) : $\quad \lambda_2 \geq 1$
From B(i,j,k) : $\quad \lambda_1 \geq 1$

On the other hand the periodicity constraint is $\lambda_1 + \lambda_2 \geq 1$. Assuming that α_C, α_P, α_A, and α_B are non-negative integers, the convex set defined by the above constraints has one vertex given by $\lambda_1 = 1$; $\lambda_2 = 1$; $\lambda_3 = 2$; $\alpha_C = 5$; $\alpha_P = 3$; $\alpha_A = \alpha_B = 0$. This solution is therefore optimal with respect to the number of delays. The corresponding systolic array is depicted in figure 7.

4.2. Bit-serial operators

Let us implement the same systolic array using the bit-serial multiplier and adder described in section ?.? We assume that the numbers have 16 bits. Therefore, the period of both adder and multiplier is 32. For the multiplier, the offset of A is 0, the offset of B and of the reset is n-1 = 15, and the offset of P is n = 16. On the other hand, the offset of the inputs of the adder is 0, and that of the output is 1. Finally, the skew of both operators is 2. In order to implement the systolic array using these operators, it is necessary to modify the equations so that the reset signals appear. Equation (4) is rewritten as follows, after introducing a reset signal rc_{ij} that follows c_{ij} and a reset signal rb_{kj} associated with b_{kj} :

$$
\begin{aligned}
C(i,j,k) &= \quad \text{if } k = 1 \text{ then } 0 \\
&\quad \text{if } k \neq 1 \text{ then } Add \ [C(i,j,k-1) , P(i,j,k) , RC(i,j,k-1)] \\
RC(i,j,k) &= \quad \text{if } k = 1 \text{ then } rc_{ij} \\
&\quad \text{if } k \neq 1 \ RC(i,j,k-1) \\
P(i,j,k) &= \quad Mult \ [A(i,j,k) , B(i,j,k) , RB(i,j,k)]
\end{aligned}
$$

$$
\begin{aligned}
A(i,j,k) &= \quad \text{if } j = 1 \text{ then } a_{ik} \\
&\qquad \text{if } j \neq 1 \text{ then } A(i,j-1,k) \\
B(i,j,k) &= \quad \text{if } i = 1 \text{ then } b_{kj} \\
&\qquad \text{if } j \neq 1 \text{ then } B(i-1,j,k) \\
RB(i,j,k) &= \quad \text{if } i = 1 \text{ then } rb_{kj} \\
&\qquad \text{if } j \neq 1 \text{ then } RB(i-1,j,k) \\
c_{ij} &\equiv \quad C(i,j,N)
\end{aligned}
\tag{5}
$$

Applying the method gives the following latency constraints :

From $C(i,j,k)$:
$$\lambda_3 \geq \omega(C) = 1$$
$$\alpha_C - \alpha_P \geq \omega(C) = 1$$
$$\lambda_3 + \alpha_C - \alpha_{RC} \geq \omega(C) = 1$$

From $RC(i,j,k)$:
$$\lambda_3 \geq 1$$

From $P(i,j,k)$:
$$\alpha_P - \alpha_A \geq \omega(P) - \omega(A) = 16$$
$$\alpha_P - \alpha_B \geq \omega(P) - \omega(B) = 1$$
$$\alpha_P - \alpha_{RB} \geq \omega(P) - \omega(RB) = 1$$

From $A(i,j,k)$:
$$\lambda_2 \geq 1$$

From $B(i,j,k)$:
$$\lambda_1 \geq 1$$

From $RB(i,j,k)$:
$$\lambda_1 \geq 1$$

On the other hand, the periodicity constraint is $\lambda_1 + \lambda_2 \geq 32$. Again assuming that the α's are non-negative, the convex set defined by the above inequalities has the following vertices :

$$\lambda_3 = 1$$
$$(\lambda_1 = 1 \text{ and } \lambda_2 = 31) \text{ or } (\lambda_1 = 31 \text{ and } \lambda_2 = 1)$$
$$\alpha_C = 17; \ \alpha_{RC} = 0 \text{ or } 17; \ \alpha_P = 16; \ \alpha_A = 0; \ \alpha_B = 0 \text{ or } 16; \ \alpha_{RB} = 0 \text{ or } 15$$

In order to minimize the number of delays, it is clear that $\alpha_{RC} = 16$, $\alpha_{RC} = 17$, and $\alpha_{RB} = 15$ must be chosen. Therefore, the number of extra delays is given by the expression :

$$(\lambda_2 - 1) + 2 (\lambda_1 - 1)$$

It can be seen that the best solution is obtained when $\lambda_2 = 31$ and $\lambda_1 = 1$, which results in the design of figure 8. The intuitive explanation of this fact is simply that the design is not symmetric in A and B, as the reset signal follows B. Therefore, it is better to minimize the number of delays along the B path.

5. Conclusion

We have described the principle of the DIASTOL system. Its purpose is to allow systolic arrays to be designed at the functional level. The method underlying DIASTOL is based on the dependence mapping procedure. The detailed design of the cells is made from a library of operators that are modelized in terms of simple timing parameters called the period, the skew and the offset of the inputs and outputs. It has been shown that a refinement of the dependence mapping makes it possible to obtain automatically the exact timing of the data, and to compute automatically the number of delays to be inserted. Moreover, the number of delays can be optimized using integer linear programming. A first version of DIASTOL implementing the first step of the dependence mapping procedure has been implemented[10]. We are currently implementing the refinement step.

References

[1] S. Y. Kung, "On Supercomputing with Systolic / Wavefront Array Processors," *Proceeding IEEE*, July 1984, pp. 867 - 884.

[2] P. Quinton, The Systematic Design of Systolic Arrays, IRISA Research Report 193, 1983.

[3] P. Quinton, "Automatic Synthesis of Systolic Arrays from Uniform Recurrent Equations," in *Proc. 11th Annual Intern. Symp. Computer Architecture,* IEEE, June 1984, Ann Arbor, pp. 208 - 214.

[4] J. V. McCanny, J. G. McWhirter, K.W. Wood, "Optimised Bit Level Systolic Array for Convolution," *IEE Proc. F,* 131, pp. 632 - 637, 1984.

[5] P. Quinton, P. Gachet, "Automatic Design of Systolic Chips," in *Future Trends in Computing*, F. Robert and C. Di Crescenzo (eds), Masson - Wiley, 1985.

[6] R. F. Lyon, "Two's Complement Pipeline Multipliers," *IEEE Trans. Comm.*, Vol. COM-24, pp. 418 - 425, Apr. 1976.

[7] R. M. Karp R. E. Miller, S. Winograd, "The Organization of Computations for Uniform Recurrence Equations," *JACM,* Vol 14, No 3, july 1967, pp. 563 - 590.

[8] H. T. Kung, "Why Systolic Architectures ? ," *IEEE Computer,* 15, pp. 37 - 46, Jan. 1982.

[9] G. Goossens, R. Jain, J. Vandewalle, H. De Man, "An Optimal and Flexible Delay Management Technique for VLSI," in *Computational and Combinatorial Methods in System Theory,* C. I. Byrnes and A. Lindquist (eds), Elsevier Science Publishers B. V. (North-Holland), 1986.

[10] P. Quinton, P. Gachet, DIASTOL user's Manual : Preliminary version, IRISA Research Report, Aug. 1984.

CHAPTER 10

Hierarchical Design of Processor Arrays Applied to a New Pipelined Matrix Solver.

P. Dewilde and J. Annevelink

Department of Electrical Engineering
Delft University of Technology
Delft, the Netherlands

ABSTRACT

The computer-aided design of dedicated pipelined processors for numerical applications and signal processing requires design tools that support system refinement, partitioning strategies and the transformation of behavioral descriptions into structure. In this paper we present an hierarchical design methodology, and apply it to the design of a pipelined matrix solver.

1. INTRODUCTION

Digital signal processing often requires very fast handling of data in a small physical area and with low power consumption. A good example is that of bit-serial digital filters which are capable of achieving high throughputs with a very limited area, taking into account numerical considerations like stability and accuracy. In such very dedicated applications, the algorithms used can be optimally adapted to the requirement of hardware minimization within the speed constraints. A prototype example of this technique can be found in [1]. In the present paper we wish to take a more general point of view where we develop the design path from a numerical algorithm down to hardware with the expressed goal of realizing a structure that is as regular as possible, however without impairing performance. Our interest will go to a braod class of algorithms, such as a matrix solver or a system to compute eigenvalues. We shall concentrate on the problems associated with mapping such algorithms on regular, dedicated hardware: hierarchical refinement, partitioning and the generation of control structures. The emphasis will be on placing all the components of the design system in one consistent framework. As we shall see, such a consistent framework is obtained by combining and modifying classical concepts so as to fit the general design situation. Foremost is the notion of *signal flow graph* which is classical in signal processing, but needs modification as will be explained further. Next are the notions of *functional calculus* and *applicative state transitions*, which are also (to a good observer) classical to signal processing and have been rediscovered (of course with a different terminology) in computer science [2]. In the meantime both notions have obtained considerable attention especially in the computer science literature [3,4,5]. On the other hand, the development of description languages for signal processing has also been considered - a prime example of which can be found in [6]. Needless to say, we have used freely some ideas already present in the literature, while devoting most of our attention to the development of a consistent (and novel) design system – the design system HIFI that is in development at Delft University [7].

The first sections will be devoted to the description of the design methodology, while the latter sections will treat a specific example: a new type of matrix solver with special pipeline properties.

2. SYSTEM DESIGN BY REFINEMENT

One may define systems' design as the successive refinement from behavioral descriptions to structure, a process that finally ends in the complete definition of the hardware. This is especially true for the design of dedicated concurrent systems, like signal processors or matrix solvers as one has in speech coding, image coding or simulation [8,9]. For these applications, it often turns out that the complexity of the overall solution is strongly dependent on the organization of the dataflow and not only on the computational complexity. Examples can be found in [10,11,12].

Conceptually, the design of a system can be split in two phases:

1. the specification of the algorithm and

2. the mapping of the algorithm onto a particular hardware architecture.

The gist of the HIFI design methodology is the translation from a *semantic level* of direct I/O functional specifications to a *structural level* where states become explicit and the algorithm gets decomposed in structural parts (construction and composition). This decomposition can then be hierarchically refined without modification of the basic ideas, all the way to a hardware description. The *semantic level* is represented by the concept of *"node"* while the *"structural level"* is described by a Signal Flow Graph (SFG) which itself then consists of nodes. In this fashion a consistent *orthogonalization* of semantics (behavior) and structure is obtained. One of the main requirements of the system is that *any* deterministic algorithm must allow for a hierarchical design in this fashion. In particular decision making on the basis of data presented must be possible at all times. Major problems with the design of a (concurrent) system are the assignment of tasks to processors and the partitioning of the algorithm. These problems can be tackled however, given a specification of the order in which the tasks should be computed, e.g. in the form of a dependence graph. In our method, the description of an algorithm essentially amounts to the specification of a dependence graph. For a review of some of the methods that can be used to transform dependence graphs to e.g. systolic or wavefront arrays, we refer to [13,14]. Partitioning methods are described in [15,16].

The overall design procedure is as shown in figure 1.

2.1 Behavioral Description in Nodes

At a high level of abstraction, any deterministic system can be described by an I/O map in the form of "formal semantics" or as a behavioral description in the form of an Applicative State Transition system [2]. We capture this with the concept of a node. A node is characterized by :

1. $F = \{f_\alpha\}$ – a set of functions which the node is able to execute (e.g. any internal state is captured in a functional way). Each f_α is a partial function: it maps some inputs to some outputs.

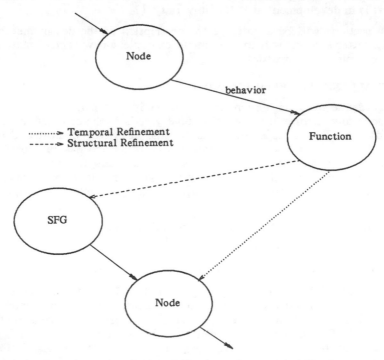

Figure 1. Hierarchical Design Procedure: from semantics to structure

2. I a collection of input ports characterized by a (fixed) type.

3. O a collection of output ports also characterized by a (fixed) type.

In any of its histories, the node will execute a sequence of functions:

$$f^0, f^1, f^2, \cdots$$

with each $f^i \in F$. When f^i is executed, we say that we stand at "event i".

Let $V_I(i)$ be the values (ev. empty) of the input ports at event i, and let $V_O(i)$ be the values (ev. empty) that the output ports will obtain at event i, then the Applicative State Transition (AST) mechanism is characterized by the map:

$$f^i = Z * V_I \rightarrow V_O * F : (i, V_I(i)) \mapsto (V_O(i), f^{i+1})$$

It can be shown that the above AST description holds for any *deterministic* system. Although the description given appears very abstract, the example given later will show that this is not the case. In fact, it can be used throughout the subsequent design refinements all the way down to the electrical network description.

2.2 The Structure of Concurrent Systems

At any given level of the design hierarchy, the system structure is represented by a Signal Flow Graph (SFG) which makes functional relations and state that are relevant at this level explicit. The communication mechanisms used in our SFG's must be consistent with our previous AST description, and are as follows:

1. Self-timing by a single token pass discipline. For each node in the SFG, an actual partial function f_i may fire when the the two following conditions are satisfied: (a) tokens are present on relevant inputs; (b) relevant outputs are free from tokens – see figure 2.

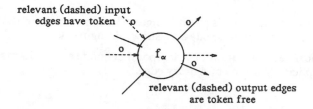

relevant (dashed) input
edges have token

relevant (dashed) output edges
are token free

Figure 2. Node Firing Discipline

2. The state that is explicit for this level is denoted by edges that are marked as *"delays"* and whose sole function is to make the delayed data available to the next event. Edges can implement an arbitrary number of delays. Except for these delays, edges can't accumulate data.

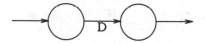

Figure 3. State Representation

This SFG model requires, however, timing verification because nodes are in no way synchronous, and mismatches between inputs and outputs are possible. We consider this a desired property, because it catches the relevant design problem at the present level of abstraction. It is easy to develop a theory of correctness for this situation based on edge trace theory, and it can be ascertained that an SFG which is trace correct is actually an AST machine. In this way the method proposed here is internally consistent.

Example 1: compute square root of an integer.

A simple algorithmic implementation is obtained as follows:

To start with, we consider a node denoting the computation of the square root of an integer. The result is again an integer value. A graphical representation of the node is shown in figure 4.

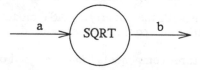

Figure 4. SQRT: Top Level

A node, like the one shown in figure 4, has to be created by a designer. In the prototype system, the designer does so interactively. Sitting behind a graphics workstation, the designer selects a number of commands, as a result of which a design object is created and stored in a design database [17]. The design object corresponding to the node shown in figure 4 merely specifies the datatypes of the input and output ports of the node. It also binds the node to a function (from a function library), to specify its behavior. The implementation of design objects will not be considered here.

An example of refinement is given in figure 5. Here we show the specification of a temporal refinement of the node *sqrt* shown in figure 4.

The graphical representation of a temporal refinement highlights the state, which is made explicit by means of a *delay* edge.

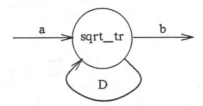

Figure 5. SQRT: First Refinement

The design object associated with this refinement will specify the initial value for the state, as well as the behavior of the node. The state introduced by the temporal refinement is a record consisting of three components: (*data, root* and *count*) The initialization of these components is described as follows:

> (state
> > (data (init a))
> > (root (init (pow 2 (/ n 2))) (final b))
> > (clock (init n)))

In addition to initializing them, we can also specify what we call a final state transfer. Here we have specified that the value of root is to be send to port b.

The behavior specified by the temporal refinement is as follows: for each data item presented on a, the node goes into a loop and produces a sequence of successive approximations to the square root. The node will output on port b, when sufficient precision has been obtained (as a result of the final state transfer). The behavior of node sqrt_tr is described as follows:

> (behavior
> > ((while (> : (root (abs (- data (* root root)))))
> > (a_sqrt : (data root count) → (root))
> > (= : (- count 1) → (count)))))

To highlight the relation with the underlying AST model for a node, remark that at this level of specification, we could have introduced

1. a data state, with three components (*data, root* and *count*)

2. a controller with four states, say T_0, ... T_3.

The controller could be implemented as a Finite State Machine (FSM). In state T_0, the node inputs a data value from port a. The next state is T_1. In state T_1 the controller determines whether the computed value satisfies the convergence criterion. If so, the next state is T_3, otherwise the next state is T_2. In state T_2 the node computes the next approximation. In state T_3 the final value is output via port b. The next state is T_0.

The next example is included to show the usage of structural refinement. The example is taken from the factorization of a matrix, and specifies a function that projects a vector on its first component. A well known way of doing this is by using the Householder reflection method [18]. A disadvantage of this method, from the VLSI point of view, is that it requires the computation of a square root, so as to compute the length of the vector. The method used here uses instead a sequences of rotations, that are applied to subsequent pairs of elements of the vector. The basic operation is a so-called GIVENS rotation that can be implemented directly in hardware [19].

To see how this is done in HIFI, we proceed as follows: First we define a node corresponding to the projection function. Next we indicate that we want to specify a structural refinement of this node. The refinement is then further specified as an array of nodes, each of which represents a CORDIC, i.e. its behavior is specified as a single function named fCORDIC. The procedure is illustrated in figure 6.

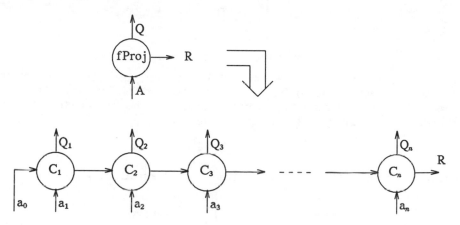

Figure 6. Structural Refinement

3. SOLVING SETS OF EQUATIONS IN A PIPELINED FASHION

Classical algorithms for solving large systems of linear equations of the type A x = b compute the factorization of the matrix A to produce an upper triangular system which is then solved by a procedure called "backsubstitution". The resulting dataflow is very unfavourable for parallel processing because the backsubstitution step needs the data outputed by the factorization step in reverese order. To overcome this problem we present an algorithm which solves the system in one pass, thereby avoiding the backsubstitution step. The resulting algorithm does not require any intermediate accumulation of data, and is ideally suited for implementation on a dedicated array of processors. We also show how the algorithm is mapped to the VLSI array, after further partitioning.

Thus, given is a system of linear equations $A x = b$ where the matrix A is n*n and typically a large banded matrix (we shall treat the problem in full generality). The traditional method of solving the system is by factoring A as A=QR where Q is a transformation matrix which we choose to be orthogonal for numerical accuracy and R is uppertriangular. If b is likewise transformed to $\beta = Q^t b$, then the system of equations is transformed to $Rx = \beta$ and x is found by backsubstituting on β. The latter operation starts with the last row in R, while the factorization produces the first row first. A conceptual architecture representing these operations is shown in figure 7.

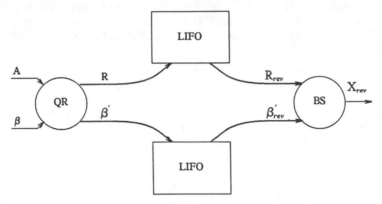

Figure 7. Architecture of the Classical Matrix Solver

By a clever arrangement of the data it is, however, possible to restrict the operations to factorization only. Inspired by the work of Faddeeva [20]. who presented a Gaussian algorithm which incorporated the backsubstitution, and following [21]. we propose to factorize the matrix

$$A' = \begin{vmatrix} A^t & I \\ -b^t & 0 \end{vmatrix}$$

The factorization, with appropriate partitioning of the matrices, gives:

$$\begin{vmatrix} U_{11} & u_{12} \\ u^t_{21} & u_{22} \end{vmatrix} \cdot \begin{vmatrix} A^t & I & 0 \\ -b^t & 0 & 1 \end{vmatrix} = \begin{vmatrix} R' & U_{11} & u_{12} \\ 0 & x^t u_{22} & u_{22} \end{vmatrix}$$

where $\begin{vmatrix} U_{11} & u_{12} \\ u^t_{21} & u_{22} \end{vmatrix}$ is an orthogonal matrix and R' is an upper triangular matrix.

The operations performed during the factorization follow the classical Householder algorithm for which we refer to [18]. The resulting parallel architecture, duly partitioned, and with the necessary datatransport and storage shown, is given in figure 8.

In figure 8 two methods of partitioning have been used, resulting in two partitioning parameters:

— LSGP (local sequential, global parallel), the grouping of c consecutive columns of the A_1 and b.

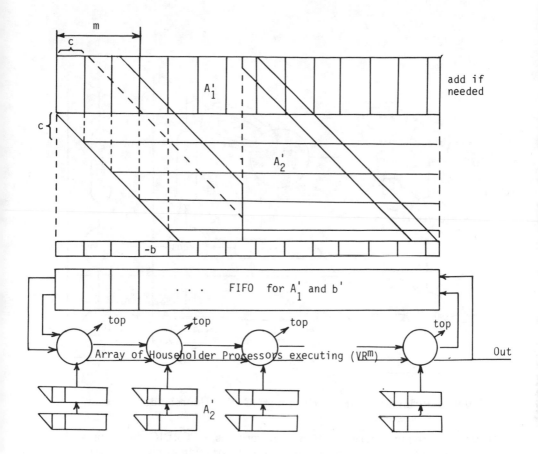

Figure 8. The Partitioned Architecture of the Single Pass Solver

— LPGS (local parallel, global sequential), whereby m parallel processors execute the first m cycles in parallel, but are then scheduled to execute the next m cycles.

Each of the processors executes a sequence $(VR^m)^*$, and consists of an Inner Product engine together with some control which takes care of the switch between vectoring and reflection, and of implementation of the different steps of the Householder algorithm. A further refinement of the design given so far is possible using the same HIFI methodology, which has been presented earlier.

3.1 Further refinement to local processing.

A further refinement can be obtained by decomposing the Householder nodes into an array of Givens nodes. Just as before, using partitioning of either the LSGP or LPGS type, new architectures are obtained with different storage properties as shown in figure 9. A systematic procedure to obtain partitioned arrays from full size artrays has been described in [16] and will be published elsewhere.

Each of the processors shown in figure 9 is of very simple type. Either it contains a complex multiplier, together with an iteration that performs the necessary

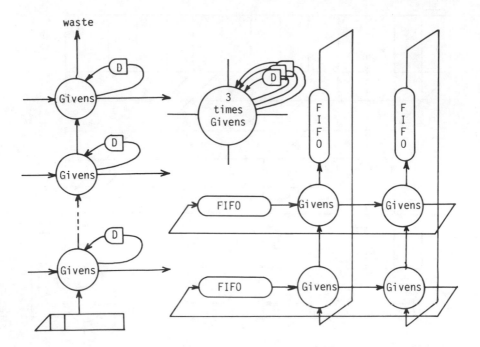

Figure 9. Refinement and Partitioning of the Householder node with Givens Processors.

square rooting and normalization or it contains a CORDIC processor which does all these operations at once and turns out to provide a more attractive solution. Again, each of these processors can be refined to the bit level, and converted into VLSI hardware [22].

4. DISCUSSION

The methodology presented here is indeed capable of refining and transforming any deterministic algorithm into dedicated hardware in which space or timing constraints are satisfied. Although the present paper has given a rather loose description of the procedure, it has been formalized and is becoming available as a prototype design system. Two features are thereby of great importance: the ability to execute simulations at any level of the design hierarchy, and the ability to do the systems' definition both automatically and interactively. Both are being supported.

References

1. J. van Genderen, H. de Man, F. Cathoor, and S. Beckers, "A Design Methodology for Compact Integration of Wave Digital Filters," *Proc. 10th Eur. Solid-State Circuits Conf.* , pp. 210-213 (Sept. 1984).

2. J.Backus, "Can Programming be Liberated from the Von Neumann Style? A Functional Style and its Algebra of Programs," *Comm. ACM* 21 pp. 613-641 (August 1978).

3. R. Milner, "Flowgraphs and Flow Algebras," *Journal of the ACM* 26(4) pp. 794-818 (Oct. 1979).

4. A. Izawa and T.L. Kunii, "Graph Based Design Specification of Parallel Computation," *Techn. Rept. Dept. of Information Science, Univ. of Tokyo*, (Dec. 1983).

5. E.A. Ashcroft and W.W. Wadge, "Lucid: A Non-Procedural Language with Iteration," *Comm. of the ACM*, pp. 519-526 (July 1977).

6. P. Le Guernic, A. Benveniste, P. Bournai, and T. Gautier, "SIGNAL: A Data-Flow Oriented Language for Signal Processing," Publication # 246, IRISA , Rennes, France (January 1985).

7. J. Annevelink, "A Hierarchical Design System for VLSI Implementation of Signal Processing Algorithms," *Proc. Intl. Conf. on Mathematical Theory of Networks and Systems MTNS-85*, (June 1985).

8. T. Kailath, *Modern Signal Processing*, Springer Verlag (1985).

9. S.Y. Kung, H.J. Whitehouse, and T. Kailath, *VLSI and Modern Signal Processing*, Prentice Hall, New Jersey (1985).

10. H.T. Kung, B. Sproull, and G. Steele, *VLSI Systems and Computations*, Computer Science Press (1981).

11.. K. Hwang and F. Briggs, *Computer Architecture and Parallel Processing*, Mc. GRaw Hill (1984).

12. K. Jainandunsing and Ed. F. Deprettere, "Design and VLSI Implementation of a Concurrent Solver for N Coupled Least Squares Fitting Problems," *IEEE Journal on Selected Areas In Communication* SAC-4(1) pp. 39-48 (Jan. 1986).

13. S.Y. Kung, "VLSI Array Processors," *Intl. Workshop on Systolic Arrays*, (July 1986).

14. S.Y. Kung, *VLSI Array Processors*, (to appear) Prentice Hall (1986).

15. K. Jainandunsing, "Optimal Partitioning Schemes for Wavefront/Systolic Array Processors," *Proc. IEEE Intl. Conf. on Circuits and Systems*, pp. 940-943 (May 1986).

16. H. Nelis, K. Jainandunsing, and Ed. F. Deprettere, "Automatic Design and Partitioning of Systolic Arrays," *Tech. Report, Dept. of EE, Delft Univ. of Technology*, (August 1986).

17. J. Annevelink and P. Dewilde, "Object Oriented Data Management for VLSI Design based on Abstract Datatypes," *Tech. Report, Dept. of EE, Delft Univ. of Technology*, (April 1986).

18. J.H. Wilkinson and C. Reinsch, *Linear Algebra*, Springer Verlag, New York (1971).

19. E.F. Deprettere, P. Dewilde, and R. Udo, "Pipelined cordic architectures for fast VLSI filtering and array processing," *Proc. IEEE Int. Conf. Acoust., Speech, Signal Processing*, pp. 41A6.1–41A6.4 (March 1984).

20. V.N. Fadeeva, *Computational Methods in Linear Algebra*, Dover publ. , New York (1959).

21. K. Jainandunsing and Ed. F. Deprettere, "A novel VLSI System of Linear Equations Solver for real–time Signal Processing," *SPIE Symp. on Optical and Optoelectronic Applied Science and Engineering*, (Aug. 1986).

22. J. Bu, Ed. F. Deprettere, and F. de Lange, "On the Optimization of Pipelined Silicon CORDIC algorithms," *European SIgnal Processing COnference'*, (Sept. 1986).

CYCLO-STATIC SOLUTIONS: OPTIMAL MULTIPROCESSOR

REALIZATIONS OF RECURSIVE ALGORITHMS

D. A. Schwartz and T. P. Barnwell III
School of Electrical Engineering
Georgia Institute of Technology
Atlanta, GA 30332
(404) 894-7346

ABSTRACT

This paper presents a set of techniques for the automatic generation of provably optimal cyclo-static multiprocessor realizations for recursive algorithms which are represented by cyclic shift-invariant flow graphs. Cyclo-static implementations are a broad class of deterministic multiprocessor schedules which include systolic and SSIMD implementations as special cases. The resulting multiprocessor compiler takes a graph representation of the algorithm as input and is capable of generating optimal solutions for a broad class of target multiprocessors. The emphasis in this paper is on the presentation of the basic concepts and their illustration with examples.

INTRODUCTION

The development of *cyclo-static* solutions for multiprocessor realizations of digital signal processing algorithms grew out of an attempt to understand why SSIMD and PSSIMD implementations [1,2] could be *rate-optimal* and *processor-optimal* (optimally fast and optimally efficient) while the systolic implementations for the same algorithms could not. There are three basic reasons for the shortcomings of systolic arrays. The first is the fact that systolic processors are static pipelines. This means that any particular operation in an algorithm is assigned to a particular processor in the systolic array, and that operation is performed by that processor on every iteration. Hence, the operations are static and only the data moves through the multiprocessor. In contrast cyclo-static processors (SSIMD and PSSIMD are special cases of cyclo-static solutions) are dynamic pipelines in which both the operations and the data move through the multiprocessor.

The second reason is the global transfer clock (tick). Indeed, this clock tick was the basic characteristic for which systolic arrays were named, giving the whole system it's 'pumping' action. There is no fundamental requirement that all the pipeline registers in the system be clocked simultaneously or that each processor must perform I/O on every clock tick. A consequence of the clock tick of systolic solutions is that it treats all operations as having effectively the same operational latency. This often leads to poor processor utilization. In contrast, the input-output operations in cyclo-static solutions move in parallel, non-overlapping wavefronts (with a periodic pattern to the spacing of successive wavefronts). The limitations of the clock tick led S. Y. Kung [3] to the development of the alternate approach of *wavefront array* processor solutions.

The third problem with systolic solutions is that in order to meet the constraints imposed by *nearest neighbor communications* it is usually necessary to generate *interleaved* realizations [4,5]. While interleaved realizations can potentially process multiple data streams in parallel, it is often the case that only one data stream is available. The existence of most systolic realizations for recursive systems require two or three-way interleaving [4]; therefore systolic realizations in the presence of one data stream have a processor efficiency bounded above by 50% or 33%. Cyclo-static realizations can meet the same type of communications constraints as systolic solutions without similar penalties in efficiency.

Cyclo-static implementations are a new class of multiprocessor solutions which can be used for the optimal realization of iterative or recursive algorithms on synchronous multiprocessors. A processor which realizes cyclo-static implementations can be considered to be a synchronous generalization of systolic processors, wavefront array processors and SSIMD processors. Cyclo-static realizations differ in that they are provably optimal with respect to multiple criteria.

Cyclo-static implementations are a broad class of deterministically scheduled synchronous MIMD schedules which include systolic and SSIMD implementations as special cases. As previously noted, the primary feature of a cyclo-static implementation which distinguishes it from a systolic implementation is that, viewed from the reference point of a single iteration of the algorithm, a systolic implementation is static where as a cyclo-static implementation is dynamic and cyclic. To state it simply, if a systolic implementation can be considered to be an array of processors in which the instructions are fixed in space and the data travels through space, then a cyclo-static implementation is one in which both the instructions and data travel through space. This extra degree of freedom in a cyclo-static implementation is very important in generating optimal realizations. As its name implies, if viewed from points in space-time separated by an appropriate period, cyclo-static implementations can be considered static.

Cyclo-static solutions can be effectively found that achieve a subset of the following optimality criteria: *rate optimality* (maximally parallel, minimum iteration period), *processor optimality* (maximum processor efficiency) and *delay optimality* (minimum throughput delay) while subject to adjacent processor communications constraints. In the next section, the optimality criteria will be formally presented. The procedure for finding solutions is a combinatorial optimization method which is efficient for typical realizations that are rate optimal. This problem was previously considered computationally intractable. In particular, previous researchers who considered optimal deterministic scheduling of flow graphs took the approach of transforming the original directed cyclic (containing loops) flow graph to a directed acyclic graph (loop free) [6]. Unfortunately the optimal solution to the acyclic graph can only fully exploit the parallelism of the original graph for a few special cases. The original aspects of this work are the direct utilization of the original cyclical graph, the concept of applying graph bounds as part of the multiprocessor compilations procedures and the introduction of a new approach to periodic scheduling.

FLOW GRAPH BOUNDS

This section presents, and reviews, the unified framework used for studying synchronous multiprocessor realizations of algorithms described by a class of shift-invariant flow graphs. In addition, this section also develops and presents the basic definitions used to define optimal realizations. Much of the material in this section is either the outgrowth of this current research effort or a unification and extension of prior work.

Algorithm Descriptions

In this research, all algorithms are specified by a form of *shift-invariant flow graphs* [2] which are restricted to contain only data independent computations. Such shift-invariant flow graphs are capable of representing a very large class of interesting DSP algorithms. They can represent a large class of linear, non-linear, and time-varying systems. In addition, shift-invariant flow graphs can also represent many algorithms which are not typically considered digital signal processing algorithms, including a large class of matrix operations as well as algorithms specified in terms of low level logic operations (e.g. a digital multiplier structure), [2]. In brief, shift-invariant flow graphs can represent all iterative algorithms which do not include any data dependent branch operations. The vast majority of digital signal processing algorithms, as well as many other important algorithms, belong to this class.

A *fully specified flow graph* (FSFG) is a shift-invariant flow graph in which the nodal operations are additionally constrained to be the atomic operations of the underlying processors which are to be used in the realization. Thus the atomic operations represent the smallest granularity at which parallelism may be exploited. Specifically, if the underlying processor has a two input (binary) addition operation and a three input summation is required, then the FSFG would represent the three input summation as two separate nodes of two input addition in cascade.

Note additionally that if the adder were a static four stage pipelined adder, the adder would be considered indivisible and the four stage pipeline would constitute a single atomic operation (as opposed to four atomic operations). It will be shown later that this will lead to implementations in which pipelined processing elements are used in an optimal manner.

In general, there may be many FSFG's which correspond to a particular generic flow graph. The processes of converting a generic flow graph into a FSFG consists of expanding all nodes which are not already atomic operations of the constituent processor into their atomic components. This can be done by expanding all non-atomic nodes (macro nodes) into atomic operations. Typically the macro node expansion is performed in a manner that maximizes the parallelism of the resulting fully specified flow graph. For the purposes of this paper it is assumed that the algorithms of interest are specified as FSFGs.

Flow Graph Bounds

Consider a FSFG that continuously processes an infinitely long input sequence of data to produce a corresponding infinitely long output sequence of data. It is possible to determine a tightest lower bound on the iteration period (reciprocal of sampling rate), hereafter referred to as the *iteration period bound*. The iteration period bound is a property of the FSFG and of the operational times of the nodal operations of the graph and is independent of the architecture of the realization. This technique is not restricted to single time index filters, but applies to any iterative algorithm that might be characterized by a representation, in FORTRAN, containing nested 'DO' loops. However, the interpretation of the iteration period is algorithm dependent in the general case.

The notation with which the iteration period bound is presented is that of Renfors and Neuvo [8]; however this is a reformulation of a result originally due to Fettweis [9]. In a different form, a similar result for the iteration period bound of a SSIMD realization was independently reported by Barnwell and Hodges [1]. The result due to Fettweis was originally applied to asynchronous implementations. This research is concerned only with synchronous realizations and will extend Fettweis' bound to the synchronous case.

The basic assumptions are that the algorithm to be implemented is represented as a FSFG, and that the *latency* (computation times) for all the nodal operations are known. Nothing is assumed about the communications structure, I/O constraints, or other details of the realization. While specific characteristics of a given architecture may not allow a realization of a specific FSFG at its iteration period bound, the resulting bound reflects the absolute limits on computation rate for the set of all possible architectures based on processing elements with the given atomic latencies.

The *iteration period bound* (IPB), is the minimum achievable latency between iterations of the algorithm. If the algorithm is a digital filter operating on a time series, this translates into the minimum achievable sample period. The iteration period bound is simply computed as follows. First, all the loops in the FSFG are identified. Then the operational delay around each loop, D_l, is computed as

$$D_l = \sum_{j \in l} d_j \tag{1}$$

where l is the loop index and d_j is the computational delay of the jth node of the lth loop. The iteration period bound is then given by

$$T_o = \underset{l \in \text{loops}}{\text{Max}}[D_l / n_l] \tag{2a}$$

where n_l is the total number of delay elements in loop l. Any loop for which $T_l = D_l / n_l = T_o$ is considered to be a *critical loop*.

The iteration period bound is a lower bound for both asynchronous and synchronous systems. In synchronous systems all events must be synchronized with the system clock. Thus in a synchronous system the iteration period bound must be an integral multiple of the system clock (clock period = 1 unit of time (u.t.)). The iteration period bound for a synchronous system is then given by

$$T_{s_o} = \left\lceil \underset{l \in \text{loops}}{\text{MAX}[D_l / n_l]} \right\rceil \quad (2b)$$

Where $\lceil x \rceil$ denotes the 'ceiling' of x, which is the smallest integer greater than or equal to x. In this paper, it is implied that the system is synchronous and that the iteration period bound is given by equation (2b). For simplicity, the 's' subscript will be dropped.

The *periodic delay bound* is the minimum achievable time between the availability of an input and the availability of the corresponding output in a system with a constant iteration period. The *periodic delay bound* is given by

$$D_o = \underset{p_f}{\text{Max}} [d_{p_f} - n_{p_f} T] \quad (3)$$

where the Max is taken over the set of all forward, loop-free, paths (p_f) from an input to a corresponding output. In this expression, d_{p_f} is the computational latency of the path p_f, n_{p_f} is the sum of the order of the ideal delay elements along the path, and T is the iteration period. Note that the *periodic delay bound* dependents on the iteration period and that it may be negative if there are no delay free paths. In contrast, the *absolute delay bound* is the absolute minimum time between the availability of an input and the availability of the corresponding output even if the algorithm is not applied iteratively. The absolute delay bound is also given by equation (3), but with T=0. Since cyclo-static implementations are iterative, the *periodic delay bound* is the applicable delay bound.

In an iterative implementation, the delay elements have an effective latency of -T times the order of the delay element. Then the periodic delay bound is just the critical path of the original graph, with respect to the latency of all loop free paths from the input to the output.

In this computation, ideal delay elements (z^{-1}) are assumed to have zero computational delay. However, extra nodes can be introduced into the FSFG to model the processing delay of an ideal delay, or communications delay, as necessary.

If D is the total (sequential) computational delay of all the nodes,

$$D = \sum_{j \in A} d_j \quad (4)$$

where A is the set of all node indices and d_j is the computational delay of the jth node, then the lower bound on the number of processors required to achieve an iteration period of T, is given by

$$P \geq \lceil D/T \rceil \quad (5)$$

This result follows very simply. If one iteration is computed every T seconds, and a total of D seconds of computations must be performed, then the average number of parallel operations is D/T. The average number of parallel operations is the lower bound on the number of required processors. Therefore an iteration period of T can never be achieved with fewer than P processors. The ceiling function is required since fractions of processors are not meaningful in a single task environment.

However, this is not a tightest lower bound on the number of processors required to achieve an iteration period of T. For most FSFGs, realizations that achieve strict equality exist. For some FSFG's, as T approaches T_o, the minimum number of processors increases. A precise determination of the tightest lower bound for an arbitrary FSFG and for any T appears to require a combinatorial search based on postulating a lower bound and searching to determine if a compatible realization exists.   For cyclo-static realizations where strict equality holds,   P is also referred to as the *maximum parallelism* of a graph.

Since T_o is the iteration period bound,   the maximum number of processors that can be   used with optimal efficiency to realize a FSFG is

$$P_o \geq \lceil D/T_o \rceil$$

It   is   conjectured that equality holds for most graphs of interest.   Therefore,   the *processor bound* can be defined as

$$P_o = \lceil D/T_o \rceil \tag{6}$$

Note: when the equality condition does not hold, P_o is used for the initial postulation of   the tightest lower bound on the number of processors.

Optimality

In the context of this work,   any realization that achieves an iteration period equal to the iteration   period   bound   is considered to be *rate optimal*,   since no faster realization   of   the underlying graph can exist. Similarly any realization that uses the minimum number of processors, to   support   an   iteration period of T,   is considered *processor   optimal* (processor efficiency optimal). For some positive Δ,   if $T \geq T_o + \Delta$, then the minimum number of processors is $\lceil D/T \rceil$. When $D/T \neq \lceil D/T \rceil$ then on the average there is less than one processor not being 100% utilized. Since   no realization (in a single task environment) exists that can achieve an iteration   period of   T,   with   less than $\lceil D/T \rceil$ processors,   the realization is considered processor   optimal   even though   the processor efficiency may be less than 100%.   If $T_o \leq T < T_o + \Delta$, then the minimum number   of   processors, P,   is   greater than $\lceil D/T \rceil$.   In these cases the   processor   efficiency, 100%(PT/D),   although   potentially   very   small,   is still considered processor   optimal.   The difference   between   100%   and   the true processor efficiency   is   called   the   *inherent processor inefficiency*.   However, as previously mentioned, it appears that for most graphs Δ is zero.   When Δ is non-zero it is usually small relative to T.   Therefore choosing $T = T_o + \Delta$ results in ideal processor efficiency and a   near   rate optimal realization.   This   will be illustrated later in this text with an example.

If   a   realization achieves an 'input to output' delay (throughput time) equal to the   delay bound, then it is considered *delay optimal*.

Of   course,   many   things   besides   the   structure of the   algorithm   and   the fundamental operational capability of the processors may limit DSP implementations.   Clearly, issues such as I/O   bandwidths,   external resource availability,   the number of available   processors,   and   the communications architecture may impact the achievable rate, delay, and processor efficiency of an algorithm.   But   in its own way,   each of these aspects can be addressed and corrected.   For   a particular   multiprocessor system (or a particular VLSI cell library) and a particular FSFG,   the bounds described above are fundamental. Hence, if implementations can be developed which achieve these bounds,   then it is clear that no other implementations exist which can operate at a higher rate,   with less delay,   or with higher efficiency.   It is this class of optimal implementations which will be generated for cyclo-static systems.

CYCLO-STATIC SOLUTIONS

A   cyclo-static   system   is   a   type   of synchronous,   multiprocessor system,   that   is deterministically   scheduled.   Cyclo-static   solutions   apply   to   iterative   problems   and   are characterized by a schedule (or program) for one iteration that is periodically repeated.   Cyclo-static   schedules   are   an extension of periodic scheduling,   or classical   multifunction   static pipeline scheduling with a constant latency (fixed period).   The schedule of an iteration repeats every iteration period.   In a static, or classical schedule, the initiation of the next iteration of computation can be represented by the initiation schedule of the previous iteration, shifted in time by one iteration period. In a cyclo-static schedule the next iteration can be represented by the schedule of the previous iteration, shifted in time by the iteration period and additionally by a fixed shift, or displacement, in the processor space.

Since the schedule of the processors is deterministic, the schedule automatically handles all precedence requirements, eliminating the need for synchronization, semaphore mechanisms or a costly real-time scheduler. In a cyclo-static system the processors are only performing direct operations of the defining algorithm and are therefore capable of ideal processor speedup.

In order to understand the basic principle of cyclo-static systems it is necessary to introduce some definitions. A *processor schedule* will refer to a complete description of the operations each physical processor will perform at each time instance. In contrast, a *schedule* will be a list of all operations that must be performed at each time instant independent of the physical processors. In other words, a schedule is a processor schedule without processor assignment. Since the problems under consideration are iterative with a constant iteration period, it is only necessary to specify the processor schedule information for one iteration period. The processor schedule for one iteration could be diagrammed in a manner similar to a reservation table or Gantt chart. Fig. 2a illustrates the processor schedule for one iteration of a task on a nine processor system for the fourth order filter of Fig. 1. Referring to the Fig. 2a, note that the processor schedule can be represented as a two dimensional space. One dimension is the index of the processor and the other dimension is time. Since the system is synchronous, all time events are integer multiples of the period of the system clock. For simplicity, all of the operations in this example were chosen to be of unit length, but in general the length of the operations can be any number of clock cycles. Let the system clock period be one unit of time (1 u.t.). This allows the time dimension of the processor schedule to be represented by a discrete time index. Since the processor index and the time index are discrete, the two dimensional space is a lattice which will be denoted as $\mathbf{P} \times \mathbf{T}$. A processor schedule is a pattern in the processor-time space $(\mathbf{P} \times \mathbf{T})$.

If the processors of the system are arranged as a two dimensional array then the processor space, $\mathbf{P}$, is two dimensional. In general, the processor space is multi-dimensional. The indexing of processors is performed modulo the cardinality of each processor dimension. For example, let $\mathbf{P}$ be one dimensional and of cardinality nine (nine physical processors), then the processor indexed by one is the same processor as that indexed by ten. For a one dimensional processor space, the processor-time lattice is an infinitely long 'cylinder.'

While the processor schedule of Fig. 2a fully specified one iteration of a task, it does not convey any information on the iteration period, or on how successive iterations are scheduled. This additional information is conveyed by a vector, $\mathcal{L}$, termed the *principal lattice vector*. The concept of the principal lattice vector is also illustrated in Fig. 2a. Fig. 2a represents a cyclo-static processor schedule for a system with nine processors and a one-dimensional processor space. Two iterations of the algorithm are illustrated with the detail of the 'zeroth' iteration specifying the order of specific operations of the underlying algorithm. Note that the 'first' iteration is identical to the 'zeroth' iteration, but is displaced in time by 2 u.t. and in processor space by one processor. Similarly this holds for each successive iteration. Recall that the processors are indexed modulo the processor space cardinality, $(|\mathbf{P}| = 9)$, and, therefore, iteration ten is also displaced by one processor from the previous iteration. Since every 2 u.t. a new iteration starts, then the system realizes an iteration period of 2 u.t., and has a processor displacement of one. Therefore the principal lattice vector for the cyclo-static schedule in Fig. 2a is, $\mathcal{L} = (1, 2), \epsilon \, \mathbf{P} \times \mathbf{T}$.

In terms of the concepts above, a cyclo-static system can now be more extensively defined. As previously noted, a *cyclo-static* system is a synchronous multiprocessor system that is deterministically scheduled. The scheduling of the system is characterized by its periodicity, in processor space and time, with a period related to the iteration interval. One iteration is a pattern (processor schedule) in the processor-time lattice, $\mathbf{P} \times \mathbf{T}$. The spatial displacement, in the lattice, between successive iterations is denoted by the *principal lattice vector*, $\mathcal{L}$.

The term cyclo-static, connotes an idea similar to cyclo-stationarity of random processes.

In a cyclo-stationary process, the statistics of observations separated by an integer multiple of the period of the process are stationary. In a cyclo-static system, any two operations in the processor-time lattice, separated by an integer multiple of the principal lattice vector, represent the exact same operation of the algorithm for different iterations.

Returning to the example of Fig. 2, several important properties of cyclo-static solutions can be introduced. Fig's 2b and 3c are also possible processor schedules to realize the same underlying algorithm as the processor schedule in Fig. 2a. Fig. 2a has only two iterations illustrated in order to simplify the complexity of the figure due to the apparent 'holes' in the schedule. The processor schedules of Fig's 2b and 2c have an iteration period of two, (every two units of time the computation of a new iteration starts). Fig's 2b and 2c have corresponding principal lattice vectors of $\mathcal{L} = (0,2)$, and $\mathcal{L} = (1,2)$.

The processor schedule in Fig. 2a belongs to the general class of cyclo-static schedules. The processor schedule in Fig. 2c belongs to the special class of SSIMD solutions [1,2]. This is because the program for each processor is identical but skewed (shifted) in time by two units of time. The processor schedule in Fig. 2b is a degenerate class of cyclo-static schedule that is also called a *static-schedule* since there is no processor component to the principal lattice vector.

All of the processor schedules in Fig. 2 have several properties in common. They all realize the same algorithm, have an iteration period of two, have identical program storage requirements and are processor optimal. The program storage requirement is whatever is necessary to represent an infinite loop program of a single iteration of the processor schedule, plus the components of the principal lattice vector (assuming a global controller). For all of the processor schedules, the processor schedule of one iteration will 'tile,' or periodically extend, along the direction of the principal lattice vector to completely cover the lattice. Therefore all of the schedules exhibit 100% processor utilization efficiency.

Fig. 1 is a FSFG of a cascaded fourth order IIR filter. Given an addition and multiplication time of 1 u.t., the iteration period bound of the graph is $T_0 = 2$ u.t. and the processor bound is $P_0 = 9$. All of the processor schedules of Fig. 2 are cyclo-static processor schedules for the filter in Fig. 1. Since all of the schedules have an iteration period equal to the iteration period bound, they are all rate optimal solutions. In fact, it turns out that the processor schedule in Fig. 1a is supported by a bi-directional ring of processors. The processor schedule of Fig. 2b is static and delay optimal, but requires an irregular complex communications support. Finally the schedule of Fig. 2.c is supported on a uni-directional ring of processors. These are just a few of a countably infinite set of different processor schedules that can realize essentially equivalent systems!

It is important to note that at any time instant, for any of these possible processor schedules, the computation of more than one iteration is proceeding in parallel. This is in contrast to classical CPM scheduling methods [6,7]. The primary flaw of the CPM approach to scheduling iterative problems is that it optimally schedules only one iteration, without a global consideration of possibly overlapped execution of iterations. The cyclo-static scheduling approach directly considers the problem as cyclic and folds in the effects of overlapped iterations.

Determination of Processor Schedules

To find a processor and rate optimal cyclo-static schedule, it is first necessary to determine the flow graph bounds of the defining FSFG. All loops are analyzed to determine the iteration period bound, to identify all critical loops, and to determine the slack time of all other loops. All loops that are not critical have spare time in the computational deadlines of all operations in a loop. This spare time is called *slack time*, and for a loop l_i is:

$$t_{s_i} = n_i T_0 - \sum_{j \in l_i} d_j \tag{7}$$

where d_i is the number of unit delays in loop l_i, T_0 is the iteration period bound, and d_j is the computational delay of operation j in loop l_i.

For a solution to exist, all operations in critical loops must be scheduled sequentially without gaps. All non-critical loops must be scheduled sequentially with a maximum total gap equal to the slack time of the loop. Non-loop operations can be scheduled at any time after their precedence requirements have been met. Of course, no operation can be scheduled before its precedence requirements have been met. The other key element to finding cyclo-static solutions is the requirement that the processor schedule tile or periodically extend along the direction of the principal lattice vector. The requirement can be simply stated as a constraint that no two points in the processor schedule for one iteration (lattice) may be separated by an integer multiple of the principal lattice vector ($k_o\mathcal{L}$). More precisely, no two entries in the complete processor schedule of one iteration can be congruent modulo $\mathcal{L}$.

Based on these simple principles, three compilers that take FSFGs as input and produce cyclo-static processor schedules have been developed at Georgia Tech. Further information on how to find cyclo-static solutions can be found in [10,11]. The remainder of the paper will present a few examples to further illustrate some fundamental properties of FSFG definitions of algorithms and how that relates to the existence of desirable multiprocessor solutions. It should be noted that all of the cyclo-static solutions presented in this paper can be determined from the material in this paper and are sufficiently simple that they can be determined with pencil and paper.

EXAMPLES

Returning to the fourth order filter in Fig. 1, let us once again examine the problem of finding a rate optimal solution. This time assume that the processor elements are inhomogeneous and are composed of multipliers and adders. Let the addition and multiplication have computational latencies of $t_a = 1$ u.t. and $t_m = 2$ u.t. Under these constraints, the critical loops of the FSFG are the loops 4-2-4 and 13-11-13. Therefore the iteration period bound, T_o, is equal to $(t_m + t_a)/1 = 3$ u.t. The processor bound concept can easily be extended to the inhomogeneous case by considering the different processor element types separately. The lower bound on the number of adders and multipliers is therefore given by

$$P_{a_o} = \lceil D_a/T_o \rceil = \lceil 8(1 \text{ u.t.})/3 \text{ u.t.} \rceil = 3$$

$$P_{m_o} = \lceil D_m/T_o \rceil = \lceil 10(2 \text{ u.t.})/3 \text{ u.t.} \rceil = 7$$

For this problem the processor bound is obtainable. However there exists no static pipeline realization that can obtain it (for the unmodified FSFG)! Referring to Fig. 3, the processor schedule indicates that the lattice vectors for the adder and multiplier subspaces are $\mathcal{L}_a = (0,3)$ and $\mathcal{L}_m = (1,3)$. In general, (there are special exceptions), if the latency of all operations are not of unit duration then static realizations that are rate and processor optimal do not exist. If all the operations have a latency of unit duration then a static realization always exists. However, different lattice vectors result in different interprocessor communications support requirements and the choice of a lattice vector is a function of the FSFG and of a processor mapping requirement.

Fig. 3 is the FSFG of a fourth order orthogonal lattice filter [12] implemented with *rotators* (i.e. CORDIC elements). This filter has the special property that it has excellent finite precision characteristics and also that all of the processing elements in the FSFG are in critical loops. In fact all of the loops except for the terminator section loop are critical. The processor schedule for all iterations is unique except for relabeling of processor names.

By symmetry it is clear that all the critical loops are similar to the loop 1-2-3-1, therefore the iteration period bound is $T_o = 3 t_r$. Given that $t_r = 1$ u.t., then $T_o = 3$ u.t. Similarly there are eight rotations, so that the processor bound is $P_o = \lceil 8/3 \rceil = 3$ (the inherent processor inefficiency is 1/9). Except for equivalent schedules, as soon as one rotation has been scheduled, all others follow immediately. Fig. 4a is a processor schedule for the filter that is rate optimal, processor optimal, delay optimal, and requires a bi-directional ring communications support. Even though it is processor optimal there is one idle processor time slot. This idle slot can be used to great advantage. If a simple dummy operation, d, is introduced in the terminator loop of the filter and scheduled in the idle slot, then the filter only requires a uni-directional communications ring. The uni-directional ring is the minimum possible communications support for this filter, therefore it is *communications optimal*. It should be noted that a systolic realization of this filter would require eight processors to achieve the same processing rate and approximately eight times as many communications links!

If a pipelined three stage rotator were the processing element, then the design would proceed as follows. Three u.t. of the pipelined clock is equal to 1 u.t. of the clock in the previous example. The iteration period bound is re-determined using the pipeline latency (not the stage latency). This results in an iteration period bound of $T_o = 9$ u.t. Similarly the processor bound is re-evaluated using the stage latency to determine the total latency of the FSFG. This yields $P_o = \lceil 9(1 \text{ u.t.})/9 \text{ u.t.} \rceil = 1$. For this graph, the minimum number of pipelined processors that can support a period of 9 u.t. is three. This is because all operations are in critical loop and no freedom is available to schedule operations in any of the other pipeline slots. This, of course, is a very inefficient implementation (33%).

If the period is increased from $T = 9$ u.t. to $T = 11$ u.t., then $P = \lceil 9(1 \text{ u.t.})/11 \text{ u.t.} \rceil = 1$ is realizable. An optimal processor schedule of the single processor is shown in Fig. 4b. Note that a small increase in the period resulted in the ability to exploit the full parallelism of the pipelined processor. This method readily extends into a powerful and helpful tool in microcoding pipelined array processors.

Fig. 5 is a FSFG of a fourth order orthogonal lattice filter where the atomic operations are addition and multiplication. The only difference between this filter and the previous filter is a different choice of atomic processing elements. All of the recursive operations, except for the termination, are in critical loops. Let $t_a = t_m = 1$ u.t., and assume a homogeneous processing element. The iteration period bound is $T_o = 3(t_m + t_a) = 6$ u.t. The processor bound is $P_o = 2(32t_m + 16t_a)/T_o 3 = 8$. As in the previous pipelined case, this is not realizable. This is because all of the critical loops require that all of the multiplies in a rotator be performed in parallel. This FSFG can be easily analyzed to determine that the minimum number of processors to support an iteration period of 6 u.t. is twelve. (In general it is very difficult to determine the true lower processor bound). Fig. 6a illustrates a static, twelve processor schedule that achieves the iteration period bound. The processor utilization is 67%.

Increasing the iteration period from 6 u.t. to 7 u.t. a static seven processor schedule can be found that achieves an ideal processor utilization of 98%. One such processor schedule is shown in Fig. 6b. This illustrates the general principal that increasing the iteration period from T_o to $T_o + \Delta$, for some small positive Δ results in the ability to find realizations with ideal processor efficiency.

DISCUSSION

This paper has presented the fundamental concepts underlying the generation of optimal cyclo-static schedules from fully specified flow graphs. The emphasis has been on the presentation of a set of simple examples which illustrate the concepts. It is important to note, however, that it is generally not reasonable to try to generate cyclo-static solutions by hand. Fortunately, all of the basic constraints (rate, efficiency, delay, and communications) can be formulated in a form which allows them to be incorporated into a constrained combinatoric search algorithm for finding optimal cyclo-static implementations. Such multiprocessor compilers are the correct approach for finding the solution to most interesting problems.

REFERENCES

[1] T. P. Barnwell III, C. J. M. Hodges, M. Randolf, "Optimal Implementation of Single Time Index Signal Flow Graphs on Synchronous Multiprocessor Machines," *ICASSP'82*, Paris, France, May, 1982.
[2] T. P. Barnwell III and D. A. Schwartz, "Optimal Implementation of Flow Graphs on Synchronous Multiprocessors," *Proc. 1983 Asilomar Conf. on Cir. and Sys.*, Pacific Grove, CA, Nov. 1983.
[3] S. Y. Kung, K. S. Arun, R. J. Gal-Ezer, D. V. Bhaskar Rao, "Wavefront Array Processor: Language, Architecture, and Applications," *IEEE Trans. on Comp.*, Nov. 1982, pp. 1054-1065.
[4] D. A. Schwartz and T. P. Barnwell III, "A Graph Theoretic Technique for the Generation of

Systolic Implementations for Shift-Invariant Flow Graphs," *ICASSP'84*, San Diego, CA, March 1984.

[5] C. E. Leiserson, "Optimizing Synchronous Circuitry by Retiming," in *Proc. 3rd Caltech Conf. Very Large Scale Integration*, edited by R. Bryant, Computer Science Press, 1983, pp. 87-116.

[6] Jan Zeman and G S. Moschytz, "Systematic Design and Programming of Signal Processors, Using Project Management Techniques," *IEEE Trans. on ASSP*, Dec. 1983, pp. 1536-1549.

[7] J. P. Brafman, J. Szczupak and S. K. Mitra, "An Approach to the Implementation of Digital Filters Using Microprocessors," *IEEE Trans. on ASSP*, Oct. 1978, pp. 442-446.

[8] Markku Renfors and Yrjo Neuvo, "The Maximum Sampling Rate of Digital Filters Under Hardware Speed Constraints," *IEEE Trans. on Cir. and Sys.*, March 1981, pp. 196-202.

[9] A. Fettweis, "Realizability of Digital Filter Networks," *Arch. Elek. Ubertragung.*, Feb. 1976, pp. 90-96.

[10] D. A. Schwartz, "Synchronous Multiprocessor Realizations of Shift Invariant Flow Graphs," Elec. Eng., Georgia Inst. of Tech, DSPL-85-2, July, 1985.

[11] D. A. Schwartz and T. P. Barnwell III, "Cyclo-Static Multiprocessor Scheduling for the Optimal Implementation of Shift-Invariant Flow Graphs," *ICASSP'85*, Tampa, FL, March 1985.

[12] S. K. Rao, T. Kailath, "Orthogonal Digital Filters for VLSI Implementation," *IEEE Trans. on Cir. and Sys.*, Nov. 1984, pp. 933-945.

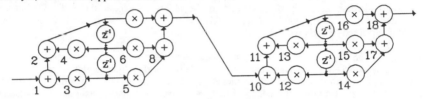

Figure 1 -- Fourth order cascade of second order sections.
Rate Bound = 2, Delay Bound =8, Processor Bound = 9

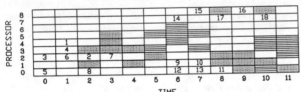

Figure 2a -- Cyclo-static schedule for bi-directional ring. L = (1,2)

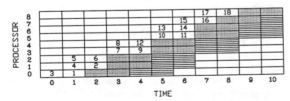

Figure 2b -- Static, delay optimal schedule. L = (0,2)

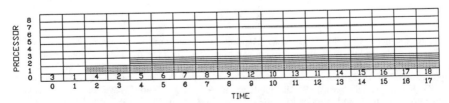

Figure 2c -- SSIMD schedule for uni-directional ring. L = (1,2)

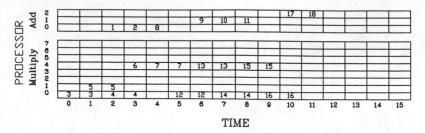

Figure 2d -- Cyclo-static schedule for inhomogeneous processors. $L_a = (0,3)$ and $L_m = (1,3)$.

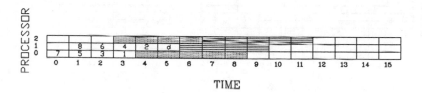

Figure 3 -- Fourth order othogonal filter implemented using Givens rotators. Rate Bound = 3, Delay Bound = 1, Processor Bound = 3

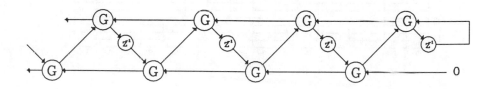

Figure 4a -- Cyclo-static, delay optimal schedule for a bi-directional ring. $L = (2,3)$

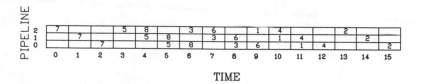

Figure 4b -- Single processor schedule using a three-stage processor.

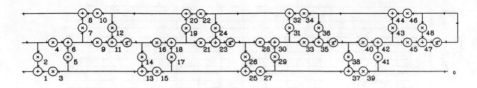

Figure 5 —— Fourth order othogonal filter.
Rate Bound = 4, Delay Bound = 4, Processor Bound = 8

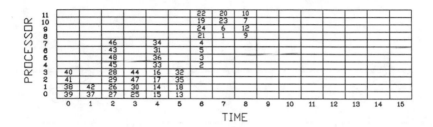

Figure 6a —— Static schedule for orthogonal filter using 4—multiply rotator.
L = (0,6)

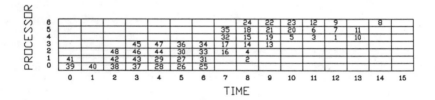

Figure 6b —— Near rate optimal schedule for a fourth order orthogonal filter
using 4—multiply rotators.   L = (0,7)

CHAPTER 12

SYSTOLIC ARRAY ARCHITECTURE FOR THE VITERBI DECODING ALGORITHM

C. Y. Chang and K. Yao
Electrical Engineering Department
University of California
Los Angeles, Ca 90024

ABSTRACT

New results on efficient forms of decoding convolutional codes based on the Viterbi algorithm using systolic array architecture are presented. A technique called strongly connected trellis decoding is introduced to increase the efficient utilization of all the systolic array processors. The issues dealing with the composite branch metric generation, survivor updating, overall system architecture, throughput rate, and computations overhead ratio are also investigated. It is shown that as the constraint length becomes large, the systolic Viterbi decoder maintains a regular and general structure as well as moderate throughput rate gain over the sequential Viterbi decoder.

INTRODUCTION

The Viterbi algorithm [1, 2] is well known to be an efficient method for the realization of maximum likelihood decoding of convolutional codes [3]. It can be viewed as a technique which finds the shortest path in the trellis diagram [4] to match the received sequence by a dynamic programming technique [5]. Presently, there are at least two commonly used techniques in the implementation of Viterbi algorithm. One is based on a sequential processor, and the other one is based on the motivation of utilizing many fully parallel processors [6, 7]. For convenience, they are respectively named as the sequential Viterbi decoder and the fully parallel Viterbi decoder. Since the error probability of convolutional codes decreases exponentially in proportion to the constraint length [2], longer convolutional codes are of great interest for high performance systems. However, as the constraint length increases, the throughput rate of the sequential Viterbi decoder decreases exponentially. On the other hand, the fully parallel Viterbi decoder suffers from the fact that the topological issue of efficiently interconnecting all the processors (or the global communication issue among all the processors) makes the design difficult for VLSI implementation.

A systolic array [8-11] is a network of processors that rhythmically process and pass data among themselves. It provides pipelining, parallelism, and simple adjacent neighbor cell interconnection structures so that it is suitable for VLSI implementation. In this paper, we shall consider the implementation of the Viterbi algorithm by using the systolic array architecture. The basic element in the systolic array design of Viterbi algorithm is that the Viterbi algorithm can be formulated in terms of general matrix vector operations. In this paper, a technique called strongly connected trellis decoding is introduced to increase the efficient utilization of all of the systolic array processors. Issues dealing with the corresponding techniques for branch metric generation and survivor updating are also investigated.

VITERBI DECODING BY SYSTOLIC ARRAY

In general, the Viterbi decoding process can be divided into three parts: branch metric generation, path metric updating, and survivor updating. The crucial factor in the implementation of Viterbi algorithm depends on how the path metric updating is performed. Different types of internal path metric updating and organization lead to different types of decoder realizations. In this section, we propose a new scheme to perform path metric updating by using the linear systolic array architecture. In later sections, we discuss its branch metric generation and survivor updating.

There are several ways to consider the Viterbi decoding process. Perhaps the simplest way is to consider it from the point of view as solving the shortest path problem in a multistage diagram [4], which can be formulated simply as a general matrix-vector multiplication [12]. Let us consider a convolutional code of code rate $1/n$ and constraint length K [7, 13]. Let N be the total number of states, i.e., $N = 2^{K-1}$. Let P be an $1 \times N$ row vector whose i^{th} element is denoted by P_i, representing the accumulated path metric from initial state 0 to state i. Let B be the $N \times N$ adjacency matrix, whose i-j^{th} element is denoted by b_{ij}, representing the branch metric from state i to state j between two adjacent stages in the trellis diagram. Then the Viterbi decoding process can be formulated as the repetitions of

$$P \leftarrow P \times B. \tag{1}$$

For example, under the maximum likelihood criterion,

$$P_j \leftarrow (P_0 x b_{0j}) + (P_1 x b_{1j}) + \ldots + (P_{N-1} x b_{N-1,j}), \tag{2}$$

where the x operator denotes conventional addition and the $+$ operator denotes the operation of taking maximum. Thus, the above expression means

$$P_j \leftarrow \max_{0 \leq i \leq N-1} \{P_i + b_{ij}\} \tag{3}$$

or

$$P_j \leftarrow \max \{P_0 + b_{0j}, P_1 + b_{1j}, \ldots, P_{N-1} + b_{N-1,j}\}. \tag{4}$$

(3) and (4) are used only to update the path metric. To determine from which state, say $\hat{i}$, the path of maximum metric (the most likely path) came from, it has the form of

$$\hat{i} = \max_{0 \leq i \leq N-1}^{-1} \{P_i + b_{ij}\}, \tag{5}$$

where $\max^{-1}$ means to take the i that yields the maximum value of $P_i + b_{ij}$. Equation (5) is used to keep track of the most likely path (the survivor). Its form is very similar to that of (3) and usually both equations are processed together. For most convolutional codes, some b_{ij} may not exist, which can be temporarily solved by making b_{ij} be a very large negative number so that this branch will never be chosen by the decoding process. We shall look more closely at this issue later. Also note that if the minimum distance decision rule is used, P_i will represent the shortest path from state 0 to state i, b_{ij} will represent the distance from state i to state j between two adjacent stages in the trellis diagram, and the $\max$ and $\max^{-1}$ in equations (3) and (5) will become $\min$ and $\min^{-1}$ respectively, where $\min$ means to choose the minimum number and $\min^{-1}$ means to pick up the path which yields the minimum distance.

From the previous discussion, we see that the Viterbi decoding process can be formulated in terms of general matrix-vector multiplication. However, it is known that systolic arrays [8] can be used efficiently for matrix-vector computations. As an example, consider a 4-state convolutional code of rate 1/2, whose trellis diagram and state transition table is shown in Figure 1. Specifically, Figure 2 shows the Viterbi decoding process as implemented on a linear systolic array. In this systolic array, R_j stays inside each processor while the P_i is moving to the right and the b_{ij} is moving down. All data movements are synchronized. It turns out that after four steps, R_j will successively contain the new updated path metric P_j. Figure 2 only describes one step of equation (1). As the Viterbi decoding process is the repetitions of equation (1), what really happens is that the next set of branch metrics is fed right on top of the current set of branch metrics and once a new path metric is obtained, (i.e., R_j contains the new path metric P_j), it is fed back into the systolic array from the left so that there are no idle periods in the whole process.

$I_{t-2}I_{t-1}$ $I_{t-1}I_t$

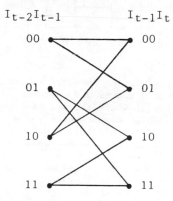

	00	01	10	11
00	0/00	1/11		
01			0/10	1/01
10	0/11	1/00		
11			0/01	1/10

Figure 1: The trellis diagram and the state transition table of a 4-state convolutional code of code rate 1/2.

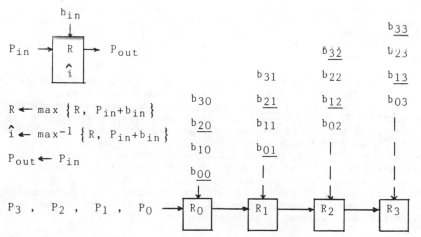

$R \leftarrow \max \{ R, P_{in}+b_{in} \}$

$\hat{i} \leftarrow \max^{-1} \{ R, P_{in}+b_{in} \}$

$P_{out} \leftarrow P_{in}$

Figure 2: The R-stay path metric updating array for the convolutional code shown in Figure 1.

131

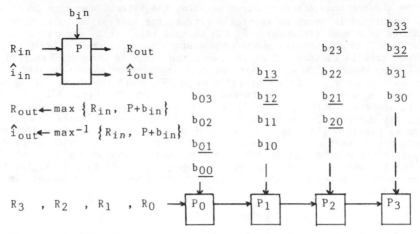

Figure 3: The R-move path metric updating array for the convolutional code shown in Figure 1.

The systolic array in Figure 2, can be considered as the "R-stay" scheme according to the behavior of R, which stays inside each processor. Another possibility is the "R-move" scheme as shown in Figure 3, where the b_{ij} is still moving downward but now the P_i stays inside each processor while the R_j is moving to the right. Of course, all data movements are also synchronized. Once R_j emerges from the right end processor, R_j will contain the new updated path metric P_j and is fed into the jth processor. The new set of branch metrics is fed right on top of the current set of branch metrics as before to avoid any idle period in the decoding process.

STRONGLY CONNECTED TRELLIS DECODING

In Figures 2 and 3, some branch metrics are underlined to indicate that there exists a connectivity between two specific states in the trellis diagram. Note that only half of the branch metrics are underlined, which means that at any time instant, only half of the processors are doing meaningful work. This low efficiency in the utilization of array processors is due to the low connectivity in the trellis diagram, i.e., the adjacency matrix B is a sparse matrix. Now, we show how the Viterbi decoding process can be modified so that it assumes the dense matrix property.

It can be seen that in the Viterbi decoding process each systolic array processor does meaningful work only when the trellis diagram is strongly connected, i.e., only when there exists a connectivity between any two states. Now, let us consider Figure 4, where every state can be reached from any other states in this two-stage trellis diagram. If each 2-branch metric is combined to be a single "composite" branch metric, we will have a strongly connected trellis diagram. For decoding, the Viterbi algorithm can work on the strongly connected trellis diagram instead of the original low connectivity trellis diagram. In other words, for the purpose of efficiently using all of the processors, several stages of the trellis diagram are combined until strongly connected. An efficient and simple scheme of generating all these composite branch metrics will be presented in the next section.

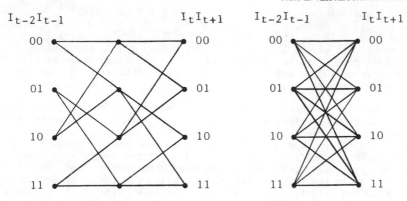

Figure 4: The 2-stage trellis diagram and the strongly connected trellis diagram of the convolutional code shown in Figure 1.

For any convolutional code, it turns out that the original trellis diagram can always be transformed to be strongly connected. The reason is that the connectivity in the trellis diagram has a very high degree of regularity. This result is stated in the following lemma and theorem.

Lemma 1: Consider a convolutional code of code rate 1/n and constraint length K. Let $\{I_t\}$ be the input binary sequence to the encoder, where t is an integer representing the time index. Let the state i at time t be denoted by the binary sequence $I_{t-K+1}I_{t-K+2}\ldots I_{t-1}$, which ranges in value from 0 to $2^{K-1}-1$. The connectivity in the trellis diagram shows that state i can only reach state $(2i \bmod 2^{K-1})$ and state $(2i+1 \bmod 2^{K-1})$, where $0 \leq i \leq 2^{K-1}-1$. #

Proof: Let $\{I_t\}$ be the input bit sequence to the encoder, where t is an integer. Supposed the state at time t is $I_{t-K+1}I_{t-K+2}\ldots I_{t-1}$. Since the current input I_t at time t can only be either 0 or 1, if the current input I_t is 0 (or 1), the next state will become $I_{t-K+2}\ldots I_{t-1}0$ (or $I_{t-K+2}\ldots I_{t-1}1$). This means that state $I_{t-K+1}\ldots I_{t-1}$ can only reach state $I_{t-K+2}\ldots I_{t-1}0$ and state $I_{t-K+2}\ldots I_{t-1}1$. Let
$$i = I_{t-K+1}I_{t-K+2}\ldots I_{t-1}$$
$$= I_{t-K+1}2^{K-2} + I_{t-K+2}2^{K-3} + \ldots + I_{t-1},$$
then it can be shown that
$$I_{t-K+2}\ldots I_{t-1}0 = 2i \bmod 2^{K-1},$$
and
$$I_{t-K+2}\ldots I_{t-1}1 = (2i+1) \bmod 2^{K-1}.$$
This proves that state i can only reach state $(2i \bmod 2^{K-1})$ and state $(2i+1 \bmod 2^{K-1})$. #

Theorem 1: For any convolutional code of code rate 1/n and constraint length K, the smallest number of stages needed to obtain a strongly connected trellis diagram from the original trellis diagram is K-1. Furthermore, in the (K-1)-stage trellis diagram, there is only one path between any two states. #

Proof: Let $\{X_n\}$ be a sequence of random numbers of two possible values, 0 or 1, where n is an integer. For convenience, when we use the term "state 2i+X" we mean "state 2i+0" or "state 2i+1". Lemma 1 shows that state $(2i + X_1) \bmod 2^{K-1}$ is reachable from state i in one step. Applying Lemma 1 n times, state $(2^n i + 2^{n-1}X_1 + 2^{n-2}X_2 + \ldots + X_n) \bmod 2^{K-1}$, 2^n states in total, will be reachable from state i in n steps. Note that if n = K-1, state $(2^{K-1}i + 2^{K-2}X_1 + \ldots + X_{K-1})$ mod 2^{K-1}, i.e., state $(2^{K-2}X_1 + 2^{K-3}X_2 + \ldots + X_{K-1})$, is reachable from state i in K-1 steps. However, $2^{K-2}X_1 + 2^{K-3}X_2 + \ldots + X_{K-1}$ uniquely represents all of the 2^{K-1} states. Obviously, K-1 is the minimum value of n to have such a property. This shows that the (K-1)-stage trellis diagram is strongly connected and also has only one path between any pair of beginning and ending states. #

There are two simple criteria that determine whether a decoder is really a maximum likelihood decoder for convolutional codes. The first criterion is that the diagram used to describe the decoding process must contain exactly all of the possible paths. The second criterion is that the decoding algorithm must be able to obtain the shortest path (or the most likely path) in this diagram. For the systolic Viterbi decoder, the diagram that describes the decoding process is the strongly connected trellis diagram. The decoding algorithm is still the Viterbi algorithm but now it is working on the strongly connected trellis diagram. The Viterbi algorithm will give us the shortest path in the strongly connected trellis diagram, since it is an algorithm to solve the shortest path problem. The remaining question is whether the strongly connected trellis diagram really contains exactly all of the possible paths. This issue can be easily clarified by comparing the strongly connected trellis diagram with the original low connectivity trellis diagram. We observe that the strongly connected trellis diagram can be considered as a partial expansion of the original trellis diagram, while the tree diagram is known to be the full expansion of the trellis diagram. This shows that the systolic Viterbi decoder is also a maximum likelihood decoder.

COMPOSITE BRANCH METRIC GENERATION

In Section 2 we proposed a way to update the path metric by linear systolic arrays which have simple and regular interconnection structure. Unfortunately, since we are dealing with sparse matrix operation, the processors are not efficiently utilized. In Section 3 we proposed an approach to increase the efficiency by letting the Viterbi algorithm work on the strongly connected trellis diagram instead of the original low connectivity trellis diagram, which can be done in theory. Now the question is in practice how to efficiently transform the low connectivity trellis diagram into one which is strongly connected. In other words, how do we efficiently evaluate all of the composite branch metrics? There are certainly many ways to generate all of the composite branch metrics. In this section, one simple and efficient approach is proposed.

Let b_{ij} be the composite branch metric from state i to state j in the strongly connected (K-1)-stage trellis diagram, whose value depends both on the code structure and the input data to the decoder (or the output data from the coding channel). Let C_{ij} be a binary vector of (K-1)xn elements, representing the codeword from state i to state j in the strongly connected trellis diagram. Let r be a real-valued vector of (K-1)xn elements, representing the current input data to the decoder, which is of length one composite branch in the strongly connected (K-1)-stage trellis diagram. It is known that under the maximum likelihood criterion the evaluation of b_{ij} can be expressed in the form of [2, 7],

$$b_{ij} = d(r, C_{ij}), \tag{6}$$

where d represents a distance function. For instance, under the soft decision d evaluates the Euclidean distance between C_{ij} and r, and under the hard decision it computes the Hamming distance between C_{ij} and the quantized value of r.

In order to generate all of the N^2 composite branch metric b_{ij}'s, all of the codeword C_{ij}'s need to be known. Supposed that all of the codewords were stored in the memory, a huge memory size, $N^2 \times (K-1) \times n$ bits in total, would be required. In the following discussion, by using the linear time invariant property of the convolutional code, we consider one technique for composite branch metric generation which does not require the storage of all of the codewords. Actually only a small and special part of the codewords need storage.

Theorem 2: Given any convolutional code, let v_{ij} be the encoded sequence (or codeword) when the initial encoder state is i and the input sequence is j, where i and j are defined in binary representation form, then

$$v_{ij} = v_{i0} + v_{0j}, \tag{7}$$

where + means bitwise modulo 2 addition. Note that the index 0 in v_{i0} represents an all 0's input sequence which has the same bit length as the index j and the index 0 in v_{0j} represents the 0 encoder state which has the same bit length as the index i. #

Proof: Recall the definition of v_{ij}, which represents the encoded sequence when the initial encoder state is i and the input sequence is j, where i and j are in binary representation form. Note that the initial encoder state is formed by the previous input of bit length K-1. Now let us form i&0 and 0&j, where & means bit concatenation. The 0 in i&0 has the same bit length as j and the 0 in 0&j has the same bit length as i. Since i&j = (i&0) + (0&j), where + means bitwise modulo 2 addition, by using the linear property of convolutional codes, it can be shown that $v_{0,(i\&j)} = v_{0,(i\&0)} + v_{0,(0\&j)}$. Furthermore, since the convolutional encoder is time invariant, this means $v_{ij} = v_{i0} + v_{0j}$. #

Here let us consider a more heuristic argument of Theorem 2. It is known that the convolutional encoder can be regarded as a special case of linear time invariant discrete systems [2]. For a linear time invariant system, it is well known that the total output response can be decomposed into two parts, the zero input response and the zero state response, so that the total response is equal to the summation of zero input response and zero state response [14]. Note that as mentioned in Theorem 2, v_{ij} means the encoded output sequence of the given convolutional code when the initial encoder state is i and the input bit sequence is j. Thus v_{i0} can be interpreted as the zero input response and v_{0j} can be interpreted as the zero state response. Equation (7) simply reconfirms the fact that the total response is equal to the summation of zero input response and zero state response.

Now let us see how the results of Theorem 2 can be applied to efficiently generate all of the composite branch metrics for rate 1/n code.

Corollary 1: Consider any convolutional code of code rate 1/n and constraint length K. Let C_{ij} be a binary vector of $(K-1) \times n$ elements, representing the codeword from state i to state j in the strongly connected (K-1)-stage trellis diagram. Then

$$C_{ij} = C_{i0} + C_{0j}, \tag{8}$$

where + means bitwise modulo 2 addition. #

Proof: Given state i and state j in the strongly connected trellis diagram, let i be the initial encoder state and j be the input sequence, then $C_{ij} = v_{ij}$, $C_{i0} = v_{i0}$, and $C_{0j} = v_{0j}$. Applying Theorem 2, we have $C_{ij} = C_{i0} + C_{0j}$. #

Example 1: Consider the same 4-state convolutional code of rate 1/2, whose trellis diagram is shown in Figure 1. From the codeword information given in the trellis diagram and the state transition table it can be shown that

$$C_0^0 = 0000 \qquad C_{00} = 0000$$
$$C_{10} = 1011 \qquad C_{01} = 0011$$
$$C_{20} = 1100 \qquad C_{02} = 1110$$
$$C_{30} = 0111 \qquad C_{03} = 1101$$

Corollary 1 states that the other C_{ij}'s can be derived from these C_{i0}'s and C_{0j}'s. For example, $C_{12} = C_{10} + C_{02} = (1011) + (1110) = 0101$ and $C_{32} = C_{30} + C_{02} = (0111) + (1110) = 1001$, etc. #

The consequence of Corollary 1 is that only a small part of the codewords need storage. In fact, one of the minimum sets of codewords which can be used to generate all of the codewords is the set consisting of C_{i0} and C_{0j} for $0 \leq i, j \leq N-1$. The other codewords can be simply derived from these C_{i0}'s and C_{0j}'s.

Figure 5 shows a linear systolic array used to generate all of the composite branch metric b_{ij}'s for the R-stay path metric updating array shown in Figure 2. Here, C_{0j} is preloaded and stored in the jth processor, and C_{i0} is used as a fanned-in input sequence which is moving to the right and will meet with the proper C_{0j} to generate the correct C_{ij} at the proper time. The processor design is also shown in Figure 5. Similarly, Figure 6 shows a linear systolic array used to generate all of the composite branch metric b_{ij}'s for the R-move path metric updating array shown in Figure 3. Here, C_{i0} is preloaded and stored in the ith processor, and C_{0j} is used as a fanned in input sequence which is moving to the right and will meet with the proper C_{i0} to generate the correct C_{ij} at the proper time.

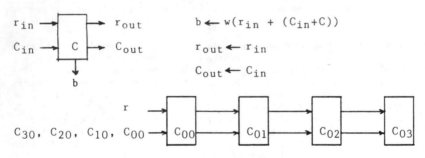

Figure 5: The composite branch metric generator for the R-stay path metric updating array in Figure 2.

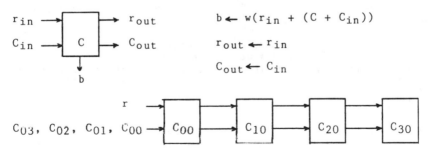

Figure 6: The composite branch metric generator for the R-move path metric updating array in Figure 3.

SURVIVOR UPDATING

Now, let us consider how in theory the survivor is updated in a systolic Viterbi decoder. Let $sur(j)$ be the survivor of state j and also let u_{ij} denote the information bits that drive the encoder from state i to state j in the strongly connected $(K-1)$-stage trellis diagram, where $0 \le i, j \le N-1$. For convenience, both $sur(j)$ and u_{ij} are represented by binary sequences. Recall equation (3) and (5) in Section 2. It is easy to see that

$$sur(j) = sur(\hat{i}) \ \& \ u_{\hat{i}j}, \tag{9}$$

where & means bit concatenation. Since $sur(\hat{i})$ represents the old survivor of state $\hat{i}$, which is known and stored in the memory, the remaining question is what $u_{\hat{i}j}$ really looks like.

For rate $1/n$ code, let the input information bit sequence to the encoder be denoted by $\{I_t\}$, where t is an integer representing the time index. The state at time t can then be denoted by $I_{t-K+1}I_{t-K+2}\cdots I_{t-1}$ as shown in Lemma 1. However, as previously defined, $u_{\hat{i}j}$ simply represents the information bit sequence of bit length $K-1$ that drives the encoder from state $\hat{i}$ to state j in the strongly connected trellis diagram. Supposing that state j is at time t and state $\hat{i}$ is at time $t-K+1$, the previous arguments simply mean that

$$\begin{aligned} u_{\hat{i}j} &= I_{t-K+1}I_{t-K+2}\cdots I_{t-1} \\ &= j. \end{aligned} \tag{10}$$

Hence, for rate $1/n$ code,

$$sur(j) = sur(\hat{i}) \ \& \ j. \tag{11}$$

SYSTEM OVERVIEW

So far we have separately considered the systolic array architecture for path metric updating and composite branch metric generation. The issue dealing with the survivor updating was also investigated in theory. Now let us consider the systems of systolic Viterbi decoders based on the R-stay and the R-move linear systolic arrays for path metric updating, which are shown in Figures 7 and 8 respectively. Each system basically consists of two linear systolic arrays cascaded together, one for composite branch metric generation and the other for path metric updating. In addition, a path memory is needed to store the survivors, which is the major part of the survivor updating process. Descriptions of the data flow and the capability of each processor in the linear systolic arrays appear in Sections 2 and 4.

In the whole system only the composite branch metric generator has the external input connection for receiving the input data to the decoder (or the output data from the coding channel), which is denoted by r. Its responsibility is to efficiently generate all of the composite branch metrics so that they can be properly fed into the path metric updating array. It does not store all of the codewords. In fact it only stores a small part of the codewords known as C_{i0} and C_{0j}, for $0 \le i, j \le N-1$, as defined in Section 4, which are sufficient to generate all of the codewords. Its outputs contain b_{ij}, the composite branch metric.

The path metric updating array receives b_{ij} from the composite branch metric generator. Its responsibility is to efficiently update the path metric in the strongly connected trellis diagram. It will tell the survivor updating device which path has been chosen as the currently most likely path (or the shortest path) for any state in the trellis diagram. Initially it stores P_i, the old path metric, which together with b_{ij} will be used to evaluate the new path metric. A variable called $\hat{i}$ is also generated and used to identify which state the best path came from. The outputs of path metric updating array contain two variables $\hat{i}$ and j which are sent to the survivor updating device, where the survivors stored in the path memory are updated according to these two pieces of information. Finally, since in reality the memory is of finite size, if the memory fills, the decoder will be forced to make a decision and send out decoded bits having the length of one composite branch.

Since the Viterbi decoding process is the repetitions of the stage transitions through the trellis diagram, the new output data from the coding channel will keep flowing into the decoder, where the composite branch metrics are generated, the path metrics and the survivors are updated, and some new decisions are made. It is clear that these procedures can also be pipelined so that every processor is busy during the decoding process.

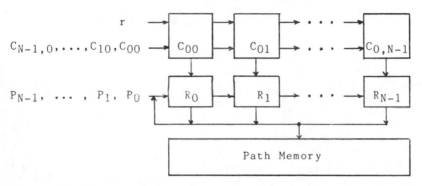

Figure 7: The system overview of the systolic Viterbi decoder based on the R-stay path metric updating array.

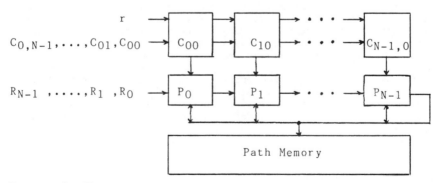

Figure 8: The system overview of the systolic Viterbi decoder based on the R-move path metric updating array.

THROUGHPUT RATE AND COMPUTATIONS OVERHEAD RATIO

In this section we like to consider the throughput rate of the systolic Viterbi decoder and the computations overhead ratio introduced by the strongly connected trellis decoding technique. Let R be the throughput rate of the systolic Viterbi decoder in bits/T, where "bits" means the number of input data to the decoder during the time T, the time needed for each processor's operations. Also let R' be the throughput rate of the sequential Viterbi decoder in bits/T', where T' is the time needed to update one path metric. Let G be defined as the ratio of R/R'. Since it is reasonable to expect that the duration of T is less than that of T', the actual throughput rate advantage of the systolic Viterbi decoder versus the sequential Viterbi decoder should be greater than G. Finally, let COR be the computations overhead ratio defined as the ratio of the number of addition/comparison in a systolic Viterbi decoder to that of a sequential Viterbi decoder over any time duration.

Corollary 2: Given any convolutional code of code rate 1/n and constraint length K, the systolic Viterbi decoder has

$$R = ((K-1)xn)/2^{K-1} \quad \text{bits/T}$$
$$G = 2x(K-1)$$
$$COR = 2^{K-2}/(K-1). \tag{12}\#$$

Proof: It is shown in Theorem 1 that it needs K-1 stages for a trellis diagram to become strongly connected. Note that a composite branch of length K-1 consists of (K-1)xn number of input data to the decoder. Also note that it takes $2^{K-1}xT$ to update all of the path metrics for one stage of transition in the strongly connected trellis diagram. This means that the throughput rate R is $((K-1)xn)/2^{K-1}$ bits/T. For the sequential Viterbi decoder R' is $n/(2x2^{K-1})$ bits/T', this shows that G = R/R' = 2x(K-1). The number of computations (addition followed by comparison) in the strongly connected trellis diagram is $2^{K-1}x2^{K-1}$. The number of computations involved in the K-1 stages of trellis diagram is $2x2^{K-1}x(K-1)$. Hence COR = $2^{(K-2)}/(K-1)$. #

Example 2: Given any convolutional code of code rate 1/2 and constraint length K

K	2	3	4	5	6	7	8
R	2/2	4/4	6/8	8/16	10/32	12/64	14/128
G	2	4	6	8	10	12	14
COR	1	1	4/3	8/4	16/5	32/6	64/7

#

Note that both G and COR shown above are very similar to those of systolic DFT over FFT. Actually, there are close philosophical similarities between systolic Viterbi decoder and systolic DFT. In both problems, more computations are introduced in comparison to the original problems (i.e., sequential Viterbi decoder and FFT algorithm) so that the problems have more well-behaved structures and can be performed by many pipelining and parallel processors to speed up the throughput rate.

CONCLUSIONS

The Viterbi algorithm has been formulated in terms of general matrix-vector multiplication so that it can be implemented by systolic array architecture. However, since the trellis diagram usually has only low degree of connectivity, the array processors are not efficiently utilized. A technique called strongly connected trellis decoding was then introduced to increase the efficiency, based on the fact that the original trellis diagram can always be transformed into one which is strongly connected. Results on the efficient evaluation of all the composite branch metrics in the strongly connected trellis diagram were also discussed. The presented scheme is good for both hard and soft decision cases. The corresponding survivor updating was also shown. Two forms of overall system architecture were then proposed. Both of them were shown to have a regular and general interconnection structure for any convolutional codes. The corresponding throughput rate was also calculated and shown to have a moderate throughput rate gain over the sequential Viterbi decoder. Even though we only considered the decoding processes of rate 1/n codes, the results can also be easily generalized to the decoding of higher rate codes and other applications of the Viterbi algorithm in communication and information processing systems.

ACKNOWLEDGEMENTS

This work was partially supported by NASA/JPL Contract PK-6-2437 and NASA/AMES Contract NAG-2-304.

REFERENCES

[1] A.J. Viterbi, "Error Bounds for Convolutional Codes and an Asymptotically Optimum Decoding Algorithm," IEEE Trans. Inf. Theory, IT-13 (1967), pp 260-269.

[2] A.J. Viterbi, and J. K. Omura, Principles of Digital Communication and Coding, McGraw-Hill, 1979.

[3] G.D. Forney, "The Viterbi Algorithm," Proc. IEEE, Vol. 61 (1973), pp 268-278.

[4] G.D. Forney, "Convolutional Codes II: Maximum Likelihood Decoding," Inf. Control, 25, pp. 222-266, July 1974.

[5] J.K. Omura, "On the Viterbi Decoding Algorithm," IEEE Trans. Inf. Theory, IT-15 (1969), pp 177-179.

[6] J.B. Cain and R.A. Kriete, "A VLSI R=1/2, K=7 Viterbi Decoder," IEEE Proc. of the 1984 NAECON, May 1984.

[7] S. Lin and D.J. Costello, Jr., Error Control Coding: Fundamentals and Applications, Prentice-Hall, 1983.

[8] H.T. Kung and C.E. Leiserson, "Systolic Arrays (for VLSI)," Proc. Symp. Sparse Matrix Computations and Their Applications, Nov. 1978, pp. 256-282.

[9] H.T. Kung, "Let's Design Algorithms for VLSI Systems," Proc. of Caltech Conf. on VLSI, 1979.

[10] H.T. Kung, "Why Systolic Architectures?," IEEE Computer Magazine, Jan. 1982.

[11] L.J. Guibas, H.T. Kung and C.D. Thompson, "Direct VLSI Implementation of Combinational Algorithms," Proc. of Caltech Conf. on VLSI, 1979.

[12] M. Gondran and M. Minoux, Graphs and Algorithms, Translated by S. Vajda, John Wiley & Sons, 1984.

[13] G.C. Clark and J.B. Cain, Error-Correction Coding for Digital Communications, Plenum Press, New York, 1981.

[14] L.A. Zadeh and C.A. Desoer, Linear System Theory, McGraw-Hill, 1963, page 144, Sec. 3.5. K. Zigangirov, "Some Sequential Decoding Procedures," Probl. Peredachi Inf., 2, pp. 13-25, 1966.

CHAPTER 13

THE THEORY AND VLSI IMPLEMENTATION OF STACK FILTERS

Edward J. Coyle

School of Electrical Engineering

Purdue University

West Lafayette, IN 47907

ABSTRACT

Stack filters are a class of digital filters that are defined by a weak superposition property known as the threshold decomposition and an ordering property known as the stacking property. These two properties lead to a VLSI architecture and implementation for these filters which is a massively parallel interconnection of very simple binary filters. They also lead to a new estimation theory based on stack filters and the mean absolute error criterion.

Considered together, this new optimality theory and the VLSI implementation of rank order filters suggest that stack filters are to the discrete Hilbert space ℓ_1 what linear filters are to ℓ_2: both linear filters and stack filters are defined by superposition properties; both classes of filters are implementable; and, both have tractable procedures for finding the optimal filter.

In this paper, we summarize both this optimality theory and the recently completed 5μ NMOS VLSI rank order filter chip which can perform all recursive and nonrecursive rank order filters of window widths 3,5,7, and 9.

1. INTRODUCTION

Stack filters are a new class of nonlinear filters that have recently been developed [1-4]. The defining properties of these new filters are two of the fundamental properties of rank order filters: *the threshold decomposition and the stacking property [5,6]*.

The threshold decomposition is a weak superposition property [5,6] which holds for all rank order filters in any dimension. This property, which is illustrated in Figure 1 with the stack filter known as the median filter, states that filtering a nonnegative multi-level signal with a rank order filter $f(\cdot)$ is the same as:

1. Decomposing the signal into many binary signals by thresholding at each possible value the signal can take on, as shown on the left side of Figure 1;
2. Filtering each binary threshold signal with the filter $f(\cdot)$; and then,
3. Adding up the outputs of all the filters, as shown on the right side of Figure 1.

Because of this property, rank order operators can be completely characterized by their effects on binary signals and can be implemented using a highly parallel architecture consisting of stacks of simple binary operators.

The second property used to define stack filters is the ordering property known as the stacking property [5,6]. A set of binary signals, all of which have the same length, possesses this property if they can be piled one on top of another so that each column of the resulting binary array looks like a stack of 1's supporting a stack of 0's. For instance, the set of binary signals obtained by thresholding a multi-level signal at each value it takes on possesses this property, as can be seen on the left side of Figure 1. A bank, or "stack", of binary filters is said to possess the stacking property if the set of

This work was supported by the National Science Foundation under grant ECS83-06235.

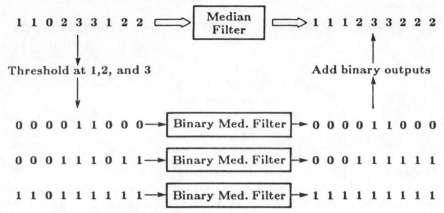

Figure 1: Ranking and Sorting vs. the Threshold Decomposition. The signal at the upper left is window width 3 median filtered the traditional way, by sorting and ranking the samples in the window, by following the heavy arrows. It is filtered by threshold decomposition by following the slender arrows. The binary signal on the bottom left is obtained by thresholding the input at level 1; the one above it is obtained by thresholding the signal at the level 2, etc.

binary output signals it produces possesses the stacking property whenever the set of binary inputs possesses this property. Thus, the binary median filters in Figure 1 have the stacking property because the binary outputs stack whenever the binary inputs stack.

Because every stack filter obeys the threshold decomposition, they all have the architecture shown in Figure 1. The difference between two stack filters lies solely in the binary operation performed on each level of this architecture. As long as this operation preserves the stacking property, the filter is called a stack filter. A necessary and sufficient condition for the preservation of this property is that the binary operation be a positive [1] Boolean function [12]. Thus, a new stack filter is obtained each time a different positive Boolean function is substituted for the binary median filter shown in Figure 1.

As a simple example, consider a positive Boolean function having a window width of three. Let the three bits appearing in the window be called x_1, x_2, and x_3. Then the binary median filter is obtained by performing the operation $x_1 \cdot x_2 + x_2 \cdot x_3 + x_3 \cdot x_1$ ("$+$" is the logical OR, "$\cdot$" is the logical AND.) on these bits [1,2]. If the positive Boolean function $x_2 + x_1 \cdot x_3$ is substituted for the binary median filter in the architecture in Figure 1, the resulting stack filter is one we have termed the asymmetric median filter [1,2]. This filter preserves edges and positive going impulses, while removing negative impulses.

There are exactly as many stack filters as there are positive Boolean functions. Thus, there are 20 stack filters of window width three, 7581 of window width five, and more than 2^{35} of window width seven, etc [13]. The exact number of positive Boolean functions for any window width is not yet known; however, the number is known to grow faster than $2^{\binom{n}{n/2}}$ [13]. Included in this large class of filters are all the rank order operators, all compositions of dilations and erosions [3,4,7], morphological filters [3,4], and, with the generalization of randomized outputs, linear filters. Furthermore, since every positive Boolean function has a minimum sum of products expression involving no complements of the input bits, every stack filter is a composition of rank order filters applied to subsets of the input bits to the positive Boolean function [3,4]. Thus, this class of filters includes many that are already widely used in practice.

This creation of a new class of digital filters, and the unification of many widely used filtering techniques under this one class, has led to significant practical and theoretical advances in signal processing.

The most significant practical advance to date concerns the new VLSI implementation of rank order filters [8-10]. This new chip promises to be extremely fast and flexible and has a much simpler architecture than other implementations [14,15] of these filters. Furthermore, since each stack filter is a particular composition of rank order filters [3,4], this chip provides a means for implementing any stack filter. The design of this VLSI chip [9,10], which was carried out by Ron Harber and Prof. Steve Bass of Purdue University, is discussed in Section 3.

The most significant theoretical advance to date as a result of this new class of filters is the technique, reported for rank order filters in [11] and for all stack filters in [17], for determining the optimal filter in any given filtering situation. This theory allows us to determine which stack filter provides the minimum mean absolute error estimate of a signal obscured by noise. This theory is outlined in Section 2.

2. CHOOSING THE OPTIMAL STACK FILTER

2.1. The Definition of Randomizing Stack Filters:

Stack filters are a subset of all window operators of a fixed window width. Defining them requires notation for the sequences that will be observed in the window of the filter. A sequence of length b will be denoted $\vec{S}_b = (S_1, S_2, \cdots, S_b)$ in which each S_i is any real number. If the sequence is a binary sequence, lower-case letters will be used. When the dimension of the vector is clear from the context in which it appears, the subscript n will be eliminated. Equality between sequences $\vec{X}$ and $\vec{Y}$ of the same length exists if and only if $X_i = Y_i$ for all i, in which case we write $\vec{X} = \vec{Y}$. We say $\vec{X} \leq \vec{Y}$ if $X_i = x$ implies $Y_i \geq x$; if we also have $\vec{X} \neq \vec{Y}$, we write $\vec{X} < \vec{Y}$.

Definition 2.1: An ordered set of sequences $\vec{S1}, \vec{S2}, \cdots, \vec{Sk}$, is said to obey the **stacking property** if

$$\vec{S1} \geq \vec{S2} \geq \cdots \geq \vec{Sk} \tag{1}$$

Thresholding operations create binary sequences which stack. Suppose that $\vec{S}$ is a sequence taking on M values, then define the new binary sequence t_ℓ, produced by thresholding $\vec{S}$ at the level ℓ, as follows:

$$t_{\ell,i} = \begin{cases} 1 & \text{if } S_i \geq \ell \\ 0 & \text{if } S_i < \ell \end{cases} \tag{2}$$

The sequence of binary threshold signals $\vec{t}_1, \vec{t}_2, \cdots, \vec{t}_M$ possesses the stacking property.

This thresholding operation is formally called the operator $T_\ell(\cdot)$, so that $\vec{t}_\ell = T_\ell(\vec{S})$ is the same statement as Equation (2). The thresholding operation and the binary threshold signals it produces may be seen on the left side of Figure 1.

The window operators we consider will be constructed from window operators with binary inputs. For now, consider any b-input Boolean operator $B: \{0,1\}^b \to \{0,1\}$.

Definition 2.2: A Boolean operator B is said to possess the stacking property if

$$B(\vec{x}) \geq B(\vec{y}) \quad \text{whenever } \vec{x} \geq \vec{y} \tag{3}$$

A necessary and sufficient condition for B to stack is that it be a *positive Boolean function [12]*.

For the purposes of the optimization procedure developed later in the paper, we extend the definition of positive Boolean functions to allow for randomization of its output.

Definition 2.3: Let the 2^b possible binary sequences of length b be ordered in some fashion as $\overline{x}_1, \overline{x}_2, \cdots$. A window-width b *randomizing Boolean function* $f(\cdot)$ is defined by the vector $\overline{p}_f$ with 2^b elements, in which the i'th element is

$$P_f(1 \mid \overline{x}_i) = \text{Prob}(f \text{ outputs } 1 \mid \overline{x}_i \text{ in its window }) \quad i = 1,2,...,2^b \qquad (4)$$

Also define $P_f(0 \mid \overline{x}) = 1-P_f(1 \mid \overline{x})$.

Definition 2.4: A randomizing Boolean function is then said to be positive if it satisfies the probabilistic stacking constraint:

$$E(f(\overline{x})) \geq E(\overline{f}(y)) \quad \text{whenever} \quad \overline{x} \geq \overline{y} \qquad (5)$$

In which $E(\cdot)$ is the expectation operator as defined on a suitable probability space.

A randomizing stack filter is defined as a "stack" of randomizing positive Boolean functions:

Definition 2.5: Let B be a randomizing Boolean window operator which possesses the probabilistic stacking property. Then the operation of the randomizing **stack filter** $\mathbf{S}_B(\cdot)$ applied to the M-valued sequence $\vec{S}$ is defined by

$$\mathbf{S}_B(\vec{S}) = \sum_{i=1}^{M} B(T_i(\vec{S})) = \sum_{i=1}^{M} B(\vec{t}_i) \qquad (6)$$

in which $T_i(\cdot)$ is the thresholding operation defined earlier, and $\vec{t}_i$ is the i'th binary threshold

Deterministic stack filters, as defined in [1,2], are simply a subset of the randomizing stack filters; the difference being the certainty with which decisions are made for each possible window vector. The class of stack filters includes all the multi-level rank order operators and all morphological operators. For instance, if we define the deterministic binary window operator

$$MF3(\overline{x}) = MF3(x_1,x_2,x_3) = x_1x_2+x_2x_3+x_3x_1 \qquad (7)$$

in which multiplication is the logical AND and addition is logical OR, then $MF3(\cdot)$ is a binary median filter of window width 3. The deterministic stack filter $\mathbf{S}_{MF3}$ is then the (multi-level) median filter as shown in its stack filter form in Figure 1.

The proof of the **threshold decomposition property** of median filters [5,6] is thus the proof that regular (i.e. multi-level) median filtering is equivalent to the operation $\mathbf{S}_{MF3}$. This shows that the threshold decomposition is a superposition property for these nonlinear filters. This property is shared by all other rank order operators [6], and, by definition, by all stack filters [1,2].

It is appropriate at this point to summarize what will follow. The class of filters over which we will optimize is called $\mathcal{F}$ and contains all randomizing stack filters. The error criterion will be an absolute error criterion. The primary result will be that a non-randomizing stack filter is always optimal, and that there exists an efficient algorithm to find the optimal filter.

2.2. The Signal and Noise at the Filter's Input:

The signal S(t) to be estimated by the stack filter is chosen, for simplicity only, to be a Markov Chain. It is corrupted by a noise signal N(t) which is Markovian and independent of S(t). The two signals are combined in a memoryless nonlinearity. The output of this nonlinearity is a process defined on the product of the state spaces of the signal and noise in order to preserve the simple Markovian nature of the process. This gives the one sample at a time evolution of the process that the filter sees as its input.

The window of the filter holds b samples so this one sample representation must be converted to a b sample representation. The resulting vector Markov process has state space Q_W and is called the *multi-level window vector process* $\vec{W}(t)$. If the state transition matrix for the signal is P_S and the one for the noise is P_N, then the state transition matrix producing the observed window vector process is

$$P_{\vec{W}} = P_S \otimes P_N \otimes \vec{I}_m \otimes I \otimes \vec{I}_b^T \tag{8}$$

in which $\vec{I}_x$ is a row vector with x 1's, m is the number of states obtained by crossing the state spaces of the signal and the noise, I is the identity matrix, and T denotes the transpose operation.

The limiting probabilities of the multi-level window process are arranged in a vector $\bar{\pi}$ whose entry for the i'th state $\vec{W}_i$ of the window process is denoted as

$$\pi(\vec{W}_i) = \lim_{t \to \infty} (\vec{W}(t) = \vec{W}_i) \tag{9}$$

2.3. The Optimization Problem for a Single Threshold:

The following are the costs a randomizing stack filter incurs in its operations on level ℓ of the threshold decomposition architecture. The cost is based on the decisions the filter makes on each length b *binary* window vector it observes. Recall that this filter produces binary output, but can choose randomly between 0 and 1 for each observed window vector.

For each binary window vector it observes, the filter on level ℓ must decide whether the desired signal S(t) is less than ℓ or not. The real valued costs associated with these decisions are as follows:

$C_\ell(\vec{W},0,0)$ The cost of deciding the signal S(t) is below level ℓ when it is actually below ℓ and the (multi-level) window process is in state $\vec{W}$.

$C_\ell(\vec{W},0,1)$ The cost of deciding the signal S(t) is equal to or greater than ℓ when it is actually below ℓ and the window process is in state $\vec{W}$.

$C_\ell(\vec{W},1,1)$ The cost of deciding the signal S(t) is greater than or equal to level ℓ when it is actually greater than or equal to ℓ and the window process is in state $\vec{W}$.

$C_\ell(\vec{W},1,0)$ The cost of deciding the signal S(t) is below ℓ when it is actually greater than or equal to ℓ and the window process is in state $\vec{W}$.

The goal is to determine the filter which minimizes the following cost function.

Definition 4.2: The cost of using the randomizing binary filter f on threshold level ℓ is a function $C(f \mid \ell)$ defined as

$$C(f \mid \ell) = \sum_{\vec{W} \in Q_W} \left[(C_\ell(\vec{W},0,0)\pi(0 \mid \vec{W},\ell) + C_\ell(\vec{W},1,0)\pi(1 \mid \vec{W},\ell))\pi(\vec{W}) \right] P_f(0 \mid \vec{w}_\ell) + \tag{10}$$

$$+ \left[(C_\ell(\vec{W},0,1)\pi(0 \mid \vec{W},\ell) + C_\ell(\vec{W},1,1)\pi(1 \mid \vec{W},\ell))\pi(\vec{W}) \right] P_f(1 \mid \vec{w}_\ell)$$

in which the sum is over all possible states $\vec{W}$ of the multi-level window process, $\vec{w}_\ell$ is shorthand for the binary window vector on level ℓ produced by thresholding the window process state $\vec{W}$ at level ℓ; thus, $\vec{w}_\ell = T_\ell(\vec{W})$. Also,

$$\pi(0 \mid \vec{W},\ell) = 1 - \pi(1 \mid \vec{W},\ell) = \text{Prob(correct output on level } \ell \text{ is 0} \mid \text{state } \vec{W}). \tag{11}$$

which will be 0 or 1 only if complete information about the window process is available.

If the window process is binary, so that there is only one level to be filtered, then choosing $C(\vec{W},0,0) = C(\vec{W},1,1) = 0$, and $C(\vec{W},1,0) = C(\vec{W},0,1) = 1$ yields the absolute error criterion for filtering binary signals.

The cost function in Equation (10) takes on a particularly simple form if we rearrange the sum, taking advantage of the fact that several states of the multi-level window process states may produce the same binary window vector $\vec{w}$ when they are all thresholded at the same level.

$$C(f \mid \ell) = \sum_{j=0}^{2^b} \left[P_f(0 \mid \vec{x}_j) E(\text{Cost} \mid \vec{x}_j, 0, \ell) + P_f(1 \mid \vec{x}_j) E(\text{Cost} \mid \vec{x}_j, 1, \ell) \right] \quad (12)$$

in which

$$E(\text{Cost} \mid 0, \vec{x}_j, \ell) = E(\text{Cost} \mid 0 \text{ output } \& \ \vec{x}_j \text{ observed on level } \ell) = \quad (13)$$

$$\sum_{\vec{W}: \ \vec{w}_\ell = \vec{x}_j} (C_\ell(\vec{W}, 0, 0) \pi(0 \mid \vec{W}, \ell) + C_\ell(\vec{W}, 1, 0) \pi(1 \mid \vec{W}, \ell)) \pi(\vec{W})$$

$$E(\text{Cost} \mid 1, \vec{x}_j, \ell) = E(\text{Cost} \mid 1 \text{ output } \& \ \vec{x}_j \text{ observed onlevel } \ell) =$$

$$\sum_{\vec{W}: \ \vec{w}_\ell = \vec{x}_j} (C_\ell(\vec{W}, 0, 1) \pi(0 \mid \vec{W}, \ell) + C_\ell(\vec{W}, 1, 1) \pi(1 \mid \vec{W}, \ell)) \pi(\vec{W})$$

which are the expected costs of deciding zero or one for each window vector observed by the filter.

Since only the binary window vectors obtained by thresholding the window vector process are observed by the filter on that threshold level, the details of the underlying (multi-level) window process are unimportant to the filter. It is sensitive only to the limiting expected costs of its decisions. This is the most likely source of the robust behavior observed when these filters are used in many applications.

With the set $\mathcal{F}$ of randomizing binary functions and the cost function introduced above, we can now state the optimization problem.

Optimization Problem: The window width b randomizing binary filter $g \in \mathcal{F}$ is said to be optimal for *window width b stack filtering* the binary threshold signal on level ℓ in the threshold decomposition architecture if it achieves

$$C(g \mid \ell) = \min_{f \in \mathcal{F}} C(f \mid \ell) \quad (14)$$

subject to the probabilistic stacking constraint

$$E(f(\vec{x})) \geq E(f(\vec{y})) \quad \text{whenever} \quad \vec{x} \geq \vec{y} \quad (15)$$

for any pair of length b binary vectors $\vec{x}$ and $\vec{y}$.
□

The probabilistic stacking constraint still ensures that if the optimal threshold function were used on other levels as well, that it would stack. All that would be necessary would be that decisions be coupled on the different levels. The long term average error would still be the same.

2.4. Solving the Optimization Problem on One Level:

It is proven in [17] that optimization problem stated in the previous subsection is equivalent to the following linear program:

Linear Program: Find the decision probability vector

$$\vec{p}_f = \left[P_f(1 \mid \vec{w}_1) \ \cdots \ P_f(1 \mid \vec{w}_{2^b}) \right] \quad (16)$$

which achieves

$$\min_{g} C(g \mid \ell) = \min_{g} \sum_{j=0}^{2^b} \left[P_g(0 \mid \overline{x}_j) E(\text{Cost} \mid \overline{x}_j, 0, \ell) + P_g(1 \mid \overline{x}_j) E(\text{Cost} \mid \overline{x}_j, 1, \ell) \right]$$

subject to:

For each of the 2^b binary vectors $\overline{x}$ of length b:

$$P_g(1 \mid \overline{x}) + P_g(0 \mid \overline{x}) = 1 \qquad (17)$$

For each pair of binary vectors $(\overline{x}, \overline{y})$ differing in only one bit position, with $\overline{x} < \overline{y}$:

$$P_g(1 \mid \overline{y}) = P_g(1 \mid \overline{x}) + \nu(\overline{x} \to \overline{y}) \qquad (18)$$

$$P_g(0 \mid \overline{x}) = P_g(0 \mid \overline{y}) + \nu(\overline{y} \to \overline{x})$$

in which all the variables are nonnegative. The variables $\nu(\cdot)$ are slack variables.
□

This linear programming formulation leads to:

Theorem 1: (Proven in [17]) The optimal window width b randomizing stack filter for binary input is a window width b deterministic stack filter. The linear program to find the optimal filter has $O(b2^b)$ variables and constraints.

The linear program nearly has the minimum possible complexity. It must examine the output of the randomizing stack filters at each of the 2^b possible binary window vectors just to determine the cost of using that particular filter. Thus, the size of the linear program is not much worse than the size of the problem of computing the cost of a single positive Boolean function.

2.5. Solving the Optimization Problem for Multi-level Input:

In the previous section it was shown that for the filtering of binary input, deterministic binary stack filters minimize the error criterion $C(f \mid \ell)$ introduced in Equation (12). In this section we define an error criterion for filtering multi-level signals by exploiting the threshold decomposition and the stacking property, the defining properties of the class of filters over which we are optimizing. The optimal filter will again be a deterministic filter. We will illustrate the technique with the absolute error criterion, which is a special case of the one level error criterion introduced earlier.

Let $C(\mathbf{S}_f)$ be the expected absolute error incurred by filtering the window process $\overrightarrow{W}(t)$ with the randomizing stack filter $\mathbf{S}_f$, which is obtained by using the randomizing binary filter f on each level of the threshold decomposition architecture shown in Figure 1. Then with $S(t)$ as the signal we are estimating

$$C(\mathbf{S}_f) = E(\mid S(t) - \mathbf{S}_f(\overrightarrow{W}(t)) \mid) = E(\mid e(t) \mid) \qquad (19)$$

Then using the threshold function defined in Equation (2), and assuming without loss of generality that the signal $S(t)$ and each element of the window vector process $\overrightarrow{W}(t)$ are M valued and nonnegative,

$$C(\mathbf{S}_f) = E(\mid \sum_{i=1}^{M} T_i(S(t)) - \sum_{i=1}^{M} T_i(\mathbf{S}_f(\overrightarrow{W}(t))) \mid) \qquad (20)$$

in which the dimension of each argument of the threshold operation is one. Since $\mathbf{S}_f$ is a stack filter constructed from the randomizing binary filter $f \in \mathcal{F}$,

$$T_i(\mathbf{S}_f(\overrightarrow{W}(t))) = f(T_i(\overrightarrow{W}(t))) = f(\overrightarrow{w}_i(t)) \qquad (21)$$

in which the binary vector $\overrightarrow{w}_i$ is the binary signal obtained by thresholding the multi-level window vector $\overrightarrow{W}(t)$ at level i. Note that the dimension of the argument of the threshold function changes on passing it inside the stack filtering operator. Equation (20) now becomes

$$C(\mathbf{S}_f) = E(\mid \sum_{i=1}^{M} T_i(S(t)) - \sum_{i=1}^{M} f(\overline{w}_i(t)) \mid) \tag{22}$$

Regroup the sum, and note that, because of the stacking property of thresholding and because of the coordination among levels of the randomizing stack filter, that all the elements of the sum are either nonnegative or nonpositive, so that the summation and the absolute value operations may be interchanged:

$$\mid \sum_{i=1}^{M} (T_i(S(t)) - f(\overline{w}_i(t))) \mid = \sum_{i=1}^{M} \mid T_i(S(t)) - f(\overline{w}_i(t)) \mid \tag{23}$$

which leads to

$$C(\mathbf{S}_f) = \sum_{i=1}^{M} E\left[\mid T_i(S(t)) - f(\overline{w}_i(t)) \mid \right] \tag{24}$$

The sum in the preceding equation is over the absolute error incurred on each threshold level. This corresponds to choosing the costs defined for filtering on level i as $C_i(\overrightarrow{W},0,0) = C_i(\overrightarrow{W},1,1) = 0$ and $C_i(\overrightarrow{W},1,0) = C_i(\overrightarrow{W},0,1) = 1$ for each threshold level i. With these costs specified in the error criterion $C(f \mid i)$ in Equation (10), we find that the absolute error associated with using the stack filter $\mathbf{S}_f(\cdot)$ is

$$C(\mathbf{S}_f) = \sum_{i=1}^{M} C(f \mid i) \tag{25}$$

the absolute error on each level is simply added together to produce the overall multi-level absolute error. Using the costs functions of the decisions of the filters for each level of the threshold decomposition architecture:

$$C(\mathbf{S}_f) = \sum_{j=1}^{2^b} \left[P_f(0 \mid \overline{x}_j) \left(\sum_{i=1}^{M} E(\text{cost} \mid 0, \overline{x}_j, i) \right) + P_f(1 \mid \overline{x}_j) \left(\sum_{i=1}^{M} E(\text{cost} \mid 1, \overline{x}_j, i) \right) \right] \tag{26}$$

in which the costs are those experienced when making decisions on each of the M levels.

The multi-level cost function now looks exactly like the single level cost function in Equation (12). The optimization procedure developed in Section 2.4 can therefore be applied without change, yielding a positive Boolean function as the optimal filter. Of course, this time the filter is being used on each level of the multi-level randomizing stack filter.

Theorem 2: (Proven in [17]) The optimal multi-level window width b randomizing stack filter is a deterministic stack filter; no randomization of outputs is necessary; the filter need only produce binary outputs on each level. The optimal filter can be found via a linear program with $O(b2^b)$ constraints and variables.

This theorem shows that deterministic stack filtering is optimal for filtering under the absolute error criterion for multi-level signals as well as for binary signals. Many generalizations are possible; for instance, it is possible to have different filters on each level of the threshold decomposition architecture while still preserving the stacking property. It is also possible to feed the binary threshold signals of several levels into a filter on just one level. This just a different type of superposition, with each filter now operating on a plane of 0's and 1's instead of just a stream of them. This is the usual approach in morphology, but all morphologists seem to have missed the significance of the stacking property in these operations -- and they certainly haven't recognized the fact that the filters they use obey a superposition property.

3. THE VLSI IMPLEMENTATION OF RANK ORDER FILTERS [8-10]

The threshold decomposition architecture shown in Figure 1 appears perfect for VLSI implementation since it is a massively parallel set of very simple pieces. By demonstrating this architecture to Prof. Steve Bass of Purdue, we enticed him to design such a chip. The challenge was to prove that the architecture did indeed have an efficient VLSI implementation. The result is a 5μ NMOS VLSI chip which can be programmed to perform any recursive or nonrecursive rank order operation of window width 3,5,7, or 9. The design produced by Prof. Bass and graduate student Ron Harber is reported in full in [10]. We only summarize the results in these papers.

Figure 2: 5μ NMOS Threshold Decomposition Rank Order Filter Chip [9,10]
This chip can perform any recursive or nonrecursive rank order operation of window width 3,5,7, or 9 on 6 bit input. There are 63 binary rank order filters on the chip; 32 on the left, 31 on the right.

As mentioned above, the threshold decomposition appears perfect for VLSI implementation because it is a massively parallel connection of very simple pieces; however, the sheer number of filters that must be placed in parallel could create space problems. For n-bit input, there are 2^n thresholding operations and binary rank order filters. It is thus very important that these operations be designed to take up as little chip area as possible. The thresholding operations cause little trouble in this regard; each threshold operation requires one-and-a-half transistors. Care must be taken with the programmable binary rank order filters, though, since so many different operations are to be designed into filter.

The novel solution to the space problem posed by these filters was the use of analog circuitry [9,10]. A binary rank order filter simply adds together all of bits in its window and compares the result against a threshold to decide whether to put out a 0 or a 1. This summing operation can be implemented in a space efficient fashion with a voltage divider circuit. This circuit helped make it possible to fit 63 of the programmable binary rank order filters onto the chip.

The final piece of the implementation is the summing of the outputs of the 2^n rank order operators. This is a large number of addition operations and could require the entire chip area by itself. The solution, discovered by Steve Bass, is to exploit the stacking property. The sum of the outputs of the filter is the same as the value of the threshold level of the highest 1 in the stack of ones at the output. A binary tree search can thus be executed to find this threshold level. The circuit implementing this search is the cascade structure in the middle of the chip; the complete design of which is shown in Figure 2. This chip takes 6-bit input; there are 32 filters on one side and 31 on the other side of the chip. Their outputs feed the tree search circuitry in the middle of the chip.

Without the stacking property there would be neither a VLSI implementation of these filters nor the multi-level optimization theory of Section 2.5.

4. CONCLUSIONS

1) The results in Section 2 are the first comprehensive optimality theory for stack filters (and therefore for gray-level morphological filters);

2) The results in Section 2 show that an appropriate way to perform optimization under the absolute error criterion is to the best possible *consistent* decision on each threshold level about whether the signal being estimated is above that threshold level or not. By consistency, we mean that we don't decide the signal is below level ℓ and then decide it is above level $\ell + i$ for some positive integer i. The consistency of decisions is guaranteed by the stacking property. Estimation under the absolute error criterion is thus massively parallel detection with a consistency condition added.

3) By quantizing the signal more and more finely, and at the same time sampling faster and faster, one could obtain a continuous time version of this paper. In such a case, it would then be filters like the analog median filter [16] that would be optimal under the absolute error criterion.

4) the proof of optimality of stack filters in higher dimensions is exactly the same as that developed in this paper for one dimensional signal processing.

In summary, the results in this paper should point out the proper use of stack filters in estimation theory; namely to reduce noise when the error is measured via an absolute error criterion.

5. REFERENCES

[1] P.D. Wendt, E.J. Coyle, and N.C. Gallagher, "Stack filters: their definition and some initial properties," *Proc. of the XIX Conf. on Information Science and Systems,* Baltimore, MD, March 1985.

[2] P.D. Wendt, E.J. Coyle, and N.C. Gallagher, "Stack filters," *IEEE Trans. on Acoustics, Speech, and Signal Processing.* vol. ASSP-34, no. 4, August 1986.

[3] J.P. Fitch, "Software and VLSI algorithms for generalized ranked order filtering," submitted to the *IEEE transactions on Circuits and Systems,* March 1986.

[4] P.A. Maragos and R.W. Schafer, "A unification of linear, median, order statistics, and morphological filters under mathematical morphology," *Proceedings of the 1985 Int. Conf. on Acoustics, Speech, and Signal Processing,* Tampa, Fla., March 1985.

[5] J.P. Fitch, E.J. Coyle, and N.C. Gallagher, "Median filtering by threshold decomposition," *IEEE Trans. Acoust., Speech, and Signal Processing,* vol. ASSP-32, pp 1183-1189, Dec. 1984.

[6] J.P. Fitch, E.J. Coyle, and N.C. Gallagher, "Threshold decomposition of multidimensional rank order operations," *IEEE Transaction on Circuits and Systems,* vol. CAS-32, pp 445-450, May 1985.

[7] J. Serra, *Image Analysis and Mathematical Morphology,* Academic Press: New York, 1982.

[8] J.P. Fitch, E.J. Coyle, and N.C. Gallagher, "Analysis and Implementation of Median-Type Filters," School of Electrical Engineering, Purdue University, West Lafayette, IN; TR. EE 84-20, July 1984.

[9] R.G. Harber and S.C. Bass, "VLSI Implementation of a fast rank order filtering algorithm," *Proceedings of the 1985 Int. Conf. on Acoust., Speech, and Signal Processing,* Tampa, FL, March 1985.

[10] R.G. Harber, S.C. Bass, J.P. Fitch, E.J. Coyle, and N.C. Gallagher, "The VLSI implementation of rank order operators," submitted to the special issue on VLSI and Signal Processing of the *Proceedings of the IEEE,*

[11] E.J. Coyle, "On the optimality of multi-level rank order operations under the absolute error criterion," *Proceedings of the 1986 Conf. on Information Science and Systems,* Princeton, NJ, March 1986; and submitted to *IEEE Trans. on Acoustics, Speech and Signal Processing.*

[12] E.N. Gilbert, "Lattice-theoretic properties of frontal switching switching functions," *J. Math. Physics,* vol. 33, pp 57-67, Apr. 1954.

[13] S. Muroga, *Threshold Logic and its Applications,* New York: Wiley, 1971.

[14] J. A. Roskind, "A fast sort-selection filter chip with effectively linear hardware complexity," *Proc. of the 1985 Int. Conf. on Acoustics, Speech, and Signal Processing,* Tampa, FL, March 1985.

[15] K. Oflazer, "Design and implementation of a single-chip 1-D median filter," *IEEE Trans. on Acoustics, Speech, and Signal Processing,* vol ASSP-31, no. 5, Oct. 1983.

[16] J.P. Fitch, E.J. Coyle, and N.C. Gallagher, "The analog median filter," *IEEE Trans. on Circuits and Systems,* vol. CAS-33, no. 1, pp. 94-102, January 1986.

[17] E.J. Coyle, "Choosing the stack filter which minimizes the mean absolute error criterion," to be presented at the *1986 Allerton Conference on Communication, Control, and Computers,* Oct. 1-3, 1986; and submitted to *IEEE Trans. on Acoustics, Speech and Signal Processing,* July 1986.

CHAPTER 14

INDEPENDENT DATA FLOW WAVEFRONT ARRAY PROCESSORS FOR RECURSIVE EQUATIONS

Sayfe Kiaei   Uday B. Desai

The Department of Electrical & Computer Engineering
Washington State University
Pullman, WA 99164-2210

A new pipelined architecture for VLSI implementation of recursive equations is developed. The new architecture, termed independent data flow wavefront array processor (IDFWAP), combines locality, modularity, and asynchoronous data-driven features of the systolic/wavefront array processors, with the additional feature of independent data flow. The IDFWAP is used for VLSI implementation of state variable equations, in particular the Kalman filter. The new architecture is general enough to handle state variable equations with time-varying parameters.

I. INTRODUCTION

The Kalman Filter (KF) is a standard tool used for state estimation [1]. It has been successfully applied to many signal processing problems like image enhancements, target prediction, sonar detection, beam formation, etc. [2] - [4], and control problems like flight control, navigation control, guidance control, etc. [5]. The computational complexity of the Kalman filter is determined by the order of the state model. For an n-th order model the Kalman filter requires $O(n^2)$ operations for time update, and $O(n^3)$ operations for the error covariance update. Thus for large n, this could pose a severe limitation on real time applications. One way to overcome this problem is to develop parallel algorithms and/or design pipelined architectures for the Kalman filter implementation.

Hierarchically, from top to bottom, there are three design methodologies for parallel processing:
   (1) System level decomposition.
   (2) Data level decomposition.
   (3) Architectural parallelism.

A combination of the above three approaches should yield "maximum" possible speedup. In this paper we shall concentrate on the third approach. Our main objective is to develop a pipelined architecture for implementing state variable filters, and in particular Kalman filters.

1.1 System Level Decomposition

Essentially, using some specific structure possesed by the state variable system, a set of decoupled systems are obtained via a transformation of the original system. Now if the original state variables are desired, the inverse transformation has to be applied, thus making the algorithm slow and inefficient for on-line implementation. Nevertheless, for many problems system decomposition does become essential. For a discussion on system decomposition, and tradeoffs between decomposition techniques, and architectural approaches see Gilbert et al. [7].

1.2 Data Level Decomposition

An algorithmic approach for achieving speedup is to segment the data, and then process these segments in parallel. Meyer and Weinert [8] presented an algorithm for linear regression where the input sequence was broken into segments at multiples of some time instant. Such a segmentation also provides fault tolerant architectures. Based on the idea of block processing, Nikias [9] - [10] introduced a new structure for realizing IIR digital filters. Based on [8] and [9] a parallel KF algorithm was presented in [11]. Also similar segmentation ideas were used in developing a parallel hierarchical estimation algorithm [12].

1.3 Architectural Parallelism

In many real time signal processing problems, there is an increasing demand for high speed computation. Often system or data decomposition methods will not yield a parallel structure to achieve the desired speedup. Thus it seems that the only alternative left is to exploit the possibility of having parallel/pipelined architecture which can handle a higher throughput rate.

In general, current parallel architectures can be divided into three classes:
(i)   Multiprocessor systems
(ii)   Vector processors
(iii) Array processors
The development of the first two classes belongs to the realm of general purpose computer design. The third class, namely array processors, consists of special purpose architectures built specifically for the desired problem. The array processor architecture is designed such that it would maximally utilize the parallelism of the algorithm in terms of pipelined computing, and concurrent processing.

The main thrust of the paper is to design an array processor for recursive equations, specifically state variable equations.

1.4 Organization of the Paper

Section 2 deals with the description of two important array processing architectures. Systolic [19] and Wavefront [22] array processors have played a significant role in the VLSI implementation of many signal processing algorithms. Both of the architectures introduced new design principles for array processing and are discussed in Sections 2.1 and 2.2.

In Section 3 we introduce the new architecture termed Independent Data Flow Wavefront Array Processor (IDFWAP). For the purpose of illustration on the application of IDFWAP, matrix vector multiplication is discussed in section 3.1, and matrix-matrix multiplication in 3.2. Section 4 discusses the implementation of recursive equations using IDFWAP, and is applied to Kalman

filtering problem in 4.1 and 4.2. The implementation of the Kalman filter in a feedback form for LQG problems is presented in 4.3.

2.   PARALLEL ARCHITECTURE FOR SIGNAL PROCESSING

The design of specialized processing chips should take into account the constraints involved in the VLSI architecture. For large scale computational systems requiring very high throughput, locality of data flow and the controller becomes an important constraint in VLSI design. Also, this locality property simplifies the hardware design of the chip. Another important consideration is modularity of the processors. Modularity will reduce the design hardship, and the architecture could be easily expanded to fit different computational size processing. And, finally, faster throughout rate can be achieved by pipelinability of the architecture; thus the rate of processing the data will be independent of the order of the model.

2.1 Systolic Array

H.T. Kung and C.E. Leiserson [13] introduced a new class of array processors called systolic arrays which revolutionized implementation of many signal processing and matrix algorithms. Systolic arrays are a network of several interconnected identical cells which satisfy the constraints of locality, modularity, and pipelinability. The basic idea behind the systolic architecture is that each cell performs some basic operations, and passes the intermediate results to the neighboring cell. The most important feature of this array is that the data can be pumped regularly through the system, thus providing a uniform and fast throughput rate. A majority of the signal processing algorithms require the following inner product operation;

$$C \leftarrow C + A * B.$$

Consequently, the cell structure in the systolic processor performs the elementary inner product operation. Each cell receives the data, carries out the inner product, and pumps the results rhythmically to the neighboring cell. Systolic arrays have been extensively used in many signal and image processing problems, and matrix computations [13] - [14]. The reader is referred to Fisher and Kung [15] for a complete reference of application of systolic arrays.

2.2 Wavefront Arrays

One problem with systolic arrays is the global control of data movement in different cells. To assure proper timing and synchronization in systolic arrays, extra delays are needed which slows down the computation, thereby decreasing throughput rate. Moreover, for large scale arrays this synchronization could become very tedious.

The above problems can be overcome by asynchronous data driven cell. This permits local control and self-timed cells in which the inner product operation is triggered by the availability of the data. This idea was developed and applied to systolic arrays by S.Y. Kung [16], where he termed the architecture as a wavefront array.

The major difference between the two architectures is the fact that wavefront array transfers the data to the next cell asynchronously by handshaking, while systolic arrays require global timing and control of data flow. For further comparison of the two architectures and applications, the reader is referred to [17] - [18].

3. INDEPENDENT DATA FLOW WAVEFRONT ARRAY PROCESSOR (IDFWAP)

The development of the IDFWAP is motivated by the systolic array of H.T. Kung [13], and the wavefront array of S.Y. Kung [17]. The basic idea is that a processing cell does not have to wait for the data until the previous cell completes its computation. The IDFWAP has the following features.

    (i)   Modularity
    (ii)   Locality
    (iii)   Self-timed cells with asynchronous data transfer
    (iv)   Independent data flow with respect to cell operation.

IDFWAP has the basic features of the wavefront array with the exception that the data flow is independent of the processing time for each cell. To illustrate the basic principle in a simple fashion, we first describe the IDFWAP architecture for the matrix-vector, and matrix-matrix multiplication. Then these ideas are applied to develop the IDFWAP architecture for recursive filters, particularly state variable filter implementation. Special emphasis is placed on the implementation of the Kalman Filter.

3.1 Matrix-Vector Multiplication Processor (MVP)

Consider an nxn matrix A and an nx1 vector x. The objective is to perform the multiplication Ax=b. Each element of b is

$$b_i = \sum_{j=1}^{n} a_{ij}x_j$$

where it can be written in recursive form as

$$b_i^{(j+1)} = b_i^{(j)} + a_{ij}x_j \; ; \; b_i^{(1)} = 0 \; , \; j=1,2,\ldots,n.$$

The detailed architecture of IDFWAP for the multiplication of a 4x4 matrix by a 4x1 vector is shown in Figure 1. Each cell consists of 3 sections: (i) Data flow path; (ii) Arithmetic logic unit (ALU); (iii) Internal registers. The key idea is to make the data flow independent of the operation of each cell; meaning that cell P_i does not have to wait for P_{i-1} to complete the inner-product operation and then receive the data. If all the cells require the same data x_j, then it can be propogated from left to right asynchron-ously, independent of the cell operation. The data is tagged with handshaking signals and it will be referred to as "token." As the token is being transmitted through all the processors, it activates the operation of each processor, and moves on to the next processor.

Data Flow Path   The data flow path consists of asynchronous register nodes D_i driven by the token. The operation of data flow in the node D_i is as follows:
   (1) Node D_i receives the token from D_{i-1} by handshaking.
   (2) Input signal x_j is then buffered to input register R_i.
   (3) D_i informs D_{i+1} that data is available to be fetched to it.
   (4) D_i sends x_j to D_{i+1}, and awaits for the new token from D_{i-1}.

It is important to note that data transfer from D_i to R_i, or to D_{i+1} is done upon readiness of them to accept the data, and is handled asynchronously by handshaking techniques. The flow of data from D_1 to D_4 is in a wavefront fashion from left to right. An obvious question that might be raised is why not use a common data bus since all cells receive the same data. The answer is because of the locality and modularity restriction on the architecture.

The main principle behind this data transfer is that data propagates independently of the inner product operation, and thus activates all the processors with only a minor delay for data communication.

Arithmetic Logic Unit (ALU)   The ALU consists of a multiplier, and an adder. The multiplier and an adder can be built by using existing fast parallel architectures (see Hwang [20], and TRW [21]). The operation of the ALU is triggered by the availability of the data in register R_i. The multiplier M_i fetches the data, performs the multiplication, and then sends the results to the adder. Then it awaits for new data from R_i. Adder performs addition of register C_i to the multiplication result. It is worth noting that by having localized independent controllers for the multiplier and for the adder, a higher degree of parallelism can be achieved.

<u>Internal Registers</u> There are two registers in each processor, R_i and C_i. The register R_i is used as an intermediate buffer to assure faster data flow in the data path. If the multiplier is busy, R_i buffers the data and overcomes the data transfer blockage in the data path. The register C_i stores the intermediate results, and after n iterations, it contains the final value b_i. The data transfer in these registers is also handled by handshaking.

Thus, this architecture employs two levels of parallelism: (1) data flow independent of cell operation, (2) parallelism by pipelining the algorithm with the use of systolic cells. Independent data flow, which is obtained by the use of independent controllers in the data path, assures the maximum utilization of all the processors, and is the main principle behind this new architecture.

3.3 Matrix-Matrix Multiplication Processor (MMP)

Using the ideas developed in the previous section, an architecture for matrix-matrix multiplication is shown in Figure 2 for n=4. This architecture is structurally the same as the one presented by S.Y. Kung [17], with the exception of data flow. The basic operation of this array processor is as follows:
(1) Processors P_{11}-P_{14}, P_{11}-P_{41} fetch the input data b_{11}-b_{14} and a_{11}-a_{41} respectively.
(2) The first wave of input data moves across the entire array (b_{11}-b_{14} from top to bottom, and a_{11}-a_{41} from left to right).
(3) Each processor P_{ij} fetches the input data and starts the inner product operation.
(4) The new wave of input data follows the previous wave and so on.

Again it should be emphasized that the flow of data occurs independent of the cell operation, thus all n^2 processors get triggered together in a very short time.

In this architecture there are two different wavefronts involved. One is the data flow wavefront, and the other wavefront is the operation of the cells. This fact increases the efficiency of the processors, and the speed up.

4. IMPLEMENTATION OF RECURSIVE EQUATIONS

Many signal processing algorithms, particularly the Kalman filter, are recursive in nature. They can be expressed in the following state variable form as:

$$x(k+1) = Ax(k) + Bu(k) \qquad (4.0)$$

$$x(k) = [x_1(k), x_2(k), \ldots, x_n(k)],$$

The above equation requires multiplication of matrix A by vector x. Due to the recursive nature of (4.0), the matrix vector multiplication of architecture of Section 3.1 cannot be used directly. The main problem arises from the feedback required to fetch $x(k)$ for the calculation of $x(k+1)$. Figure 3 shows the architecture proposed for the above problem.

This architecture exploits most of the features of the MVP, namely independent data flow, localized distributed controllers, parallelism in ALU, and other features of IDFWAP. The differences between the two architectures are:

(1) In this architecture, the inner product operation starts with the right-most processor P_1.

(2) The input data $x_j(k)$ is fetched only once, and gets updated at the next recursion $k+1$.

(3) This processor has two data paths. Input data path flowing from left to right, and the output path flowing in the opposite direction. The data along these two paths move independent of each other, and independent of cell operation.

In Figure 4 the snapshots of the processing at different times are shown. The operation of the processor starts with x_1 entering the input path. To guarantee that x_1 is fetched at P_1, it is tagged with a signal XNEW. XNEW labels the leading edge of the new state vector x at time k. After x_1 is fetched into R_1, the processor P_1 will transmit the signal FETCH to P_2. The FETCH signal will initiate transfer of x_2 into R_2 after it reaches node D_2. The activities in the input data path are summarized by the following steps:

1. Token x_1 propogates from node D_4 to D_1. Once D_1 recognizes x_1 (by use of XNEW), data transfer from D_2 to D_1 is halted.

2. x_1 is fetched into R_1, and P_1 transmits signal FETCH to P_2.

3. Once x_2 is available at D_2, it is fetched into R_2, and similarly P_2 transmits FETCH to D_3, and so on.

This action is propagated throughout the input path from right to left in a hierarchical fashion. The external input data $u(k)$ gets fetched in a similar fashion in a separate path.

After completion of multiplication at each processor, the result is transmitted to the lower path. The structure and data transfer hierarchy in the output path is similar to the input path with the exception that each node performs an addition, and then transmits the data to the left. The addition is performed when both operands are available. Once $x_1(k+1)$ is calculated, it is fedback to the input path along with the signal XNEW, for the next recursion.

The wavefront structure of the data flow and the fetching instruction gives a natural delay in between multiplication operations. This delay is indeed required in the output path for the addition operation. Another important feature of this architecture is that it can be used for time varying parameter case. In this case the local memory which contains a_{ij} has to be updated at each time k. Generalization of this architecture to multi-input (vector $u(k)$) is shown in Figure 5. Figure 6 describes the timing and operation flow for the single input case.

Speedup and Efficiency   Referring to Figure 6, the time involved in the calculation of each recursive time step k is:

$$\tau = nM + nS + (n-1)S = nM + (2n-1)S$$

where M = time required for one multiplication

S = time required for one addition. Speedup obtained by this architecture over sequential processing is:

$$SU = \frac{Time\ sequential}{Time\ parallel} = \frac{n^2M+n^2S}{nM+(2n-1)S}$$

Now efficiency defined as speedup over number of processors required will be:

$$\varepsilon = \frac{SU}{\#\ of\ processors} = \frac{n^2M+n^2S}{n^2M+n^2S+n(n-1)S}$$

If we let one multiplication time to be equal to α summations, namely, M = αS. Then

$$SU = \frac{n^2(\alpha+1)}{n\alpha+2n-1} \cong \frac{n(\alpha+1)}{\alpha+2}$$

$$\varepsilon = \frac{n^2(\alpha+1)}{\alpha n^2+2n^2-n} \cong \frac{\alpha+1}{\alpha+2}$$

It is seen that depending on the architectures of the multiplier and the adder, efficiency can be from 67% to 97%.

4.1 Kalman Filter

In this section, the implementation of both measurement and time update equation for the Kalman filter are presented. Architecture for the case of both scalar as well as vector measurements is discussed.

The State Space Model   Consider a zero mean stationary process $y(\cdot)$ generated by

$$y(k) = Hx(k) + v(k) \tag{4.1}$$

where the nx1 state vector x(k) satisfies the recursion

$$x(k+1) = Fx(k) + Gu(k)\ ,\ x(0) = x_o \tag{4.2}$$

u(k) and v(k) are (mx1) and (px1) white processes, with zero mean processes and covariances

$$E \begin{bmatrix} u(i) \\ v(i) \end{bmatrix} [u'(j)\quad v'(j)]' = \begin{bmatrix} Q(i) & C(i) \\ C'(i) & R(i) \end{bmatrix} \delta_{ij}$$

The initial state x_o has covariance Π_o and is uncorrelated with u(k) and v(k). The parameters of the model are assumed to be known a priori, and in this case we assume the model is time invariant. Let the time update, one step predicted estimate be

$\hat{x}(k+1/k)$ = Linear least-squares estimates

$\quad\quad$ ($\ell.\ell.$s.e.) of $x(k)$ given $\{y(k-1),\ y(k-2),\ \ldots,y(0)\}$ $\quad$ (4.4)

And the measurement update (filtered estimate) be

$\hat{x}(k/k)$ = $\ell.\ell.$s.e of $x(k)$ given $\{y(k),\ y(k-1),\ldots,\ y(0)\}$ $\quad\quad\quad$ (4.5)

Then the well-known time and measurement update equations are:

TIME UPDATE EQUATION

$\hat{x}(k+1/k) = F\hat{x}(k/k)$ $\quad\quad$ (for $C(k) = 0$) $\quad\quad\quad\quad\quad$ (4.6)

$P(k+1/k) = FP(k/k)F' + GQG'$ $\quad\quad\quad\quad\quad\quad\quad\quad\quad\quad$ (4.7)

MEASUREMENT UPDATE EQUATIONS

$\hat{x}(k/k) = \hat{x}(k-1/k) + K(k)(y(k) - H\hat{x}(k/k-1)$ $\quad\quad\quad\quad$ (4.8)

$K(k) = P(k/k)H'R_\varepsilon^{-1}(k)$ $\quad\quad\quad\quad\quad\quad\quad\quad\quad\quad\quad\quad$ (4.9)

$R_\varepsilon(k) = HP(k/k)H' + R(k)$ $\quad\quad\quad\quad\quad\quad\quad\quad\quad\quad\quad$ (4.10)

$P(k/k) = P(k/k-1) - P(k/k-1)H'R_\varepsilon^{-1}(k)HP(k/k-1)$ $\quad\quad$ (4.11)

4.2 Implementation of Measurement and Time Update Equations

We assume that the Kalman gain vector $K(k)$ is available either by an offline or an online calculation at each step k. There have been several architectures proposed for online calculation of the Kalman gain, and the reader is referred to [22] for a discussion of these architectures. The measurement update equation (4.8) can be written as

$\hat{x}(k/k) = A(k)\hat{x}(k/k-1) + K(k)y(k)$ $\quad\quad\quad\quad\quad\quad\quad$ (4.12)

where
$\quad\quad A(k) = I - K(k)H$
and is available at every recursion step k.

The structure of the time and measurement update processor is shown in Figure 7. The processor consists of two rows. The top row computes the measurement update, which is then fed into the lower row, where the time update value $\hat{x}(k+1/k)$ is computed. The architecture for each row is identical to the processor developed for recursive equation (4.0), and preserves the independent data flow and wavefronting principle. The time required for the calculation of both the measurement and the time update at each step k is 2τ, where τ = time required for matrix-vector multiplication at each row.

4.3 Implementation of Non-Stationary KF with an "External Feedback Loop"

In many situations, particularly in control system applications, one wants to implement the Kalman filter in the form:

$\hat{x}(k+1/k) = F\hat{x}(k/k-1) + K(k)\varepsilon(k)$ $\quad\quad\quad\quad\quad\quad\quad$ (4.13)

where $\varepsilon(k) = y(k) - H\hat{x}(k/k-1)$ is the innovations process. Note that (4.13) has two feedback loops, one inner loop due to the recursive nature of the equation, and an outer loop due to the term $y(k) - H\hat{x}(k/k-1)$. The above filter requires calculation of 3 variables: (i) innovations $\varepsilon(k)$; (ii) feedback term $K(k)\varepsilon(k)$; (iii) $F \hat{x}(k/k-1)$

Figure 8 shows the IDFWAP architecture for implementing (4.13). This architecture has 5 cells (for n=4), where P_5 is dedicated for multiplying $K(k)$ with $\varepsilon(k)$. As shown in Figure 8, the local memory for P_1 to P_4 contains the values of $-H_i$ in the bottom row, and then F_{ij} are stored above.

The operation of the processor is as follows:
(1) Processing starts with the multiplication $-H\hat{x}(k/k-1)$
(2) The first variable calculated at the output path is $\varepsilon(k)$.
(3) Once $\varepsilon(k)$ is available (which is labelled with XNEW), it is fetched into P_5.
(4) P_5 calculates the product of $K_1(k)\varepsilon(k)$. This value is stored until the next recursive step k+1.
(5) Simultaneously with the calculation of $\varepsilon(k)$, first row of F, namely F_1 is being multiplied by $\hat{x}(k/k-1)$.
(6) After $F_1\hat{x}(k/k-1)$ is computed, the summation $F_1\hat{x}(k/k-1)+K_1(k)\varepsilon(k) = \hat{x}_1(k+1/k)$ will be performed, (K_1 = first row of K(k)).
(7) The next rows will be computed in a similar fashion, and are fed back to the input path.

This architecture has two significant advantages over the one shown in Figure 7.

(i) It requires (n+1) cells
(ii) Only the value of K(k) at P_5 has to be updated at each step k.
     Values of local memory P_1 to P_4 are not changed.

The modification of both the architectures shown in Figures 7 and 8 for "steady state stationary" Kalman filters is obvious. In both of these architectures it is assumed that the feedback gain K(k) is available at each recursion. Currently we are investigating the development of an IDFWAP for online computation of the gain vector K(k). The simulation and hardware design of IDFWAP for the algorithms discussed in the previous sections are also under investigation.

REFERENCES

[1] B.D.O. Anderson and J.B. Moore, Optimal Filtering, Prentice-Hall, Englewood Cliffs, NJ, 1979.

[2] J. Biemond, J. Rieske, J.J. Gerbrands, IEEE Trans. on Acoust., Speech and Signal Processing, Vol. ASSP-31, No. 5, 1983, pp. 1248-1256.

[3] J. Biemond, F.G. Van der Putten, Proc. IEEE Int. Conf. on Acoust. Speech, and Signal Processing, Florida, 1985, pp. 660-663.

[4] R.H. Travassos, in VLSI and Modern Signal Processing, Prentice-Hall, N.Y. 1985, pp. 375-388.

[5] P.S. Maybeck, Stochastic Models, Estimation and Control, Vol. 1-3, Academic Press, 1980-1982.

[6] A. Andrews Int. Conf. Parallel Process, 1981, pp. 216-220.

[7] B.K. Gilbert et al., in VLSI and Modern Signal Processing, Prentice-Hall, N.Y., 1985, pp. 451-473.

[8] G.L. Meyer, H.L. Weinert, in Statistical Signal Processing, Marcel Dekkar Inc., 1985, pp. 507-516.

[9] C.L. Nikias, IEEE Trans. on Acoust., Speech, and Signal Processing, 1984, pp. 770-774.

[10] H.H. Chiang, C.L. Nikias, Proc. IEEE Conf. on Acoust., Speech, and Signal Processing, Florida, 1985, pp. 1301-1304.

[11] U.B. Desai and B. Das, Proc. Amer. Cont. Conf., Boston, 1985, pp. 920-921.

[12] U.B. Desai and S. Kiaei, "Parallel Implementation of Hierarchical Estimators," Tech. Rpt. 1985, Dept. of ECE, Washington State University, Pullman, WA 99164.

[13] H.T. Kung and C.E. Leiserson, in Introduction to VLSI Systems, Mead and Conway, Addison-Wesley, 1980, Sec. 8.3.

[14] H.T. Kung, IEEE Comput. Mag., 15 (1), Jan 1982.

[15] A.L. Fisher and H.T. Kung, in VLSI and Modern Signal Processing, Edt. S.Y. Kung et al., 1985, pp. 153-169.

[16] S.Y. Kung and R.J. Gul-Ezer, Proc. SPIE Conf., Arlington, VA, May 1982.

[17] S.Y. Kung, "VLSI Array Processors," IEEE ASSP Mag., July 1985, pp. 4-22.

[18] D.S. Broomhead, et al., Proc. IEEE Workshop on VLSI Signal Processing, Los Angeles, 1984.

[19] S.Y. Kung, H.J. Whitehouse, and T. Kailath, Editors, VLSI and Modern Signal Processing, Prentice-Hall, N.Y., 1985.

[20] K. Hwang, Computer Arithmetic Principles, Architecture, and Design, John Wiley, N.Y., 1979.

[21] TRW, VLSI Data Book, La Jolla, CA, 1985.

[22] J.M. Jover and T. Kailath, Proc. ICASSP, 1984, pp. 8.5.1-8.5.4.

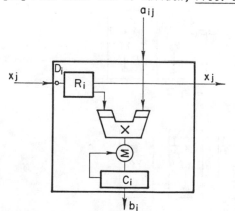

Figure 1.a   Internal Structure of MVP

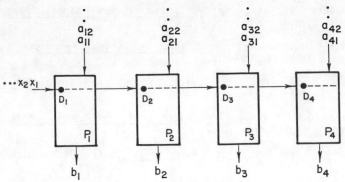

Figure 1.b Matrix Vector Multiplication Using IDFWAP (n=4)

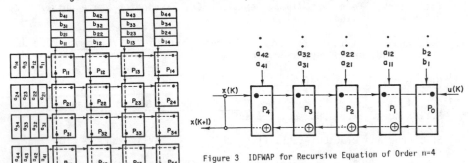

Figure 2 Matrix Matrix Multiplication Using IDFWAP (n=4)

Figure 3 IDFWAP for Recursive Equation of Order n=4

STEP 1: $x_1(K)$ ENTERS THE ARRAY PROCESSOR

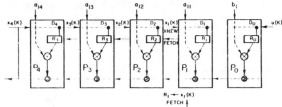

STEP 2: $x_1(K)$ AWAITS TO BE FETCHED AT P_1,
$x_1(K)$ IS FETCHED INTO R_1,
P_0 IS COMPUTING $b_1 = u(K)$

STEP 3: $x_2(K)$ IS FETCHED TO P_2, P_1 IS COMPUTING $x_1(K) = a_{11}, \ldots$

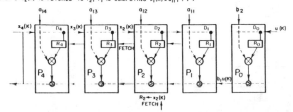

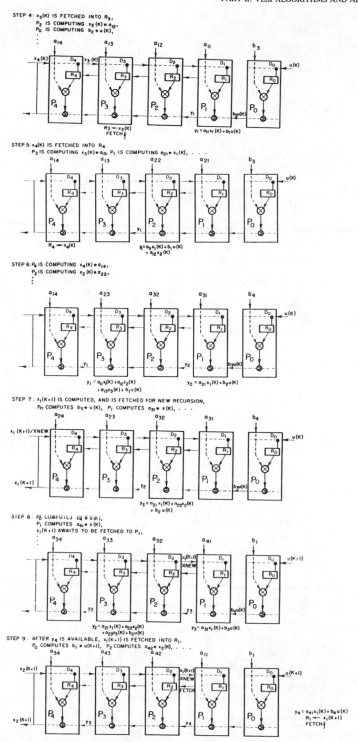

STEP 4: $x_3(K)$ IS FETCHED INTO R_3,
P_2 IS COMPUTING $x_2(K) * a_{12}$,
P_0 IS COMPUTING $b_2 * u(K)$,

$R_3 \leftarrow x_3(K)$
FETCH

$y_1 = a_{11}x_1(K) + b_1u(K)$

STEP 5: $x_4(K)$ IS FETCHED INTO R_4
P_3 IS COMPUTING $x_3(K) * a_{13}$, P_1 IS COMPUTING $a_{21} * x_1(K)$, . . .

$R_4 \leftarrow x_4(K)$

$y_1 = a_{11}x_1(K) + b_1u(K) + a_{12}x_2(K)$

STEP 6: P_4 IS COMPUTING $x_4(K) * a_{14}$,
P_2 IS COMPUTING $x_2(K) * a_{22}$,

$y_1 = a_{11}x_1(K) + a_{12}x_2(K) + a_{13}x_3(K) + b_1u(K)$

$y_2 = a_{21}x_1(K) + b_2u(K)$

STEP 7 : $x_1(K+1)$ IS COMPUTED, AND IS FETCHED FOR NEW RECURSION,
P_0 COMPUTES $b_3 * u(K)$, P_1 COMPUTES $a_{31} * x(K)$, . . .

$y_2 = a_{21}x_1(K) + a_{22}x_2(K) + b_2u(K)$

STEP 8 : P_0 COMPUTES $b_4 * u(K)$,
P_1 COMPUTES $a_{41} * x(K)$,
$x_1(K+1)$ AWAITS TO BE FETCHED TO P_1,

$y_2 = a_{21}x_1(K) + a_{22}x_2(K) + a_{23}x_3(K) + b_2u(K)$

$y_3 = a_{31}x_1(K) + b_3u(K)$

STEP 9 : AFTER y_4 IS AVAILABLE, $x_1(K+1)$ IS FETCHED INTO R_1,
P_0 COMPUTES $b_1 * u(K+1)$, P_2 COMPUTES $a_{42} * x_2(K)$, . . .

$y_4 = a_{41}x_1(K) + b_4u(K)$
$R_1 \leftarrow x_1(K+1)$
FETCH

163

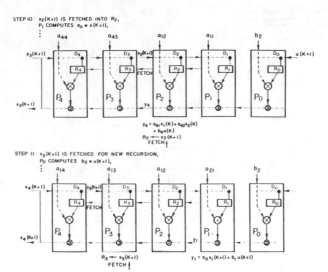

Figure 4 Snapshots of the Operation of IDFWAP for Recrusive Equation of Order n=4

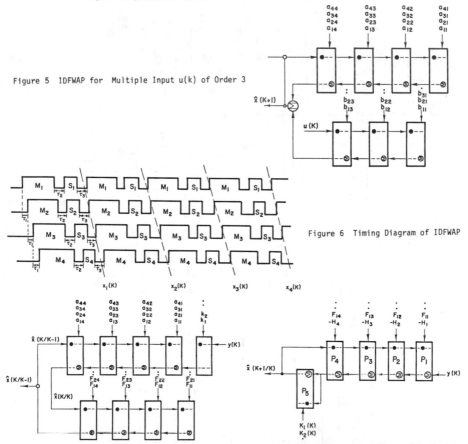

Figure 5 IDFWAP for Multiple Input u(k) of Order 3

Figure 6 Timing Diagram of IDFWAP

Figure 7 IDFWAP for Kalman Filter

Figure 8 IDFWAP for Kalman Filter with an "External Feedback Loop"

CHAPTER 15

PARALLELIZATIONS: OLD TRANSFORMATIONS AND A NEW PARALLEL PROCESSING SYSTEM

W. Holsztynski, P. Mancuso, R. Raghavan,* C. H. Ting, and P. Wong
Research & Development Division
Lockheed Missiles & Space Company, Inc.**
3251 Hanover Street
Palo Alto, California 94304

ABSTRACT

An outline of a parallel processing system currently under development is
presented. The research involves hardware, software, and studies on
computational complexity. The following new features characterize our work
in parallel processing:

1. Design and Construction of a new controller for parallel processors [1]
2. Analysis of some important translation invariant algorithms, including
    thresholding and neighborhood operations [2], resulting in the shortest
    possible programs
3. Matrix multiplications [3] and the Kalman filter [4] implemented on the
    processor
4. Segmentation of images and labeling [5, 6]
5. Rotations as an illustration of the computational complexity of
    nontranslation invariant operations [7].

The parallel processor used is the Geometric-Arithmetic Parallel Processor,
or GAPP [8]. We illustrate the above advances of the GAPP and its
controller through the use of simulators [9] and a multi-user software
system [10]. The formalism for the study of translation invariant
processes is now available; however, detailed results still remain for the
future.

INTRODUCTION

Parallel processing and VLSI fabrication have come together in recent
years. Parallel processing is a term which refers to both single
instruction multiple data (SIMD) as well as a variety of multiple
instruction multiple data (MIMD) machine architectures. Both architectures
are involved in our system. We shall indicate the range of problems that
can be mapped either naturally, or with some interpretation, onto a
geometric SIMD machine, of which the GAPP is an example. This processor
can be thought of as operating on a square lattice with nearest neighbor
shifts. The controller for the processor, on the other hand, is a MIMD
device. The entire system, consisting of more than one GAPP and one
controller, is an example of a MIMD architecture.

Conventional single instruction single datastream (SISD) processors cannot
adequately handle the computational burden of many image processing

---

*Project Leader
**W. Holsztynski is a consultant

transformations that require fast throughput rates. Currently, multiprocessor architectures that offer possibilities for significant improvement are emerging. However, conventional algorithms must be recast into a form suitable to benefit from multiprocessing environments. Such an environment was created in a simulation of GAPP [11]. M. Cayward has written UNIX shell procedures on which our software-user interaction is based [10].

The "MIMD" software development system provides the GAPP programmer with a relatively automated, universal interface to the GAPP simulator. The system relies on standardized naming conventions for flexibility established by the system's text editors. The key system command, "MIMD," searches for files according to file extension, then compares file modification times in order to translate, to compile, and to archive library objects efficiently. The script also compiles a user-specified application program, resolving differences using public and private libraries, and executes the object code. The result in a smooth incorporation of the user's GAPP program into the MIMD system, calling for little user attention to file format details, to file dependencies, archive maintenance, or to command sequences that run the application. Our primary software development system uses a simulator written in C (P. Wong, Ref. 9) and runs on a standard computer or workstation in the UNIX environment.

However, a sophisticated image processor can simulate a very large GAPP array. Since the image processor can perform logic operations on all the pixels in an image at 30 Hz, it is a much faster simulator than a single CPU computer. Another important advantage is that the progress of simulation can be observed visually as the imagery data is displayed on a CRT monitor.

We have used an IP5500 image processor made by De-Anza, Gould, Inc., to simulate a 512 by 512 element GAPP array (C. H. Ting, Ref. 9). Four image planes are used to simulate the GAPP registers, and 16 image planes are used to simulate 16 memory planes of the GAPP. Three more image planes are needed to store the output of the ALU. The ALU functions and the routing of data among registers, ALU output, and RAM memory are all performed through lookup tables associated with the image memory planes. In one TV frame time, the simulator can perform one data routing function. It generally requires one to four frame times to accomplish one GAPP instruction on the entire image.

The GAPP chip has been extensively described elsewhere [11], and our comments here will be brief.

The GAPP is a bit-serial, mesh connected array processor with nearest neighbor connections. It runs at 10 MHz. It can be shown that it is universal in the sense that, with sufficient memory for input and output, it can perform any finite window operation [12], [13]. The difficulty in its use lies in the fact that the GAPP, like several other parallel processors, provides computing power without corresponding computing intelligence.

CONTROLLING THE PROCESSOR: DMC [1] AND FORTH ENGINE [14]

The DMC is a controller, invented by one of us [1], that has unique features which enable us to overcome the bottlenecks alluded to. A patent application is being considered, and we hope to provide details on this device at the time of presentation. It addresses the crucial new problem in computing science, namely that of providing computing intelligence to powerful parallel processors. Our hardware development consists of this new controller on a GAPP Array; the demonstration system for image processing is expected to be complete toward the end of 1986.

We are complementing our system with another and simpler controller based on the NC 4000 microprocessor developed by Novix, Inc., which executes high level FORTH instructions [14].

COMPLEXITY OF PARALLEL TRANSFORMATIONS [2, 12, 13]

The object of the algebra of image processing is to study the composition of transformations of images, and the decomposability of transformations into simpler transformations. Images are arbitrary functions of a lattice into any finite set A (of colors, grey scales, etc.). The most common lattice in image processing is a square pixel array. A finite rectangular portion is usually assumed.

The complexity theory of image transformations is a quantitative refinement of the algebra of image processing. Each GAPP instruction is a finite window transformation. The GAPP complexity of a window transformation T is the minimal number n where T is a composition of n GAPP instructions. More generally, the notion of GAPP complexity applies also to transformations which can be derived from window transformations with the use of auxiliary arrays.

In the special case of image transformations, if X and Y are sets of images, then the transformation T is derived from a window transformation T' if T(x) = T'(x,b) for a fixed image b and arbitrary image x. Two examples of this will be presented in the following sections.

The GAPP complexity of a transformation T derived from a window transformation is the minimal number n of GAPP instructions where T can be derived from the composition of these n instructions. The transformations are assumed to operate entirely on contents of GAPP RAM as opposed to GAPP registers.

Three Obvious Sources of Lower Bounds

GAPP programs may be lengthy due to one or more of the following factors: the number of memory accesses, the number of shifts, and Boolean complexity. In general, two or more of the above factors may combine and have a nontrivial effect on a program. In the following, a register of the GAPP called CM, and used primarily for input-output, is disregarded for simplicity. The NS and EW registers are used to shift data in the obvious directions, respectively.

Number of Memory Accesses. At most, one GAPP address is actively involved in any one GAPP instruction. If a transformation T operates on n-bit pixel images and produces k-bit pixel images, and every bit plane of the input image is essential, and no output bit plane is identical to any of the input bit planes, then the GAPP complexity of T is at least k+n. More precisely, every GAPP program which implements T has to include n instructions which read GAPP RAM, and k instructions which write into GAPP RAM.

Number of Shifts (Translations). The flow of information between different GAPP cells is restricted to the NS and EW 1-bit latches of the nearest neighboring cells. Let W be the minimal window of a transformation T. Let points (m,y) and (x,n) belong to W. Then a GAPP program which implements T

has to include at least m GAPP instructions which involve shifts of the
EW register, and n which involve shifts of the NS register. Furthermore,
if (u,w) belongs to W then there has to be at least u + w GAPP
instructions which involve shifts of the NS or EW registers (or both).
Thus the GAPP complexity of such T is at least 2 + u + w (the summand 2
corresponds to the minimum of 2 memory accesses).

The window W decomposes into the set union of windows for each input bit
plane. When T has a multibit input, then an upper bound can be obtained in
terms of all these windows. Often T is such that all bit plane windows are
identically equal to W. For such W, if the input consists of 2*t - 1 or
2*t bit planes, and an output of k planes, then the complexity of T is at
least 2 + t*(u + w), where (u,w) belongs to W.

Boolean Complexity. If the boolean complexity of T is high, then so is the
GAPP complexity of T. The same is also true for any computing digital
device, not just for GAPP.

Transformations Dominated By Memory Accesses

GAPP is suited to perform additions and subtractions of two arrays, of a
constant and an array, of an array and its 1 step shift in any of the four
directions and of two 1-step shifts of a given array in the orthogonal
directions. All the above mentioned additions and more than half of the
subtractions have a complexity equal to the number of the necessary GAPP
RAM accesses, when shifts are not involved. Consider two other examples:

Example 1: T thresholds an n-bit image f. Therefore, T has an n-bit input
and 1-bit output. Let the constant integer v be the threshold and let
(Tf)(x,y) = 1 if f(x,y) > v, or = 0 otherwise. Since each v defines a
different threshold transformation T, the number, k, of essential input bit
planes may be smaller than n. Only k + 1 RAM accesses are necessary but
the GAPP complexity is k + 2, whenever k > 1.

Example 2: T multiplies an n-bit integer array by 3. The output is an n
+ 2 bit integer array. The least significant bit plane of the input and of
the output are equal. On the other hand the output higher bit planes truly
depend on all n-bit planes. Thus, for n > 1, n RAM reads and n + 1 RAM
writes are necessary, for a total of 2* n + 1 GAPP RAM read/write
instructions. It turns out that 2* n + 1 GAPP instructions suffice,
therefore the GAPP complexity of T is 2* n + 1.

Simple But Unsolved Problems

For the following three simple transformations the GAPP programs obtained
(by
W. Holsztynski) are perhaps the shortest possible but a proof does not
exist:

(i)   The absolute value of an n-bit integer array requires 3 GAPP
       instructions per bit (plus/minus a small constant which depends on
       the exact formulation of the transformation)
(ii) The vertical fork (n) = [(k) and C] or [(m) and ~C] requires 5 GAPP
       instructions/bit (as do so-called horizontal forks). Here (n)
       denotes the contents of the nth location in GAPP RAM, and C that of
       the C register [8, 11].
(iii) Horizontal addition of an n-bit integer array takes 3n + 1 GAPP
       instructions (as does the addition of two n-bit arrays in RAM).

MATRIX MULTIPLICATIONS AND THE KALMAN FILTER [3, 4]

The subject of systolic architectures to implement matrix operations is
well studied [15]. The architecture of the GAPP is particularly suited to
this. However, writers [15, 16] who specify the functional architecture
and not the hardware logic assume equal delays for multiplication and
addition. This is not true for the GAPP, nor is it feasible for such
high-density arrays. Both the GAPP and hypothetical "wavefront processors"
perform matrix-matrix multiplication in $O(N)$ steps. The tradeoff is
between processor area (real estate) and processing delays (speed).

For our purposes, we note the arguments that matrix-matrix multiplication
and matrix-vector multiplication are operations that map naturally on to
array processors with interconnections corresponding to nearest neighbors
on a square lattice. (The literature on "systolic" and "wavefront" array
processors covers this argument.) These arguments may be summarized as
follows.

Consider a matrix A with entries $A(i, j)$, and a column vector x with
components $x(i)$. A variety of algorithms for Ax, where juxtaposition
denotes matrix-vector multiplication, may be envisaged depending on how the
data is available to the array processor: (we assume that the array
processor consists of processing cells of yet unspecified power arranged in
a rectangular array with near-neighbor communication, as indicated above)
(1) If the data stream is arranged such that $x(j)$ reaches all the cells
where $A(i, j)$ arrives for all i, the elements of the product are formed in
one multiplication time, (2) Frequently, it is better to isolate the
problem by assuming that A is available to the processors in the
"canonical" way; i.e., each processor in the rectangular array has access
to the corresponding element of A. In this case again, by having an array
which consists of multiple copies of the vector x, one may ensure that the
product is formed in one multiplication time called Tm, say.

Matrix-matrix multiplication can be performed on $O(n)$ time. An explicit
procedure for the matrices A and B with indices from 0 to $n - 1$ is as
follows [3]: Form A_0 and B_0 where $A_0(i, j) = A[i + j, j]$, $B_0(i, j)$
$= B(i, i + j)$. Addition is considered modulo n. Let $A_k = S_N A_{k-1}$,
$B_k = S_E B_{k-1}$ where $S_N X$, $S_E X$ means shifting the array X by one
unit to the north or east, respectively.

Let $C_0 = A_0 * B_0$ where $*$ denotes pointwise mulitplication which is
done in parallel, i.e., in a fixed time T_m. It may be seen that if C_k
$= C_{k-1} + A_k * B_k$, then C_{n-1} is the matrix product AB, which thus
needs n multiplies and $n - 1$ shifts and adds.

Discussion of Kalman Filter Example [4]

The following example is the linear Kalman filter prediction and
measurement update equations for the state vector and the covariance
equations. It is shown that for an eight-dimensional state vector and a
two-dimensional measurement vector, it takes the equivalent of 30
matrix-vector multiplies to complete all calculations. This is obtained by
counting the number of such operations in a traditional set of Kalman
filter equations [16, 17]. Thus, with 24-bit precision for the eight state
vector and two measurement vector it will take ~7 ms (30 matrix-vector
multiplies times 0.24 ms per multiply) to complete all calculations. The
implementation worked out here is based on the original linear Kalman
filter equations. The more recent versions [17] of the Kalman update
equations involve square-root formulation or lower-upper LU decomposition.

These more recent versions will not help to decrease the number of required multiplications (and they may increase the number), but they have the advantage of lowering the required precision for the matrix operations.

If a large number of Kalman filter updates need to be performed in parallel, the high parallelism of such architectures may be an advantage. The GAPP is also well suited to some early processing operations, e.g., local convolutions, thresholding, track formation etc., that often precede the Kalman filter operation. This again is an advantage in a system. However, the slow multiplication of the bit-serial machine is a disadvantage when very rapid updates for a small number of objects is desired.

Since each GAPP chip has 72 processing elements in a 12 by 6 array, two GAPP chips can be combined to form a 12 by 12 array with 144 elements. Thus, six GAPP chips can do either six 8-state or three 12-state filters simultaneously.

SEGMENTATION AND LABELING [5, 6]

Image analysis and scene understanding frequently require partitioning an image into subregions defined by a common quality. Partitioning procedures often involve comparisons between adjacent pixel intensities. Computations including such comparisons are easily implemented on a single instruction multiple datastream (SIMD) processor [12]. Extended neighborhoods [13] and other complex techniques for making comparisons were not considered since the primary area of concern was the implementation of the procedure on a parallel processor. Initially, noisefree images containing features against a white background were used. In practice, images rarely contain distinct regions throughout and more complex processing is usually necessary to ensure that the following edge based image segmentation procedure will be successful.

Segmentation

The images used were eight bit graylevel images. Since the task was to separate regions of homogeneity, it is implied that relatively large changes in pixel intensity will occur at region boundaries. For this reason, edge detection was used for discerning these boundaries. Several filters can be used to detect edges. The algorithm employed here was a parallel version of the Sobel edge operator in its parallel version [11]. The coding of the Sobel operator using GAPP instructions is straightforward [5, 11, 12].

The next procedure in the segmentation process operates on the Sobel magnitude array. Initially, a simple threshold was imposed on the Sobel image to differentiate edge pixels from interior pixels. Unfortunately, this approach usually produced a binary array with disconnected edges and edges of varying width. The problem can be remedied by either lowering the threshold or using an edge augmentation procedure, such as simple dilation, to connect the edge segments. The former method was used most often here and produced good results. A skeletonization technique [6] which preserves connectivity was then used to thin the thresholded edge pixels [12]. The segmented image contained zeros in the interior and ones at the boundaries of a region. It should be emphasized that the proper labeling of an image (subsequent step) does not require an edge detection procedure as used here.

Labeling

The approach taken to solve the labeling problem exploits the parallel nature of the GAPP architecture. A method described by Holsztynski for labeling [6] was successfully tested on several images.

The method gives each non-edge pixel an initial label, such as the pixel's coordinate. Then a parallel relabeling process is invoked. Each non-edge pixel was compared with one of its neighbors and the minimum label was identified. Each pixel was relabeled with the minimum, provided the neighbor was not an edge pixel. This procedure was repeated with each pixel's immediate neighbors: east, west, north, and south until the relabeling process was completed and each pixel has a label smaller than any of its immediate neighbors. Every four-connected region has a unique label. The smallest initial label in each region has been propagated through the entire region by the relabeling process.

Parallel Segmentation and Label Performance

This segmentation and labeling technique was implemented on the GAPP simulator and tested using a series of imagery of an industrial area south of San Francisco, CA shown in Fig. 1. First the Sobel edge operator is applied. Since the Sobel operator is scene independent, each case required not more than about $31 n + 70$ instructions, where n is the number of bits in the image intensity.

Next, a threshold was applied to the Sobel image. The thresholding process takes one GAPP instruction per bit. Pixels greater than or equal to the threshold value are changed to one (white) and those less than the threshold become zero (black).

Skeletonization, being scene dependent, takes varying numbers of GAPP instructions to complete. Each cycle of the skeletonization procedure takes 10 instructions (1 µs) per direction. Figure 2 shows the results and the corresponding number of GAPP instructions needed to skeletonize each edge image. Regions are now separated by boundaries, or edges, that are one pixel wide.

Finally, the labeling process assigns a unique label to each region. Again, the number of GAPP instructions required to complete this process is dependent on the scene itself. We have not optimized the labeling program. It presently requires 170 GAPP instructions per direction. Rectangular areas take n+m neighborhood comparisons to propagate a label, where n and m are the length and width of the region. As a worst case, a narrow serpentine region will cause the performance to degrade to that of conventional SISD processing where the number of neighborhood comparisons equals or exceeds the number of pixels in the region. Figure 3 shows the results of the labeling step.

ROTATIONS [7]

An example of nontranslation invariant operations, namely attempting to rotate an array or image residing in the GAPP, provides a nice illustration of implementing operations that are neither translation invariant, nor with finite window on a SIMD machine with local interconnections. It is also an example of a "routing" problem on a SIMD processor array.

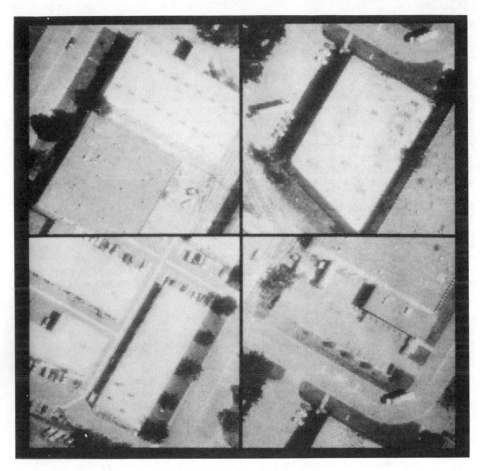

Fig. 1 Original Image for an Edge–Based Segmentation and Labeling
        on the GAPP Simulator. An industrial area of South of
        San Francisco

As indicated in the section on complexity, the problem is reformulated as
an invariant one by the use of auxiliary data. We proceed as follows.
First we divide a square lattice with unit length between neighboring
points into disjoint "circles." (We use quotes to distinguish the word
from the true circle.)

Consider a given quadrant. Let γ be the radius of a point p from the
origin. The point is connected to one of its N, NW, or W neighbor (at
radii γ_0, γ_1, γ_2, respectively) by (1) first checking if $[\gamma_0] = [\gamma]$
or $[\gamma_2] = [\gamma]$, (2) if neither of the above conditions is met, then it can
be proved $[\gamma_1] = [\gamma]$ and p connects to its NW neighbor. Here $[\gamma]$ can
denote either the closest integer, or the integer part of γ.

It is easily proved that the points of the lattice then lie on disjoint
"circles," Fig. 4(a). It is advantageous in general to have disjoint paths
for the flow of data. In general, e.g., with random source to destination

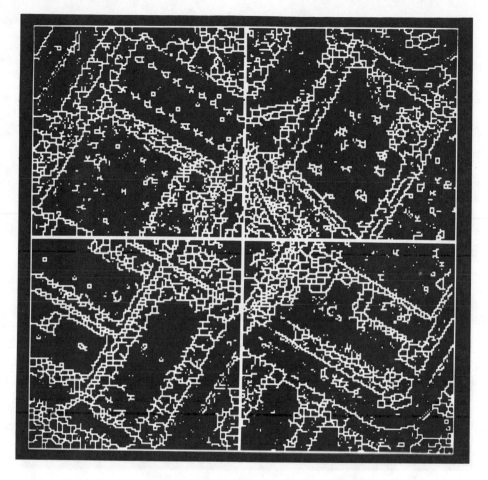

Fig. 2   Transformed Image After Edge Detection, Thresholding, and
         Skeletonization. The image is ready for the labeling process

vectors, this is impossible. The "circles" obtained here results in an
invertible transformation when a rotation is performed. The procedure
above guarantees a bounded radial error as points are moved along the arc.
However, the simplest algorithm mentioned below will result in an unbounded
error along the arc (as the radius increases), albeit a bounded angular
error.

To move a point p along the arc, we associate with each point a counter
indicating the number of steps along the arc required. Thus, the array
that needs to be pre-computed includes a flag for the neighbor on a given
"circle" and the counter indicating how far a point has to move along the
circle. With the aid of these two auxiliary arrays, the problem becomes
translation invariant: IF (counter $\neq$ O AND neighbor is correct) THEN (move
and decrement counter) ELSE (retain image). This can be easily programmed
on a SIMD machine. Since rotation is a global operation, on a NXN GAPP
array, the process takes O (NlnN) steps). The lnN comes from the

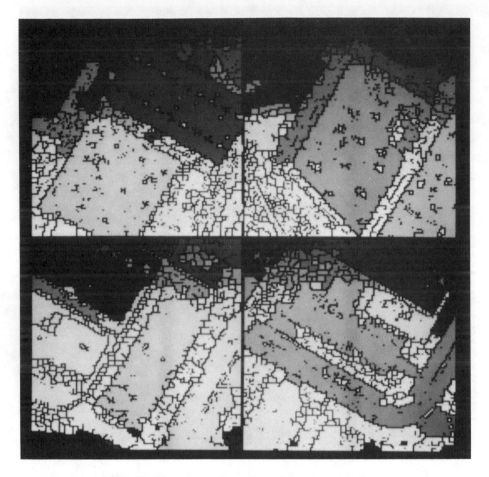

Fig. 3 Labeled Image. A color coding for the labels were used,
and a black and white photograph of the result is shown

requirement of decrementing the counter which takes 0 ($\ln_2 N$) bits for
storage. The results of the simplest such operation are shown on
Fig. 4(b). More refined algorithms are under development.

This problem is rather special, but nicely illustrates a data routing
problem and its complexity on geometric SIMD machines. The research has
many obvious directions of growth, and some of these are currently being
explored. We remark that both labeling and rotations involve the
transmission of information along the paths. In the case of rotations,
the paths are pre-determined and no information is lost. For labeling,
the paths are data dependent, and old labels are destroyed in the process
of transmission.

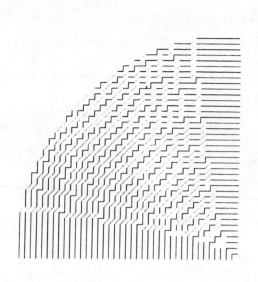 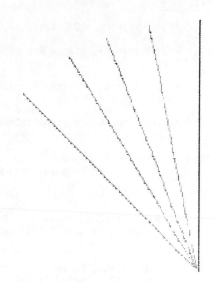

(a) The "Circles" are displayed for small radii.

(b) The simplest algorithm to move a straight line (y-axis) at angles of $10°$, $20°$, $30°$, and $45°$

Fig. 4   Illustrations of Approximate Rotations

CONCLUSION

It is apparent that a wide class of problems can be solved by using this parallel processing architecture. The Unix-based GAPP simulator and development system was not only used to explore the feasibility of parallel processing in many different applications, but it also proved to be a very useful analytical tool.

Many translation-invariant and noninvariant algorithms were implemented on this software system and their computational complexity analyzed. A significant improvement over traditional SISD implementations was generally found.

Comparisons with other parallel systems was done only for some transformations. In each case studied, our system was clearly advantageous.

The controller will serve to considerably enhance the capability of the processor array and its use in complicated real-time environments. "Smart" GAPP programs can be fetched into the new controller and executed without slowing the GAPP down. As mentioned, our demonstration system will be ready towards the end of the year 1986.

ACKNOWLEDGMENTS

This program is being supported with Lockheed Independent Research Funds.

REFERENCES

1. W. Holsztynski, "DMC tutorial," Lockheed Internal Report.

2. W. Holsztynski, "Image Transformations on the GAPP," Lockheed Internal Report.

3. W. Holsztynski, "Matrix Multiplications on the GAPP," Lockheed Internal Report.

4. Raghavan and Rauch, "The Kalman Filter on the GAPP," Lockheed Internal Report.

5. P. Wong and P. Mancuso "Segmentation and Labeling on the GAPP Simulator," Lockheed Internal Report.

6. Holsztynski, W., "Segmentation and Labeling," Lockheed Internal Report.

7. M. Johnson and R. Raghavan, "Rotations on a Parallel Machine," submitted to the Eight International Conference on Pattern Recognition, 1986.

8. W. Holsztynski, Geometric-Arithmetic Parallel Processor, Canadian Patent 1201208, priority date 19 March 1983.

9. C. H. Ting "GAPP Simulation on the DeAnza Image Processor;" P. Wong "The SIMD Simulator in C;" W. Holsztynski "DMC Simulation;" P. Mancuso "A GAPP Assembler and Disassembler," Lockheed Internal Reports.

10. M. Cayward, "MIMD Software Development System."

11. Electronic Design, cover articles Oct. 1984, Nov. 1984, and Dec. 1984.

12. E. Cloud and W. Holsztynski, "Higher Efficiency for Parallel Processors," Professional Program Session Record 14/4, Computer Vision, SOUTHCON 1984.

13. W. Holsztynski, "Theory of Translation Invariant Transformations," ERIM Internal Report.

13. Preston, K., and Duff, M., "Modern Cellular Automata," Plenum Press, C-5, 6 pp. 103-136, NY and London (1984).

14. C. H. Ting, "A Simple GAPP Controller Using the NC 4000 FORTH Engine," Lockheed Internal Report.

15. Toriwaki, J., and Fukumura, T., "Extraction of Structural Information from Digitized Grey Pictures," Computer Graphics Image Proc. 7, pp. 30-51 (1978).

16. S. Y. Kung, Whitehouse, and T. Kailath (Eds.) (1985). "VLSI and Modern Signal Processing." Englewood Cliffs, New Jersey, Prentice-Hall.

17. J. M. Jover, and T. Kailath (1986). "A Parallel Architecture for Kalman Filter Measurement Update and Parameter Estimation." Great Britain, Pergamon Press, Ltd.

18. R. E. Kalman (1960). "A New Approach to Linear Filtering and Prediction Problems," Journal of Basic Engineering, Trans. ASME, pp. 96-108.

Parallel and Pipelined VLSI Architectures Based on Decomposition

Joseph JáJá
Department of Electrical Engineering &
Institute for Advanced Computer Studies

Anil Kapoor
Department of Electrical Engineering &
Systems Research Center

University of Maryland
College Park, Md 20742.

Abstract

In this paper we introduce fully pipelined structures that are based on a novel strategy consisting of decomposing a computation into a set of subcomputations that can be executed in parallel. A problem of size n will be roughly decomposed into $\sqrt{n}$ subproblems each of size $\sqrt{n}$ such that all these subproblems can be solved in parallel on a fully pipelined bit-serial systolic architectures. The class of problems for which such decompositions exist include filtering, convolution and the Discrete Fourier Transform.

1. Introduction

Several architectures have been recently proposed for handling basic signal processing computations. In this paper, we introduce fully pipelined structures that are based on a novel strategy consisting of decomposing a computation into a set of subcomputations that can be executed in parallel. A problem of size n will be roughly decomposed into $\sqrt{n}$ subproblems each of size $\sqrt{n}$ such that all these subproblems can be solved in parallel on a fully pipelined bit-serial systolic architectures. The class of problems for which such decompositions exist include filtering, convolution and the Discrete Fourier Transform(DFT). We show that these structures can be implemented quite efficiently with compact hardware. As a matter of fact, we outline the design of a 25–MHz chip for computing the 240–point DFT that can handle up to 30,000 such pipelined DFTs per second. This chip is being currently fabricated through MOSIS. The rest of the paper is organized as follows.

The basic algorithms and the corresponding structures are introduced in the next section, while the design of each of the subcomponents is outlined in section 3. We end in section 4 with a brief description of a chip for the 240–point DFT.

2. Basic Structures

We start our description with the generic problem of computing $y = Ax$, where x is an n-dimensional input vector, A is a matrix defining the given problem, and y is an n-dimensional output vector. Our basic paradigm is to decompose x into $\approx \sqrt{n}$ vectors each of size $\approx \sqrt{n}$ and operate on these vectors simultaneously to obtain the $\sqrt{n}$ vectors that form y. The merits of this strategy will depend on whether we can carry out such a decomposition in a way that the interaction between the different subcomputations can be implemented efficiently in terms of speed and hardware. Not only do we show that such decompositions are possible, but that the overall arithmetic requirements are less than those of the standard implementations and that the hardware is quite regular and modular. We now outline such algorithms for computing the DFT and the convolution.

The Discrete Fourier Transform of n points $x_0, x_1, \ldots, x_{n-1}$ is given by:

$$y_i = \sum_{k=0}^{n-1} \omega^{ik} x_k \quad ,$$

where ω is an nth root of unity. Let $n = n_1 n_2$, where n_1 and n_2 are relatively prime. Then one can show that the above computation can be decomposed into n_2 simultaneous subcomputations each of size n_1 as follows. Let X be the $n_1 \times n_2$ matrix corresponding to the vector x stored in row-major order form. Similarly, Y can be obtained from y. Then

$$Y = A_2(A_1 X)^T$$

where A_1 and A_2 are DFT algorithms of sizes n_1 and n_2 respectively, and T is the transpose operation. Moreover, A_i can be specified with three matrices (S_i, D_i, T_i), where $A_i = S_i D_i T_i$, T_i and S_i consist of $0, 1$, or -1 , and D_i is a diagonal matrix consisting of all the multipliers, $i = 1, 2$. See [ER],[W1] for more details. Therefore the above algorithm can be implemented with the structure

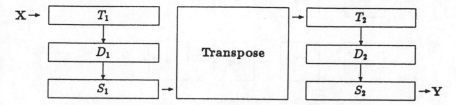

Figure 1. Structure of the DFT Hardware

If we choose $n_1 \approx n_2 \approx \sqrt{n}$, then we have potentially reduced the overall time by a factor of $\sqrt{n}$. We will show that this is indeed the case in the next section.

We now look at the convolution of two vectors h and x. The cyclic convolution $\{y_p\}$ of two length-n sequences $\{x_i\}$ and $\{y_i\}$ is defined by

$$y_p = \sum_{i=0}^{n-1} h_i x_{p-i} \quad , \quad 0 \leq p \leq n-1$$

where $p - i$ is computed modulo n. As before, let $n = n_1 n_2$, where n_1 and n_2 are relatively prime. Then if X, H, Y are the matrices corresponding respectively to the vectors x, h, y stored in row-major order form, then one algorithm to compute Y is given by

$$Y = C_1(C_2(A_2(A_1 H)^T \circ B_2(B_1 X)^T))^T$$

where (B_1, A_1, C_1), and (B_2, A_2, C_2) are the algorithms corresponding to the convolution of sizes n_1 and n_2 respectively, and o is pointwise multiplication. Again if we choose $n_1 \approx n_2 \approx \sqrt{n}$, then it is possible to achieve a speedup of order $\sqrt{n}$ over other systolic implementations. Assuming that H is known, this algorithm can be implemented with the following structure:

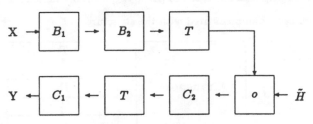

Figure 2. Convolution

We can in a similar fashion develop algorithms for filtering. Notice that the same type of components are required for all the algorithms mentioned above.

3. The Design of the Basic Components

It is clear from the previous section that the above algorithms require four basic types of hardware. The first type corresponds to the matrix operation SX, where S consists of integer constants $(0, 1, -1$ in the case of the DFT), and X is an array of inputs (or intermediate outputs) of size $\approx \sqrt{n} \times \sqrt{n}$. A typical systolic architecture can implement such an operation. However, since S is known, the elements of S can be directly embedded in the systolic array. Such a component will be called *the summation component*. The input is processed in a fully pipelined bit-serial fashion as shown in Figure (4).

The second type of hardware required corresponds to the matrix operation DX, where D is a diagonal matrix containing the precomputed multipliers (in 2's complement form) and X is the array of inputs (intermediate outputs) of size $\approx \sqrt{n} \times \sqrt{n}$. In this case, the ith row of X is *scaled* by the element D_{ii}. This can be implemented using a bit serial pipelined version of the Pezaris Array Multiplier[H]. The bit serial multiplier has p subcomponents which compute the partial products as the input moves through them serially. Figure(3) shows the block diagram of such a multiplier for $p = 4$. The sequence s has delay elements in its path at each subcomponent so that it appears to be moving at half the frequency relative to x.

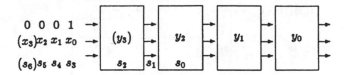

Figure 3. Bit Serial Multiplier

Figure(5) shows the input/output data flow of such a component which will be called *the scaling component*. The multiplier is followed by a set of shift registers to retain only the higher order bits of the result of the multiplication and discard the least significant bits.

Another type of hardware required corresponds to the operation $A \circ X$, where A is a matrix of predetermined constants, X is a matrix of intermediate results, and $\circ$ is the pointwise multiplication. This component is similar to the bit serial multiplier with the constant input shifted through a serial-in parallel-out register to perform the desired multiplication with the elements of X. The input/output requirements are identical to the *scaling component*.

Finally, we need a component that can perform the matrix transpose operation while satisfying the same I/O requirements as the previous components. This is similar to the *transpose component* described in [JO]. It consists of an array of shift registers that will execute a pipelined skewed transpose. At each clock, the elements in each shift register t of the ijth unit are shifted one position with a new bit coming from one of the two left neighbors of the unit. The control inputs to the transpose component at the kth clock unit is determined as follows:

if $k \geq 2(i + j) - (p - 2)n_1$

   $C_{ij} = 1$

else

   $C_{ij} = 0$

endif

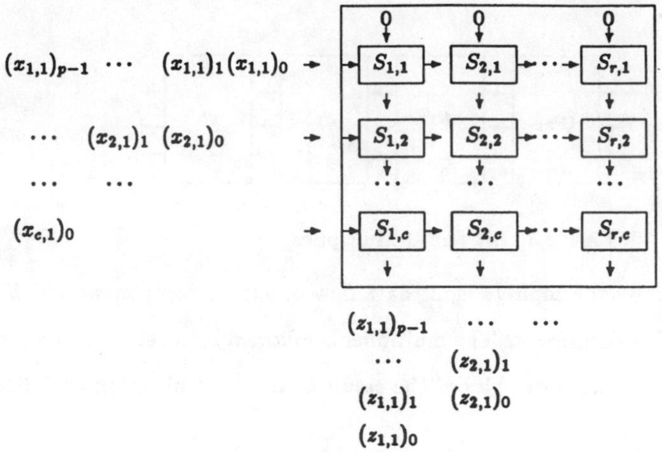

0
0
0
$(x_{1,1})_{p-1}$ ⋯ $(x_{1,1})_1$ $(x_{1,1})_0$
⋯ $(x_{2,1})_1$ $(x_{2,1})_0$
$(x_{c,1})_0$
$S_{1,1}$
$S_{2,1}$
$S_{r,1}$
$S_{1,2}$
$S_{2,2}$
$S_{r,2}$
$S_{1,c}$
$S_{2,c}$
$S_{r,c}$
$(z_{1,1})_{p-1}$
$(z_{2,1})_1$
$(z_{1,1})_1$
$(z_{2,1})_0$
$(z_{1,1})_0$

Figure 4. Summation Component.

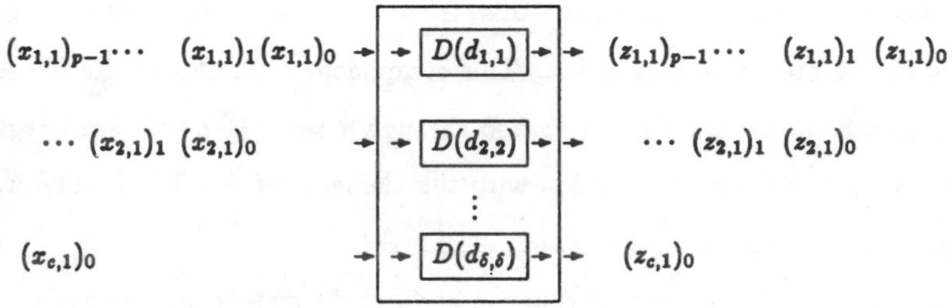

$(x_{1,1})_{p-1}$ ⋯ $(x_{1,1})_1$ $(x_{1,1})_0$
⋯ $(x_{2,1})_1$ $(x_{2,1})_0$
$(x_{c,1})_0$
$D(d_{1,1})$
$D(d_{2,2})$
$D(d_{\delta,\delta})$
$(z_{1,1})_{p-1}$ ⋯ $(z_{1,1})_1$ $(z_{1,1})_0$
⋯ $(z_{2,1})_1$ $(z_{2,1})_0$
$(z_{c,1})_0$

Figure 5. Scaling Component.

$T_{1,1}$
$C_{1,1}$
$T_{1,2}$
$C_{1,2}$
$T_{2,1}$
$C_{2,1}$
$T_{2,2}$
$C_{2,2}$
$T_{1,c}$
$C_{1,c}$
$T_{r,1}$
$C_{r,1}$
$T_{r-1,c-1}$
$C_{r-1,c-1}$
$T_{r-1,c}$
$C_{r-1,c}$
$T_{r,c-1}$
$C_{r,c-1}$
$T_{r,c}$
$C_{r,c}$

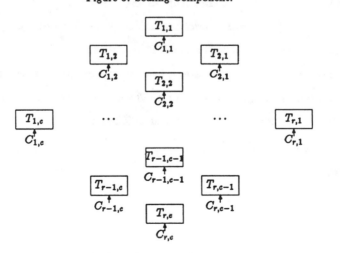

Figure 6. Transpose Component.

The operation of a shift register can itself be described formally as follows:

if $C_{ij} = 1$

$\qquad t[0] = (x_{in})$

else

$\qquad t[0] = (y_{in})$

for $i = 1, 2, \ldots, p-2$

$\qquad t[i+1] = t[i]$

end

where p is the number of bits of precision we wish to retain.

Note that the elements of the input/output array are supplied and generated in a bit skewed fashion. The scaling component retains only the higher order p bits of the result of multiplication to keep up with the flow of data. The single bit inputs, $C_{i,j}$ to a shift register of the transpose component control the flow of data through the component and moreover $C_{i,j}$ is the same for all i, j for which $i + j$ is identical. The control requirements are minimal with no control required for the Summation and Scaling components.

4. A 240–point DFT Chip

In this section we present a brief description of a DFT chip that is being currently fabricated. The basic components are laid out as given below.

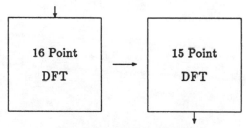

Figure 7. 240 Point DFT

The design for the 16-point DFT which is under fabrication is shown in Figure 8.

The summation components consist of 288 subcells each of size approximately $100\lambda \times 100\lambda$ with a delay of atmost 15ns. The scaling component consists of 288 subcells each of size approximately $150\lambda \times 150\lambda$ which worked correctly with a clock period of 40ns. The 15-point DFT can be realized as 5×3 DFTs. The clock period that can be used for reliable operation of each cell is 40ns, yielding a clock frequency of 25MHz. At this frequency with our fully pipelined approach, we can compute 30,000 DFTs per second with 16 bits of input/output precision.

Acknowledgements

This project was supported by the National Security Agency, contract number MDA-904-85H-0015 and by the Systems Research Center, University of Maryland, contract number 0IR8500108.

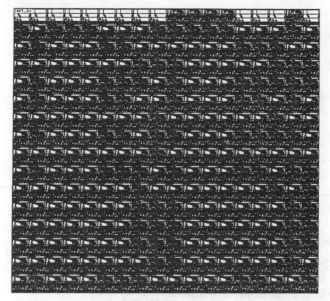

Plate 1. Scaling Component

Figure 8. 16–Point DFT Chip

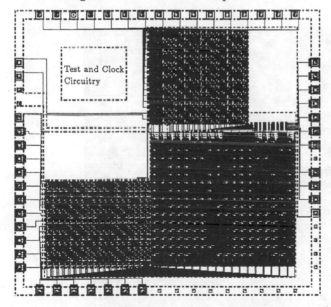

Test and Clock
Circuitry

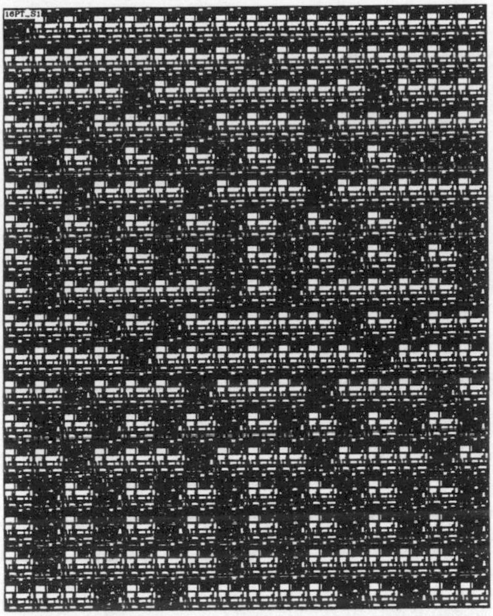

Plate 2. Input Summation Component

Plate 3. Output Summation Component

References

[AC]   Agarwal, R. and J. Cooley, *"New Algorithms for Digital Convolution"*, IEEE Trans. Acoustics, Speech and Signal Processing, 25, pp.392-410, 1977.

[CT]   Cooley, J. and J. Tukey, *"An Algorithm for the machine calculation of Complex Fourier Series"*, Math. Comp., 19, pp.297-301, 1965.

[DR]   Peter Denyer and David Renshaw,*"VLSI Signal Processing: A Bit-Serial Approach"*, Fourier Transform Machines, pp.147-197,1985.

[ER]   Elliot, D. and K. Rao, Fast Transform Algorithms, Analyses, Applications, Academic Press, 1982.

[G]   Good, I., *"The interaction algorithm and practical Fourier analysis"*, J. Royal Stat. Soc., ser. B, vol.20, pp.361-272, 1958, Addendum, vol.22, pp.372-375, 1980.

[H]   Hwang, K. Computer Arithmetic Principles, Architecture and Design pp. 175-178.

[J]   Ja'Ja', J. *"High-Speed Networks for Computing the Discrete Fourier Transform"*, Proceedings of the Advanced Research in VLSI, MIT, Jan. 1984.

[JO]   Ja'Ja', J. and R.M. Owens *"A VLSI Chip for the Winograd/Prime Factor Algorithm to Compute the DFT"*, to appear in August 1986 IEEE Trans. Acoustics, Speech & Signal Processing.

[KL]   Kung, H. T., and C. E. Leiserson *"Algorithms for VLSI processor Arrays"*, Symposium on Sparse Matrix Computations, Knoxville, Tenn.

[KP]   Kolba, D. and I. Parks, *"A Prime Factor FFT Algorithm using High Speed Convolution"*, IEEE Trans. Acoustics, Speech and Signal Processing, ASSP-25, pp.281-294, 1977

[KRY]   Kung H. T., L. M. Ruane and D. W. L. Yen *"A two-level Pipelined Systolic array for Convolutions"*, in Kung, Sproull and Steele, pp. 255-264, 1981.

[KS]   Kung H. T. and S. W. Song *"A Systolic 2-D Convolution Chip"*, CMU-CS-81-110, Dept. of C.S., Carnegie-Mellon Univ., Pittsburg, Pa.

CHAPTER 17

AN AREA EFFICIENT VLSI FIR FILTER

Robert Michael Owens and Mary Jane Irwin
Department of Computer Science
Pennsylvania State University
University Park, PA 16802

ABSTRACT

In this paper we present a technique which can be used to implement area efficient VLSI FIR filters. In a straightforward systolic approach, a FIR filter, whose lag is β, can be implemented using β processing cells. Each of these cells contains a multiplier. In the technique presented in this paper, a FIR filter can be implemented using some small number (≤ 4) of multipliers which, nevertheless, is still as fast as FIR filters implemented using β multipliers. Since the area occupied by VLSI FIR filters is directly related to the number of multipliers used, a filter based on our design will be much smaller. In fact, it appears to be possible to implement a programmable FIR filter for a moderately large β on a single chip using our technique.

INTRODUCTION

Algorithmically, a FIR computes the aperiodic convolution y_i (it's output) of a limited sequence h_i of length β with a quasi-infinite sequence x_i (it's input), where

$$y_i = \sum_{j=0}^{\beta-1} h_j \; x_{i-j}$$

For example, consider the simple single pole, low pass filter described by

$$RC \; \dot{y} + y = x$$

The discrete equivalent of this filter is described by

$$1.5 \; y_i - 0.5 \; y_{i-1} = x_i$$

and a FIR approximation ($\beta = 3$) by

$$y_i = 0.666667 \; x_i + 0.222222 \; x_{i-1} + 0.074074 \; x_{i-2}$$

For an unit step input this filter produces

	x_i	y_i
-2	0.000	0.000
-1	0.000	0.000
0	1.000	0.667
1	1.000	0.889
2	1.000	0.963
3	1.000	0.963

Note that for $i \geq 2$, the y_i's do not become increasingly closer to 1.0 as expected. This is because our low pass filter example can only be approximated by a FIR filter with a finite number of terms. However, by increasing the number of terms (increasing β) a better approximation can be obtained.

If the y_i's are computed in a straightforward systolic manner [KL], the generation of each of the y_i's requires β multiplications. To reduce the computational requirements of generating each y_i, we convert (as is usually done) the aperiodic convolution into a series of length α cyclic convolutions [N] that have the form

$$\hat{y}_{\lambda,i} = \sum_{j=0}^{\alpha-1} \dot{h}_{i-j}\, \hat{x}_{\lambda,j} \quad \text{for } i = 0, 1, \cdots, \alpha - 1$$

where for some λ, $\hat{x}_{\lambda,j} = x_{\lambda+j}, 0 \leq j < \alpha, \hat{y}_{\lambda,i} = y_{\lambda+i}, \beta\text{--}1 \leq i < \alpha$, and

$$\dot{h}_j = \begin{cases} 0 & j \geq \beta \\ h_j & \beta > j \geq 0 \\ \dot{h}_{2\beta-j} & 0 > j \end{cases}.$$

While the previous equation uses α of the x_i's, it produces only only $\alpha - \beta + 1$ of the y_i's. Hence, consecutive cyclic convolutions must be "overlapped" by at least $\beta - 1$. For a discussion concerning the choices for α and λ's for a given β see [N]. The matrix form for the low pass filter example for $\alpha = 4$ and $\lambda = -2$ is given in Figure 1.

$$\hat{\mathbf{y}} = \begin{bmatrix} \hat{y}_{-2,0} \\ \hat{y}_{2,1} \\ y_0 \\ y_1 \end{bmatrix} = \mathbf{H}\,\hat{\mathbf{x}} = \begin{bmatrix} h_0 & 0 & h_2 & h_1 \\ h_1 & h_0 & 0 & h_2 \\ h_2 & h_1 & h_0 & 0 \\ 0 & h_2 & h_1 & h_0 \end{bmatrix} \begin{bmatrix} x_{-2} \\ x_{-1} \\ x_0 \\ x_1 \end{bmatrix} =$$

$$\begin{bmatrix} 0.667 & 0.000 & 0.074 & 0.222 \\ 0.222 & 0.667 & 0.000 & 0.074 \\ 0.074 & 0.222 & 0.667 & 0.000 \\ 0.000 & 0.074 & 0.222 & 0.667 \end{bmatrix} \begin{bmatrix} 0.000 \\ 0.000 \\ 1.000 \\ 1.000 \end{bmatrix} = \begin{bmatrix} 0.296 \\ 0.074 \\ 0.667 \\ 0.889 \end{bmatrix}$$

Figure 1. Low Pass Filter

For a sufficiently large β, it is computationally more efficient to compute each of the cyclic convolutions using the fast Fourier transform. In this case, the generation of each of the y_i's requires $K \log \beta$ multiplications. However, from [AC, ER, N] we see that for a wide range of β each of the cyclic convolutions can be computed by computing: 1) an initial set of sums; 2) a set of products; and 3) a final set of sums. Algebraically, this process can be described as follows

$$\hat{\mathbf{y}} = \mathbf{C}\,(\mathbf{A}\,\dot{\mathbf{h}} \circledast \mathbf{B}\,\hat{\mathbf{x}})$$

where $\circledast$ is elementwise product, $\mathbf{A}$, $\mathbf{B}$, and $\mathbf{C}$, are matrices whose elements are fractions, and, for some λ,

$$\left[\hat{\mathbf{x}}\right]_i = \hat{x}_{i+\lambda}, \quad \left[\hat{\mathbf{y}}\right]_i = \hat{y}_{i+\lambda}, \quad \text{and} \quad \left[\dot{\mathbf{h}}\right]_i = \dot{h}_i$$

Since the $\dot{h}_i$'s are predefined, $\dot{\mathbf{h}} = \mathbf{A}\,\dot{\mathbf{h}}$ can be precomputed. Furthermore, both $\mathbf{C}$ and $\mathbf{B}$ can be written as the product of matrices whose elements are either -1, 0, or 1. While in theory the lower bound on the number of these matrices is one, the published algorithms use more than one matrix. This is because their goal was to minimize the number of additions not, necessarily, the number of matrices. For example, the matrices for the cyclic convolution of four points is shown in Figure 2.

$$\mathbf{C} = \begin{bmatrix} 1 & 0 & 1 & 0 \\ 0 & 1 & 0 & 1 \\ 1 & 0 & -1 & 0 \\ 0 & 1 & 0 & -1 \end{bmatrix} \begin{bmatrix} 1 & 1 & 0 & 0 & 0 \\ 1 & -1 & 0 & 0 & 0 \\ 0 & 0 & 1 & 0 & -1 \\ 0 & 0 & 1 & -1 & 0 \end{bmatrix}$$

$$\mathbf{A} = \begin{bmatrix} 0.25 & 0.25 & 0.25 & 0.25 \\ 0.25 & -0.25 & 0.25 & -0.25 \\ 0.5 & 0.0 & -0.5 & 0.0 \\ 0.5 & -0.5 & -0.5 & 0.5 \\ 0.5 & 0.5 & -0.5 & -0.5 \end{bmatrix}$$

$$\mathbf{B} = \begin{bmatrix} 1 & 0 & 1 & 0 \\ 1 & 0 & -1 & 0 \\ 0 & 1 & 0 & 1 \\ 0 & 1 & 0 & 0 \\ 0 & 0 & 0 & 1 \end{bmatrix} \begin{bmatrix} 1 & 0 & 1 & 0 \\ 1 & 0 & -1 & 0 \\ 0 & 1 & 0 & 1 \\ 0 & 1 & 0 & -1 \end{bmatrix}$$

Figure 2. Four Point Cyclic Convolution

In the next section an area efficient VLSI architecture to compute this equation is presented.

THE ARCHITECTURE

We will now focus on the problem of building a VLSI chip which computes

$$\hat{\mathbf{y}} = \mathbf{C} \, (\hat{\mathbf{h}} \circledast \mathbf{B} \, \hat{\mathbf{x}})$$

as defined above. To compute $\hat{\mathbf{y}}$ we propose using three different types of arithmetic cells. The word level definition of these cells is given in Figure 3.

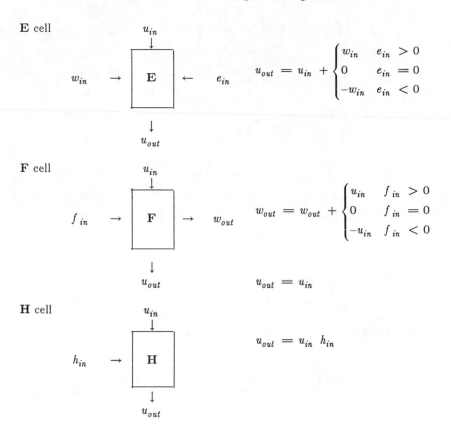

$$u_{out} = u_{in} + \begin{cases} w_{in} & e_{in} > 0 \\ 0 & e_{in} = 0 \\ -w_{in} & e_{in} < 0 \end{cases}$$

$$w_{out} = w_{out} + \begin{cases} u_{in} & f_{in} > 0 \\ 0 & f_{in} = 0 \\ -u_{in} & f_{in} < 0 \end{cases}$$

$$u_{out} = u_{in}$$

$$u_{out} = u_{in} \, h_{in}$$

Figure 3. Word Level Cell Descriptions

We will first consider how to compute $\mathbf{B} \, \hat{\mathbf{x}}$ (multiplication by $\mathbf{C}$ is similar). Recall that $\mathbf{B}$ can be written as the product of matrices whose elements are -1, 0, and 1 as follows

$$\mathbf{B} = \mathbf{E}_k \, \mathbf{F}_{k-1} \, \cdots \, \mathbf{E}_4 \, \mathbf{F}_3 \, \mathbf{E}_2 \, \mathbf{F}_1$$

We can, without loss of generality, assume k is even. Hence, we need only develop an area efficient VLSI architecture to compute

$$\mathbf{E}_2 \, \mathbf{F}_1 \, \hat{\mathbf{x}}$$

since it can then be replicated to compute $\mathbf{B} \, \hat{\mathbf{x}}$. The architecture illustrated in Figure 4 can be used to compute $\hat{\mathbf{u}} = \mathbf{E}_2 \, \mathbf{F}_1 \, \hat{\mathbf{x}}$.

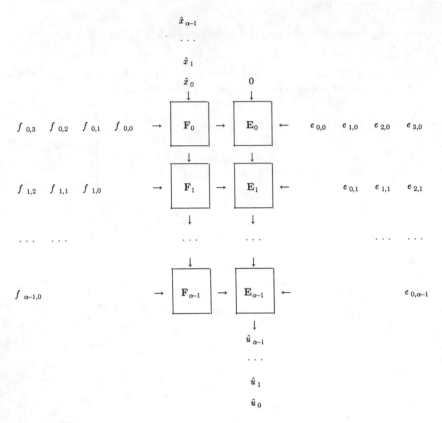

Figure 4. Sum Structure

If the sum structure above is supplied with data as indicated, it will correctly compute $\hat{\mathbf{u}}$. For example, how $\hat{\mathbf{u}} = \mathbf{B}\,\hat{\mathbf{x}}$ is computed, where $\hat{\mathbf{x}}$ and $\mathbf{B}$ are defined in Figures 1 and 2, respectively, is shown in Table 1.

Table 1. Initial Sum Example

	F				E			
	f in	u in	w out	u out	e in	u in	w in	u out
0	1.00	0.00	0.00	0.00	0.00	0.00	0.00	0.00
	0.00	0.00	0.00	0.00	0.00	0.00	0.00	0.00
	0.00	0.00	0.00	0.00	0.00	0.00	0.00	0.00
	0.00	0.00	0.00	0.00	0.00	0.00	0.00	0.00
1	0.00	0.00	0.00	0.00	0.00	0.00	0.00	0.00
	1.00	0.00	0.00	0.00	0.00	0.00	0.00	0.00
	0.00	0.00	0.00	0.00	0.00	0.00	0.00	0.00
	0.00	0.00	0.00	0.00	0.00	0.00	0.00	0.00
2	1.00	1.00	0.00	0.00	0.00	0.00	0.00	0.00
	0.00	0.00	0.00	0.00	0.00	0.00	0.00	0.00
	0.00	0.00	0.00	0.00	0.00	0.00	0.00	0.00
	0.00	0.00	0.00	0.00	0.00	0.00	0.00	0.00
3	0.00	1.00	1.00	1.00	0.00	0.00	1.00	0.00
	-1.00	1.00	0.00	0.00	0.00	0.00	0.00	0.00
	1.00	0.00	0.00	0.00	0.00	0.00	0.00	0.00
	0.00	0.00	0.00	0.00	0.00	0.00	0.00	0.00
4	0.00	0.00	1.00	1.00	1.00	0.00	1.00	0.00
	0.00	1.00	-1.00	1.00	0.00	0.00	-1.00	0.00
	0.00	1.00	0.00	0.00	0.00	0.00	0.00	0.00
	1.00	0.00	0.00	0.00	0.00	0.00	0.00	0.00
5	0.00	0.00	1.00	0.00	1.00	0.00	1.00	1.00
	0.00	0.00	-1.00	1.00	0.00	1.00	-1.00	0.00
	1.00	1.00	0.00	1.00	0.00	0.00	0.00	0.00
	0.00	1.00	0.00	0.00	0.00	0.00	0.00	0.00
6	0.00	0.00	1.00	0.00	0.00	0.00	1.00	1.00
	0.00	0.00	-1.00	0.00	0.00	1.00	-1.00	1.00
	0.00	0.00	1.00	1.00	1.00	1.00	1.00	0.00
	-1.00	1.00	0.00	1.00	0.00	0.00	0.00	0.00

	F				E			
	f in	u in	w out	u out	e in	u in	w in	u out
6	0.00	0.00	1.00	0.00	0.00	0.00	1.00	0.00
	0.00	0.00	-1.00	0.00	1.00	0.00	-1.00	1.00
	0.00	0.00	1.00	0.00	-1.00	1.00	1.00	2.00
	0.00	0.00	-1.00	1.00	0.00	2.00	-1.00	0.00
8	0.00	0.00	1.00	0.00	0.00	0.00	1.00	0.00
	0.00	0.00	-1.00	0.00	1.00	0.00	-1.00	-1.00
	0.00	0.00	1.00	0.00	0.00	-1.00	1.00	~~0.00~~ (2.00)
	0.00	0.00	-1.00	0.00	0.00	0.00	-1.00	2.00
9	0.00	0.00	1.00	0.00	0.00	0.00	1.00	0.00
	0.00	0.00	-1.00	0.00	0.00	0.00	-1.00	-1.00
	0.00	0.00	1.00	0.00	0.00	-1.00	1.00	~~1.00~~
	0.00	0.00	-1.00	0.00	1.00	-1.00	-1.00	(0.00)
10	0.00	0.00	1.00	0.00	0.00	0.00	1.00	0.00
	0.00	0.00	-1.00	0.00	0.00	0.00	-1.00	0.00
	0.00	0.00	1.00	0.00	0.00	0.00	1.00	~~1.00~~
	0.00	0.00	-1.00	0.00	0.00	-1.00	-1.00	(-2.00)
11	0.00	0.00	1.00	0.00	0.00	0.00	1.00	0.00
	0.00	0.00	-1.00	0.00	0.00	0.00	-1.00	0.00
	0.00	0.00	1.00	0.00	0.00	0.00	1.00	~~0.00~~
	0.00	0.00	-1.00	0.00	1.00	0.00	-1.00	(-1.00)
12	0.00	0.00	1.00	0.00	0.00	0.00	1.00	0.00
	0.00	0.00	-1.00	0.00	0.00	0.00	-1.00	0.00
	0.00	0.00	1.00	0.00	0.00	0.00	1.00	~~0.00~~
	0.00	0.00	-1.00	0.00	0.00	0.00	-1.00	(-1.00)

So that

$$\hat{\mathbf{u}} = \begin{bmatrix} 2.000 \\ 0.000 \\ -2.000 \\ -1.000 \\ -1.000 \end{bmatrix}$$

is computed as desired.

The architecture illustrated in Figure 6 can be used to compute $\hat{\mathbf{v}} = \hat{\mathbf{h}} \circledast \hat{\mathbf{u}}$.

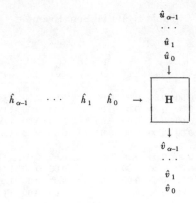

Figure 5. Product Structure

If the product structure above is supplied with data as indicated, it will correctly compute $\hat{\mathbf{v}}$. For example, how $\hat{\mathbf{v}} = \hat{\mathbf{h}} \circledast \hat{\mathbf{u}}$ is computed, where $\hat{\mathbf{u}}$ is defined in Table 1, is shown in Table 2.

Table 2. Product Example

		H	
	u in	h in	u out
0	2.00	0.24	0.00
1	0.00	0.13	0.48
2	-2.00	0.30	0.00
3	-1.00	0.19	-0.59
4	-1.00	0.41	-0.19
5	0.00	0.00	-0.41

So that

$$\hat{\mathbf{v}} = \begin{bmatrix} 0.481 \\ 0.000 \\ -0.593 \\ -0.185 \\ -0.407 \end{bmatrix}$$

is computed as desired.

The architecture in Figure 4 can again be used to compute $\mathbf{C}\,\hat{\mathbf{v}}$. For example, how $\hat{\mathbf{y}} = \mathbf{C}\,\hat{\mathbf{v}}$ is computed, where $\mathbf{C}$ and $\hat{\mathbf{v}}$ are defined in Figure 2 and Table 2, respectively, is shown in Table 3.

Table 3. Final Sum Example

	f in	u in	w out	u out	e in	u in	w in	u out
0	1.00	0.48	0.00	0.00	0.00	0.00	0.00	0.00
	0.00	0.00	0.00	0.00	0.00	0.00	0.00	0.00
	0.00	0.00	0.00	0.00	0.00	0.00	0.00	0.00
	0.00	0.00	0.00	0.00	0.00	0.00	0.00	0.00
1	1.00	0.00	0.48	0.48	0.00	0.00	0.48	0.00
	1.00	0.48	0.00	0.00	0.00	0.00	0.00	0.00
	0.00	0.00	0.00	0.00	0.00	0.00	0.00	0.00
	0.00	0.00	0.00	0.00	0.00	0.00	0.00	0.00
2	0.00	-0.59	0.48	0.00	0.00	0.00	0.48	0.00
	-1.00	0.00	0.48	0.48	0.00	0.00	0.48	0.00
	0.00	0.48	0.00	0.00	0.00	0.00	0.00	0.00
	0.00	0.00	0.00	0.00	0.00	0.00	0.00	0.00
3	0.00	-0.19	0.48	-0.59	0.00	0.00	0.48	0.00
	0.00	-0.59	0.48	0.00	0.00	0.00	0.48	0.00
	0.00	0.00	0.00	0.48	0.00	0.00	0.00	0.00
	0.00	0.48	0.00	0.00	0.00	0.00	0.00	0.00
4	0.00	-0.41	0.48	-0.19	0.00	0.00	0.48	0.00
	0.00	-0.19	0.48	-0.59	0.00	0.00	0.48	0.00
	1.00	-0.59	0.00	0.00	0.00	0.00	0.00	0.00
	0.00	0.00	0.00	0.48	0.00	0.00	0.00	0.00
5	0.00	0.00	0.48	-0.41	1.00	0.00	0.48	0.00
	0.00	-0.41	0.48	-0.19	0.00	0.00	0.48	0.00
	0.00	-0.19	-0.59	-0.59	0.00	0.00	-0.59	0.00
	1.00	-0.59	0.00	0.00	0.00	0.00	0.00	0.00
6	0.00	0.00	0.48	0.00	0.00	0.00	0.48	0.48
	0.00	0.00	0.48	-0.41	0.00	0.48	0.48	0.00
	-1.00	-0.41	-0.59	-0.19	0.00	0.00	-0.49	0.00
	-1.00	-0.19	-0.59	-0.59	0.00	0.00	-0.49	0.00

	f in	u in	w out	u out	e in	u in	w in	u out
7	0.00	0.00	0.48	0.00	1.00	0.00	0.48	0.00
	0.00	0.00	0.48	0.00	1.00	0.00	0.48	0.48
	0.00	0.00	-0.19	-0.41	1.00	0.48	-0.19	0.00
	0.00	-0.41	-0.41	-0.19	0.00	0.00	-0.41	0.00
8	0.00	0.00	0.48	0.00	0.00	0.00	0.48	0.48
	0.00	0.00	0.48	0.00	0.00	0.48	0.48	0.48
	0.00	0.00	-0.19	0.00	0.00	0.48	-0.19	0.30
	0.00	0.00	-0.41	-0.41	0.00	0.30	-0.41	0.00
9	0.00	0.00	0.48	0.00	0.00	0.00	0.48	0.00
	0.00	0.00	0.48	0.00	1.00	0.00	0.48	0.48
	0.00	0.00	-0.19	0.00	-1.00	0.48	-0.19	(0.48)
	0.00	0.00	-0.41	0.00	1.00	0.48	-0.41	(0.30)
10	0.00	0.00	0.48	0.00	0.00	0.00	0.48	0.00
	0.00	0.00	0.48	0.00	0.00	0.00	0.48	0.48
	0.00	0.00	-0.19	0.00	0.00	0.48	-0.19	(0.67)
	0.00	0.00	-0.41	0.00	0.00	0.67	-0.41	(0.07)
11	0.00	0.00	0.48	0.00	0.00	0.00	0.48	0.00
	0.00	0.00	0.48	0.00	0.00	0.00	0.48	0.00
	0.00	0.00	-0.19	0.00	0.00	0.00	-0.19	(0.48)
	0.00	0.00	-0.41	0.00	-1.00	0.48	-0.41	(0.67)
12	0.00	0.00	0.48	0.00	0.00	0.00	0.48	0.00
	0.00	0.00	0.48	0.00	0.00	0.00	0.48	0.00
	0.00	0.00	-0.19	0.00	0.00	0.00	-0.19	(0.00)
	0.00	0.00	-0.41	0.00	0.00	0.00	-0.41	(0.89)

From the example we see that

$$\hat{y} = \begin{bmatrix} 0.296 \\ 0.074 \\ 0.667 \\ 0.889 \end{bmatrix}$$

is computed as desired.

Only one multiplier is used in the array. However, because of the overlap need-ed in the consecutive cyclic convolutions and pipeline delays through the proposed array, multiple arrays are needed to obtain a throughput similar to that of the straightforward systolic linear array (i.e., where one clock produces one value). Hence, several multipliers are needed in the overall design. However, for designs of practical size ($\beta \leq 1024$) no more than four multiplers are needed.

Each of the cells in Figure 3 can be implemented in either word parallel hardware or in digit serial hardware. Because of the constraints on pin out in VLSI we have elected to build the FIR filter using digit serial hardware. Details for the implementation of **E** (**F** is similar to **E**) and **H** using a digit serial, digit pipelined approach are given below.

The **E** Cell Digit Level Description

The digit level description of the **E** cell is given in Figure 6 where b is the base and p is the precision [IO]. There is a similarity between a standard one-bit serial adder which starts adding with the least significant bits of the two operands and the adder of Figure 6 which starts adding with the *most significant* digits of the two operands. Both adders produce an internal carry and a sum. In the bit serial adder, the carry is stored and the sum is output. In the adder of Figure 6, the present sum digit, which is determined by the first digit addition operation (1), is stored in an internal latch to be used in the next cycle. Then a second digit addition operation (2) adds the present carry to the previous sum and this result is output. Most importantly this second addition is *guaranteed* to be carry free because of the restricted digit sets allowed for the carry, c_j, and the sum, s_j. The latency of this adder is one digit; i.e., the first digit of the result is generated one cycle after the first digits of the operands have been input.

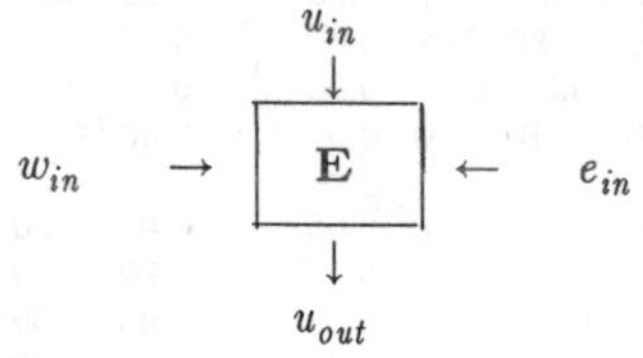

$$u_{in}$$
$$\downarrow$$
$$w_{in} \quad \rightarrow \quad \boxed{\text{E}} \quad \leftarrow \quad e_{in}$$
$$\downarrow$$
$$u_{out}$$

$$(1) \quad s_j = u_{in} + \begin{cases} w_{in} - b \ c_j & \text{if } e_{in} > 0 \\ 0 & \text{if } e_{in} = 0 \\ -w_{in} - b \ c_j & \text{if } e_{in} < 0 \end{cases}$$

$$\text{where } -1 \le c_j \le 1 \text{ and } -(b-2) \le s_j \le (b-2)$$

$$(2) \quad u_{out} = s_{j-1} + c_j$$

Figure 6. The **E** Cell Digit Level Description

For the current VLSI implementation $b = 4$ so w_{in}, e_{in}, u_{in}, $u_{out} \in \{\overline{3}, \overline{2}, \overline{1}, 0, 1, 2, 3\}$, s_j, $s_{j-1} \in \{\overline{2}, \overline{1}, 0, 1, 2\}$, and $c_j \in \{\overline{1}, 0, 1\}$ where $\overline{3} = -3$. By using a two's complement encoding for the base 4 digits and standard logic minimization techniques, a four level logic implementation of both digit addition operations can be defined. The **E** cell requires a digit complementing cell, two digit adder cells, and a one digit storage cell as shown in the block diagram of Figure 7. The **E** cell is 600λ by 600λ in custom logic, double metal CMOS. The input/output requirements are 12 data lines. A 3 micron feature size MOSIS implementation of the cell operates in less than 50 nsec including input/output pad time. Only *one* **E** cell is needed regardless of the operand precision.

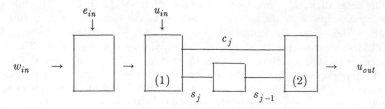

Figure 7. **E** Cell Logic Block Diagram

Examples of most significant digit (msd) base 4 addition $(c_{in} > \underline{0})$ are given in Table 4 for $w_{in} = .1223_4$ and $u_{in} = .1213_4$ followed by $w_{in} = .1232_4 \, (= .1112_4)$ and $u_{in} = .1121_4 \, (= .1113_4)$. The first sum, $u_{out} = .3102_4$, would incur a full precision carry if performed using a conventional ripple carry adder. For a sequence of additions, the msd's of the operands for the next addition can be input at the next clock after the lsd's of the operands for the present addition have been input. As long as overflow is not allowed, the interim carry formed during operation (1) for the next addition is guaranteed to be zero. This interim carry is the one used in operation (2) to complete the present addition and if it was not zero an incorrect lsd for the present addition would be output.

Table 4. **E** Cell Example $(e_{in} > 0)$

clock	1	2	3	4	5	6	7	8	9
w_{in}	1	2	2	3	1	2	-3	2	
u_{in}	1	2	1	3	1	1	2	-1	
c_j	0	1	1	1	0	1	0	0	0
s_j	2	0	-1	2	2	-1	-1	1	
u_{out}		3	1	0	2	3	-1	-1	1

The **H** *Cell Digit Level Description*

A **H** cell is built out of $p + 3$ scaling cells [IO]. A scaling cell is defined in Figure 8 which scales the input digit value, u_{in}, by the scale digit value, h_{in}.

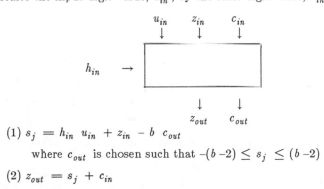

(1) $s_j = h_{in} \, u_{in} + z_{in} - b \, c_{out}$

where c_{out} is chosen such that $-(b-2) \le s_j \le (b-2)$

(2) $z_{out} = s_j + c_{in}$

Figure 8. Scaling Cell Digit Level Description

The scale digit, h_{in}, must belong to the digit set $\{\bar{2}, \bar{1}, 0, \underline{1}, \underline{2}\}$ while the digits of the input value, u_{in}, are assumed to belong to the digit set $\{3, 2, 1, 0, 1, 2, 3\}$. A mapping which covers the range $-10 \leq h_{in}\, u_{in} + z_{in} \leq 10$, must be used for operation (1). In this mapping $-2 \leq c_{out} \leq 2$ and, thus, $-4 \leq z_{out} \leq 4$. The scaling cell has been estimated at 600λ by 600λ in custom logic, double metal CMOS. The input/output requirements for the scaling cell are 18 data lines since z_{in} and z_{out} require four lines each. The scale cell has been simulated at an operational speed of less than 50 nsec. The latency of the scaler is one digit.

To build a **H** cell capable of multiplying two p digit value operands, $p + 3$ scaling cells must be interconnected as shown in Figure 9 for $p = 4$. The three extra scaling cells are necessary to allow time for the appropriate carries and sums to be combined. The three bottommost scaling cells are hardwired with $h_{in} = 0$. Thus, another function of the bottommost components is to allow for the three bit encoded, carry free formation of z_{out}. During processing the digits of the input value, msd first, are broadcast on the u_{in} line. The latency of the multiplier is two.

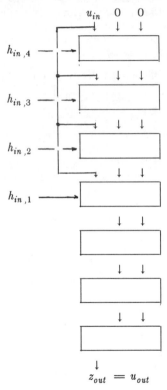

Figure 9. The **H** Cell Digit Level Description

CONCLUSIONS

If the FIR filter is implemented using conventional word parallel hardware, each of the **E** or **F** cells is implemented as a linear array of p adder cells of fixed area, where p is the precision. The **H** cell is implemented as a $p \times p$ mesh of adder cells of fixed area. Hence, the overall size of the word parallel version is $O(p^2 + p\ \beta)$. Furthermore, the word parallel version can compute β output values in time $O(\beta)$. Therefore, on the average, a single output value can be computed in time $O(1)$. The straightforward systolic approach leads to a word parallel implementation which computes a single output in time $O(1)$ but has area $O(\beta\ p^2)$. Hence our approach is as fast as the straightforward approach but uses a factor of β less area.

A more appropriate implementation for VLSI is to build the FIR filter out of digit serial hardware. In the digit serial version, each of the **E** or **F** cells is implemented as a single cell of fixed area. The **H** cell is implemented as a linear array of p cells. Hence, the overall size of the digit serial approach is $O(p + \beta)$. Furthermore, the digit serial version can compute β output values in time $O(\beta\ p)$. Therefore, on the average, a single output word value can be computed in time $O(p)$. The straightforward systolic approach leads to a digit serial implementation which computes a single word output in time $O(p)$ but instead has area $O(\beta\ p)$. Hence our approach is again as fast as the straightforward approach but again uses a factor of β less area.

Each time the array is stepped, only one new $\hat{x}_i$ is needed and only one new $\hat{y}_i$ is generated in the word parallel version. Hence, by initially having the elements of **F**, **H**, and **E** stored on the chip, the input/output requirements of the array can be reduced to only $2pq$ pins, where q is the number of bits needed to encode one digit. In the digit serial version only $2\beta q$ input/output pins are needed if the chip is again initialized with the elements of **F**, **H**, and **E**.

ACKNOWLEDGEMENTS

This work is supported in part by the Army Research Office under Contract DAAG29-83-K-0126.

REFERENCES

[KL] Kung, H.T. and C.E, Leiserson, "Systolic Arrays (for VLSI)," *Sparse Matrix Proceedings,* 1978, ed. J.S. Derff and G.W. Stewart, SIAM, 1979, pp. 256-282.

[AC] Agrawal, R. and J. Cooley, "New Algorithms for Digital Convolution," *IEEE Transactions on ASSP,* ASSP-25, pp. 392-410, 1977.

[ER] Elliot, D. and K. Rao, *Fast Transforms Algorithms, Analyses, Applications,* Academic Press, 1982.

[N] Nussbaumer, H.J., *Fast Fourier Transforms and Convolution Algorithms,* Springer-Verlag, Berlin, 1982.

[OI] Owens, R.M. and M.J. Irwin, "Area Efficient VLSI Architectures for Aperiodic Convolution," in preparation.

[IO] Irwin, M.J. and R.M. Owens, "Digit Pipelined Processors," accepted by *The Journal of Supercomputing,* July 1986.

CHAPTER 18

A HIGH SPEED VLSI COMPLEX DIGITAL SIGNAL PROCESSOR BASED ON QUADRATIC RESIDUE NUMBER SYSTEM

Magdy A. Bayoumi

The Center for Advanced Computer Studies
University of Southwestern Louisiana
Lafayette, LA 70504

ABSTRACT

The special mathematical properties of Quadratic Residue Number System (QRNS), of eliminating the interaction between the real and imaginary channels in complex arithmetic, establishes parallelism and modularity on both the functional and implementation levels for VLSI Digital Signal Processing (DSP) structures. In this paper, a high performance signal processor for complex DSP applications based on QRNS is developed. The processor design is optimized for efficient computation of Fast Fourier Transform (FFT) and signal processing algorithms based on FFT, such as Finite Impulse Response (FIR) filtering, Inverse Fast Fourier Transform (IFFT), convolution, correlation, and multiplication. This computational versatility is achieved through macroprogrammability. A high computational throughput is achieved by maximal utilization of the hardware resources (only two real multiplications are required for performing complex multiplication) and by using First-in First-out (FIFOs) for storing input, intermediate, and output data (and coefficients). The processor gains its high performance by using bit parallel processing within small (1-7 bits) computational fields. Customized look-up tables are employed for implementing the residue operations. The developed processor can be employed either as a stand-alone processor or as a peripheral processor (co-processor).

1. INTRODUCTION

The increasing demands of high speed real-time Digital Signal Processing (DSP) coupled with low-cost, high density VLSI technology has suggested that VLSI based architectures are a viable approach to achieve tremendous computation capability in a cost_effective way [1]. To achieve such performance using conventional architectures and methodologies, very complicated systems have to be designed where cost, size, power, and reliability are sacrificed. Parallel processing and custom design are the key factors in building high performance signal processors with reasonable hardware complexity. Achieving parallelism on the mathematical level has been the motivation for renewed interest in the Residue Number System (RNS) applied to the implementation of DSP architectures. In this system, the computational field is decomposed into a set of completely independent subfields, in which computation is performed in parallel in a Single Instruction Multiple Data (SIMD) like structure. Moreover, RNS has inherent high speed mathematical operations, since addition and subtraction have no inter-digit carries or borrows and multiplication does not need the generation of partial products. Several special purpose DSP systems have been successfully designed and built based on RNS [2-4].

For complex DSP applications such as Fast Fourier Transform (FFT), digital filtering, homomorphic speech processing, and radar and satellite communication, the computation requirements are doubled. The conventional implementation of these systems are based on a formation of two channels; real and imaginary, with interaction between both of them especially during complex multiplication. Recently, the algebraic properties of RNS has been rediscovered and employed for complex DSP applications in what has been widely called quadratic RNS (QRNS) [5]. This mathematical system establishes parallelism on the functional level, by eliminating the interaction between the real and imaginary channels. A high computational throughput is achieved by maximal utilization of the hardware resources; only two real multiplications are required for implementing a complex multiplication versus 4 multiplications and 2 additions in conventional arithmetic.

QRNS has the same features and properties of the standard RNS and employs the same computational kernels. The advent of VLSI technology provides an opportunity for significantly increased efficiency of both RNS and QRNS systems as they support the main VLSI design properties and features:

(1) **Concurrency**: the arithmetic operations are performed independently for each modulus, with separate channels (real and imaginary) in case of QRNS.

(2) **Regular Interconnections**: a large fraction (40-50%) of the area of a chip is typically used in providing for data paths or interconnection among various logic elements [6]. Both RNS and QRNS structures exhibit regular and simple interconnection patterns. In case of QRNS no interconnection is there between the real and imaginary channels.

(3) **Modularity**: RNS offers modularity on both functional and layout levels as the processing structure for each modulus is similar in architecture and performance.

(4) **Fault Tolerant Capabilities**: digits can be added to expand the word length, or deleted due to hardware faults, without destroying the legibility of the data. Also, errors can be isolated easily. Moreover, in QRNS if one channel fails it can be isolated and the other one performs both computation in a half data rate.

In this paper, a devoted signal processor for complex DSP applications is proposed based on QRNS. The motivation to this design is the fact that the currently available general and specialized signal processors do not have special hardware capability for complex arithmetic, but rather they employ software or firmware techniques which limit the speed performance of these processors. The proposed processor belongs to the dedicated specialized VLSI architectures family. But, it offers more computational versatility and application flexibility through macroprogrammability. The processor design is optimized for efficient computation of the FFT and signal processing operations based on FFT such as FIR filtering, IFFT, convolution, correlation, and multiplication. On the circuit level modularity is observed; look-up table modules are used as the computational kernels. Their layout design has been optimized through custom design procedure for modulo look-up tables. Selecting this type of circuit design allows bit parallel processing rather than bit serial processing, which offer high performance with simple hardware complexity.

2. QUADRATIC RESIDUE NUMBER SYSTEMS (QRNS)

RNS in general is a means of representing information in a non-weighted number system. The base of the general RNS is a set of moduli; $\{m_1, m_2, \ldots\ldots, m_L\}$, this set of elements is pairwise relatively prime [12]. In conventional (integer) RNS, the ring of integers modulo M $(M = \prod_{i=1}^{L} m_i)$ can be represented by a set of independent subrings based on the Chinese Remainder Theorem;

$$R(M) = (R(m_1), R(m_2), \ldots\ldots, R(m_L))$$

The total number of integers that may be unambiguously encoded is M defined in the following dynamic range:

$$\left\{ -\frac{M-1}{2} , \frac{M-1}{2} \right\} \qquad \text{for M odd}$$

$$\left\{ -\frac{M}{2} , \frac{M}{2} -1 \right\} \qquad \text{for M even}$$

An integer X within this dynamic range is encoded as L-tuple, $(x_1 , x_2 , \ldots , x_L)$, where $x_i = X$ modulo m_i , and is defined by:

$$x_i = |X|_{m_i} , \qquad i = 1,2,\ldots ,L$$

Arithmetic operations for each modulus are fully independent. The arithmetic operations between two numbers X and Y have the following property:

$$Z = |X \circ Y|_M$$
$$z_i = |x_i \circ y_i|_{m_i}$$

where the operator '$\circ$' can be either addition, subtraction, or multiplication. Figure 1 shows a general RNS-based system.

For complex DSP applications, complex RNS (CRNS) has been used for encoding information [13]. CRNS has the same properties and features of integer RNS, but each number is coded in a complex representation as

$$x_i = x_i^r + j x_i^i , \qquad \text{where}$$

x_i^r is the real part modulo m_i

x_i^i is the imaginary part modulo m_i

j is $\sqrt{-1}$

The residue arithmetic between two complex numbers x_i , y_i is performed as follows:

$$\text{addition} \quad z_i = x_i + y_i$$
$$= |x_i^r + y_i^r|_{m_i} + j |x_i^i + y_i^i|_{m_i}$$
$$= z_i^r + j z_i^i$$

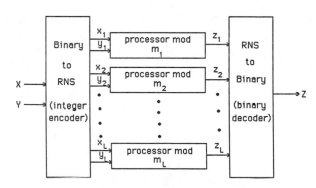

Fig.1 A General RNS_Based System

multiplication $\quad z_i = x_i * y_i$

$$= | x_i^r y_i^r - x_i^i y_i^i |_{m_l} + j | x_i^r y_i^i + x_i^i y_i^r |_{m_l}$$

$$= z_i^r + j z_i^i$$

As in the conventional RNS, CRNS is based on the isomorphism between the ring of complex integers modulo M and a set of L smaller and independent rings modulo m_i;

$$CR(M) = (CR(m_1), CR(m_2), \ldots\ldots, CR(m_L))$$

Figure 2 shows a general complex RNS based system. CRNS suffers as other conventional number systems from the high price of the hardware requirements in the case of complex multiplication as it is performed by 4 real multiplications and 2 real additions besides the necessary interaction between the real and imaginary channels. To reduce such complexity QRNS has been proposed for complex computation as it has the capability of significantly reducing the hardware complexity especially in VLSI implementation.

QRNS structure has been employed in Fermat Number Transform by Nussbaumer [14], but it has not been proposed for general complex DSP algorithms except recently by Leung [5]. The algebraic foundation for QRNS is based on the following theorem[15].

Theorem: The number -1 is a quadratic residue of all primes of the form $4n + 1$ and a quadratic non-residue of all primes of the form $4n + 3$.

In case of quadratic residue, the following equation has a solution for all moduli m_i:

$$x^2 = -1 \bmod m_i \qquad\qquad (1)$$

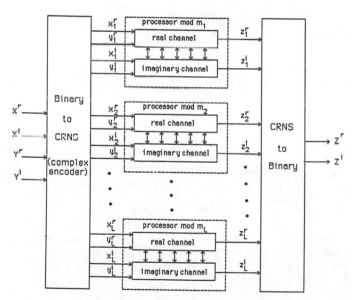

Fig.2 A General CRNS_Based Structure

If m_i expressed by its prime factors, $m_i = p_1^{k_1} \, p_2^{k_2} \, ...p_n^{k_n}$, equation (1) can be rewritten as follows:

$$x^2 = -1 \bmod p_i \qquad (2)$$

Equation (2) always has two solutions $\bar{j}_{i1}$ and $\bar{j}_{i2}$, which are both additive and multiplicate inverses of each other. The entity $\bar{j}$ is a real integer. As an example, for $m = 17$,

$$\bar{j}^2 = -1 \bmod 17$$
$$\bar{j}_1^2 = 16, \, \bar{j}_1 = 4, \text{ then}$$
$$\bar{j}_2 = 13, \text{ i.e. } \bar{j}_2^2 = 169 = -1 \bmod 17$$

In QRNS the complex numbers are represented by decoupled real and imaginary pair expressed as follows:

$$X = (x, x^*)$$

The QRNS is also a finite ring; QR(M) consisting of M pairs and has the isomorphism property based on the CRT as the conventional RNS:

$$QR(M) = (QR(m_1), QR(m_2), ..., QR(m_L))$$

The mathematical relationship is defined in QR(M) as follows [15]:

$$(x_i, x_i^*) + (y_i, y_i^*) = (z_i^1, z_i^{1*})$$
where $z_i^1 = |x_i + y_i|_{m_i} \; \& \; z_i^{1*} = |x_i^* + y_i^*|_{m_i}$
$$(x_i, x_i^*) * (y_i, y_i^*) = (z_i^2, z_i^{2*})$$
where $z_i^2 = |x_i * y_i|_{m_i} \; \& \; z_i^{2*} = |x_i^* * y_i^*|_{m_i}$

Figure 3 shows a general QRNS based system. Leung [5] and Jenkins [15] have presented an isomorphic mapping between the CRNS and QRNS such that:

$$\text{If} \quad s_i = x_i + jy_i \qquad \text{(CRNS), and}$$
$$s_i = (z_i, z_i^*) \qquad \text{(QRNS)}$$
$$\text{then} \quad z_i = |x_i + \bar{j}y_i|_{m_i}$$
$$z_i^* = |x_i - \bar{j}y_i|_{m_i}$$
$$x_i = |2^{-1}(z_i + z_i^*)|_{m_i}$$
$$y_i = |2^{-1}\bar{j}^{-1}(z_i - z_i^*)|_{m_i}$$

QRNS offers significant hardware saving and faster performance as complex multiplication only requires two real multiplies.

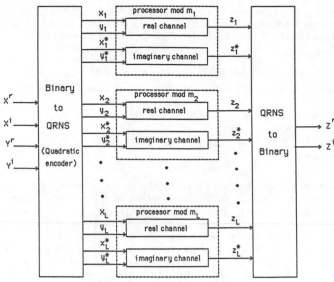

Fig. 3 A General QRNS_Based Structure

3. THE LOOK-UP TABLE IMPLEMENTATION

The implementation strategy of residue arithmetic has evolved based on the development of the state of the art of integrated circuit technology. The available technology determines the implementation approach on both circuit and system levels. Evolving from SSI to MSI and then LSI, RNS based hardware implementation went through the following approaches [9]:

(1) Look-up table implementation.

(2) Direct Logical implementation using conventional binary modules.

(3) The hybrid implementation.

The look-up table approach is based on storing the RNS functions in memory modules. In the binary number system, the table look-up approach is not feasible because of the enormous storage required; for a word length of B bits, 2^{2B} entries are required in the table. In the RNS with a comparable range, i.e., $\sum_{i=1}^{L} m_i \approx 2^B$, each modulus m_i requires m_i^2 entries in the table, hence a total of $\sum_{i=1}^{L} m_i^2$ entries are needed. For reasonable values of L and $\{m_i\}$, $\sum_{i=1}^{L} m_i^2 << \prod_{i=1}^{L} m_i^2$. Besides, the look-up table approach offers the best solution for high speed realization through pipelining [8]. However restrictions have to be placed on the size of the moduli in a system. The look-up table approach can provide great savings for hardware if some of the operands are fixed; constants can be premultiplied or added, and stored along with the binary operation being implemented. In the VLSI technology the look-up table approach is more efficient implementation than the other two approaches for moduli size up to 6 or 7 bits [9,10].

VLSI has been an opportunity to overcome several special problems experienced when using commercially available memory packages, such as the constraints on selecting the moduli set, the wasted memory locations, and the problems practiced in interconnecting large number of memory packages. Using the custom design approach the designer determines the required number of memory locations and the geometrical configuration of the layout. A procedure has been developed to determine the area and time required to implement a modulo look-up table based on a computational model reported in [11]. The main features of the layout procedure are:

(a) The word width is $\lceil \log_2 m \rceil$, where m is the modulus.

(b) It supports any modulus (not necessarily a power of 2).

(c) The storage array is not necessarily a power of 2, and can be divided into multi-subarrays.

(d) The storage array can be laid out in a hierarchical fashion which offers high speed performance.

(e) The number of decoder outputs is m_i^2; this leads to substantial saving in the required area.

The first phase of the layout procedure is to generate all the viable layout options for a modulo look-up table. Each option is defined by 5 tuples {W, H, A, T, AT} where W and H are the width and height in unit length; A is the area in unit area; T is the time in unit time; and AT is the area-time product. Each option has distinct parameters. In the second phase, the user selects the most efficient option according to his performance measure; area, time, area-time product, or any other constraints on the dimensions due to placement and routing. Figure 4 shows an experimental mod-7 look-up table adder chip implemented in $2\mu m$ NMOS technology.

4. QRNS-BASED SIGNAL PROCESSOR

The proposed processor is based on a Cooley-Tukey, power of 2, decimation in time FFT algorithm [16]. The FFT algorithm has been chosen by several authors [17,18] to be the central computational algorithm because it allows not only the frequency domain related signal processing, but, also it is a basic part of efficient algorithms for FIR filtering, convolution, and correlation. The Cooley-Tukey approach provides suitable VLSI implementation with appropriate simplicity, modularity, and hardwired programmability. The computational kernel of this processor is a radix-2 butterfly which offers simple complexity for data recording. L subprocessors (modulo processors) are used in parallel (one for each modulus). They are identical and have the same architecture.

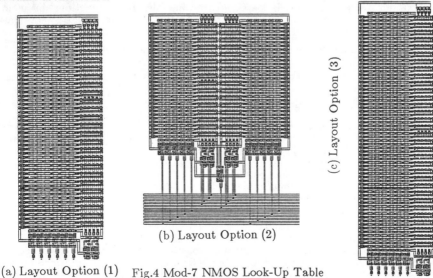

(a) Layout Option (1)   (b) Layout Option (2)   (c) Layout Option (3)

Fig.4 Mod-7 NMOS Look-Up Table

4.1. The Modulo Processing Organization

The input data and twiddle factors are assumed to be residue encoded. Only one modulo processor (for each modulus) is used for executing N-point FFT, executing butterflies in order, starting from the left column going through each column in descending direction, (see reference [16], pp.362). FIFO's (First-In, First-Out) are used for storing input, intermediate, and output data. Data shuffling among FFT passes can be implemented without additional hardware or time overhead using FIFOs compared to using RAMs. FIFO input/output data rate is faster than the RAMs, too. Two sets of FIFOs are used for input data and twiddle factors. Each set is composed of two separate and independent real and imaginary FIFOs. The modulo processor is composed mainly of a complex radix-2 butterfly, a complex multiplier and scaler. Two bidirectional busses are employed for input/output and data transfer for the real and imaginary parts, Fig. 5. This allows simultaneous access of the real and imaginary data words and coefficients to support the main feature of QRNS. Scaling has been analyzed in the context of RNS in [20], and several proposed implementations in QRNS have been reported in [19].

4.2. The Modulo Processor Functions

The main operations performed by the proposed processor are FFT, Inverse FFT (IFFT), Finite Impulse Response Filtering (FIR), convolution, correlation, and multiplication. Macroprogramming is employed to configure the data flow according to the executed algorithm. A single instruction is devoted for each operation, it is loaded into the processor and stored in an instruction register. Then the instruction is decoded by a single Programmable Logic Array (PLA), which issues the appropriate control signals. The processor stays in a certain configuration till it receives another instruction. The necessary information for each algorithm such as the number of data points to be processed and the number of FFT passes should be supplied, too. The input data are properly ordered in case of FFT while they are bit reversed ordered for IFFT. The FIR algorithm is performed by computing the FFT of the incoming data, multiplying with the coefficients, and then computing the IFFT of the product.

The control part of the processor is responsible for synchronizing the different algorithms and operations steps. It manages the control of the data transfer to and from the FIFO. The arithmetic operations and the I/O operations for a particular function are overlapped such that there is no time overhead to the overall operation. Also, pipelining can be achieved easily through the processors computational resources. This processor can be employed as a stand-alone unit controlled by another controlling chip. It can be used as a co-processor with a simple asynchronous control interface with the host computer. A large modulo processor can be built by multiplexing smaller modulo processors. For purpose of error detection, a modulo processor can function as a watchdog processor for another one.

4.3. Modulo Processor Implementation

A radix-2 butterfly can be defined using RNS as follows:

$$Y(0)_j = \left| X(0)_j + \left| \, (W_N^k)_j \bullet X(1)_j \mid_{m_j} \right|_{m_j} \right.$$

$$Y(1)_j = \left| X(0)_j - \left| \, (W_N^k)_j \bullet X(1)_j \mid_{m_j} \right|_{m_j} \right.$$

for $k = 0,1,2,3, \ldots\ldots, (N-1)$, where $X(n)_j$ and $Y(n)_j$ are the residue representation of the j'th residue of the the input and output signals, respectively, and $W = e^{\frac{-2\pi i}{N}}$. For complex signals, QRNS is employed.
Let,

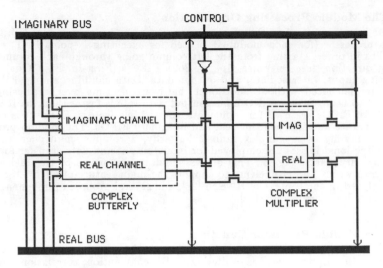

IMAGINARY BUS

CONTROL

IMAGINARY CHANNEL

IMAG

REAL CHANNEL

REAL

COMPLEX
BUTTERFLY

COMPLEX
MULTIPLIER

REAL BUS

Fig.5 Data Flow of the Computational Part of the
Modulo Processor

$$X_j\,(n) = (x_j\,(n),\, x_j^{\,*}\,(n))$$
$$Y_j\,(n) = (y_j\,(n),\, y_j^{\,*}\,(n))$$
$$z_j = w_N^k{}_j \cdot x(1)_j$$
$$z_j^{\,*} = w_N^k{}_j^{\,*} \cdot x(1)_j^{\,*}$$

Now the radix-2 butterfly is represented in QRNS as follows:

$$y(0)_j = x(0)_j + z_j$$
$$y(0)_j^{\,*} = x(0)_j^{\,*} + z_j^{\,*}$$
$$y(1)_j = x(0)_j - z_j$$
$$y(1)_j^{\,*} = x(0)_j^{\,*} - z_j^{\,*}$$

A block diagram for the QRNS based butterfly is shown in Fig. 6.

Look-up tables are used to implement the modulo operations. Their layouts are optimized using the model and procedure explained in Section 3. The various employed memory modules are connected together in a regular topology according to the following rules:

1. The data bits are routed in dedicated paths (i.e., each bit has a certain path).
2. Only one word ($\lceil \log_2 m \rceil$) can be transferred at a time.
3. The transfer time depends on the longest bit path.
4. Horizontal paths are laid out on one layer and the vertical ones are laid out on another layer.

In Fig. 6, look-up table (1) is laid out such that the layout height is larger than the width with 20% of the optimal layout option. Tables (2) and (3) are laid out based on the most efficient area_time product. Figure 7 shows a compact layout (NMOS, 2μm) for a modulo-13 processor which exhibits regular layout and very small

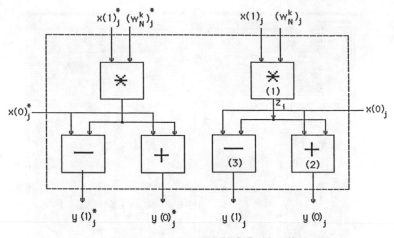

Fig.6. A Radix-2 QRNS Butterfly

interconnection area. A simple control circuitry is included for routing the output of the butterfly for configuring the processor to execute the required algorithm. The overall area is 3.96×3.83 mm^2, the time delay of one complex butterfly is around 300 nSecs. Fig. 8 shows an experimental modulo 13 NMOS chip.

5. COMMENTS AND CONCLUSIONS

The QRNS, which has been rediscovered recently, is a good candidate to support the growing need for high speed computing hardware for real-time complex digital signal processing applications. The complex arithmetics are encoded using the algebraic properties of modular number codes which leads to a complete separation between the real and imaginary channels. This feature is highly required for VLSI implementation. Besides reducing the computational hardware requirements by 50%, the interconnection area is dramatically minimized, which can lead to an overall area reduction of around 70% of the area required when using CRNS. QRNS compares favorably to the conventional arithmetic, the overall reduction depends on the overhead area required for converting to/from QRNS. The parallelism within the system is doubled which leads into more modularity and faster performance.

On the circuit level, look-up tables have been used as building block modules which are considered efficient for VLSI implementation for up to 6-7 bits moduli. Using the custom design approach, these modules can be optimized according to the overall system requirements and constraints. The proposed processor is optimized for computing the FFT algorithms and related applications. It belongs to those processors which can be employed as stand alone or as peripheral processors needed for a wide range of applications for speech processing with few KHz to radar processing at several Mhz.

The novel proposed architecture offers a good VLSI solution for DSP architectures. It represents a tradeoff between the overhead of the general purpose processors and the limitation of the dedicated ones. It is considered a contemporary processor dedicated to a specific family of algorithms and programmable within this family domain. Moreover, the processor gains its high performance from using bit parallel processing which has been possible by using the parallelism inherent in the residue arithmetic. With this coordination on several levels, mathematical, architectural and implementation, the processor is capable of computing 1024 complex point FFT in around 2 mSecs.

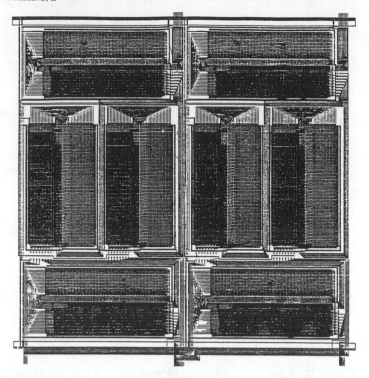

Fig.7 A Compact NMOS Mod-13 Processor

REFERENCES

[1] H.T. Kung, "Special Purpose Devices for Signal and Image Processing: An Opportunity in Very Large Scale Integration (VLSI)", SPIE Vol-241, pp. 76-84, 1980

[2] C.H. Huang, D.G. Peterson, H.E. Rauch, J.W. Teague, and D.F. Frasher, "Implementation of Fast Digital Processor using the Residue Number System", IEEE Trans. Circuits and Systems, Vol. CAS-28, No. 1, pp. 32-38, Jan. 1981.

[3] H.K. Nagpal, G.A. Jullien, W.C. Miller, "Processor Architectures for Two-Dimensional Convolvers Using a Single Multiplexed Processor Element", IEEE Trans. on Computers, Vol. C-32, No.11, pp. 989-1000, 1983.

[4] S.D. Fouse, G.R. Nudd and A.D. Cumming, "A VLSI Architecture for Pattern Recognition using Residue Arithmetic", Proc. of the 6th International Conference on Pattern Recognition.

[5] S.H. Leung, "Application of Residue Number Systems to Complex Digital Filters", in Proc. 15th Asilomar Conf. Circuits and Systems., pp. 70-74, Nov. 1981.

[6] C.A. Mead and L.A. Conway, Introduction to VLSI Systems, Addison Wesley, 1979.

[7] M.A. Bayoumi, G.A. Jullien, and W.C. Miller, "Models for VLSI Implementation of Residue Number System Arithmetic Modules", Proc. 6th Symposium on Computer Arithmetic, pp 174-182, Denmark, June 1983.

[8] G.A. Jullien, "Residue Number Scaling and Other Operations Using ROM Arrays", IEEE Trans. Computers, Vol. C-27, No.4, pp. 325-337, April 1978.

[9] M.A. Bayoumi, G.A. Jullien, and W.C. Miller, "A VLSI Implementation of RNS_based Architectures", 1985 IEEE International Symposium on Circuits and Systems, Japan 1985.

[10] Chao_Lin Chiang and Lennart Johnsson, "Residue Arithmetic and VLSI", Proc. of ICCD'83, pp. 961-968, 1983.

[11] M.A. Bayoumi, G.A. Jullien, and W.C. Miller, "A VLSI Design Methodology for RNS Structures Used in DSP Applications", Submitted to the IEEE Trans. on Circuits and Systems.

[12] N.S. Szabo and R.J. Tanaka, Residue Arithmetic and its Applications to Computer Technology, McGraw Hill, New York, 1967.

[13] W.K. Jenkins, "Complex Residue Number Arithmetic for High Speed Signal Processing", Electronic Letters, Vol. 16, No.17, pp. 660-661, Aug. 1980.

[14] H.J. Nussbaumer, "Complex Convolutions via Fermat Number Transforms", IBM Journal of Research and Development, Vol. 20, pp. 282-284, May 1976.

[15] J.V. Krogmeier and W.K. Jenkins, "Error Detection and Correction in Quadratic Residue Number Systems", Proc. of the 26th Midwest Symposium on Circuits and Systems, Mexico, pp. 408-411, Aug. 1983.

[16] L. Rabiner and B. Gold, Theory and Applications of Digital Signal Processing, Prentice Hall, 1975.

[17] E.E. Swartzlander and G.H. Hallnor, "Frequency Domain Filtering with VLSI", Proc. of the Workshop on VLSI and Modern Signal Processing, 1983.

[18] Peter P. Reusens, Richard W. Linderman, Walter H. Ku, "CUSP: A Custom VLSI Processor for Digital Signal Processing Based on the Fast Fourier Transform", Proc. of IEEE ICCD, pp. 757-760, 1983.

[19] F.J. Taylor, "A Single Modulus Complex ALU for Signal Processing", IEEE Trans. on ASSP, Vol.33, No.5, pp. 1302-1315, Oct. 1985.

[20] B. Tseng, G.A. Jullien, and W.C. Miller, "Implementation of FF Structures Using the Residue Number System", IEEE Trans. on Computers, Vol. C-28, No. 11, pp. 831-844, Nov. 1979.

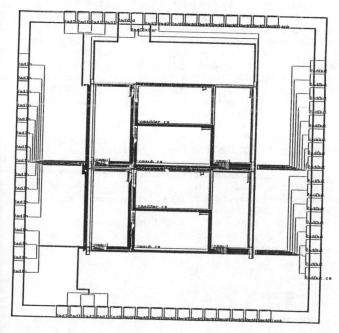

Fig.8 An Experimental
Mod-13 NMOS Chip

LINEAR CONTENT-ADDRESSABLE SYSTOLIC ARRAY FOR SPARSE MATRIX MULTIPLICATION

Robert Charng and Ferng-Ching Lin
Dept. of Computer Science and Information Engineering
National Taiwan University
Taipei, Taiwan, R.O.C.

ABSTRACT

We present a linear systolic array with content-addressable memories for fast computation of sparse matrix multiplication. Let N_A be the number of not-all-zero rows of matrix A, N_b be the number of nonzero elements of matrix B, and p be the number of cells in the linear systolic array. The system takes $N_A + \lceil N_A/p \rceil N_b - 1$ time steps to compute $A \cdot B$.

1. INTRODUCTION

Sparse matrices occur in many application areas including signal and image processing. Implementations of various algorithms on VLSI array structures for matrix multiplication have been extensively studied [1-7]. Although they are ideally suited for either dense matrices, strictly banded matrices or symmetric matrices, these designs can not effectively compute the multiplication of sparse matrices since the structure of nonzero elements in a large sparse matrix is irregular and sparse.

In this paper, we present a regular and simple linear systolic array equipped with content-addressable memories to solve the problem in high speed. Let A be an $m \times n$ matrix, B be an $n \times k$ matrix. Also let N_A be the number of not-all-zero rows of A, N_b be the number of nonzero elements of B, and p be the number of cells in the linear systolic array. The system takes only $N_A + \lceil N_A/p \rceil N_b - 1$ time steps to compute $A \cdot B$.

In a content-addressable memory (CAM), each individual memory word can be read from or written to by specifying its content or partial content. The recent advances in VLSI technology make the implementation of CAM more feasible since two major problems, implementation cost and interrogation drive, have already been solved [8]. The general characteristics of CAM realized by VLSI are discussed in [9, 10]. Special CAM-based array structures have been proposed for solving a system of sparse linear equations [11], language processing [12], image analysis [8], etc.

In Section 2, we introduce some primitive compression operations to derive efficient storage schema for sparse matrices. In Section 3, we deal with the case of sufficient array size, i.e., $p = N_A$. We first describe a linear content-addressable systolic array for sparse matrix-vector multiplication ($k=1$) and then extend its use to the general matrix-matrix multiplication. In Section 4, we treat the case of insufficient array size, i.e., $p < N_A$, by partitioning the not-all-zero rows of A and duplicating the nonzero elements of B. Some concluding remarks are given in Section 5.

2. STORAGE SCHEMA FOR SPARSE MATRICES

A matrix (vector) is sparse if most of its elements are zero. We shall call a row (column) of matrix not-all-zero if not all its elements are zero. Many methods have been used for efficient storage of sparse matrices [13-16], however, they are mainly suitable for uniprocessor system. In order to improve the storage efficiency and the execution speed of our system, we introduce some storage schema in which a matrix or vector is compressed by four primitive operations: horizontal element compression (HEC), vertical element compression (VEC), column compression (CC), and row compression (RC).

The HEC (VEC) operation condenses sparse row (column) vectors of a matrix into 'dense' row (column) vectors of the matrix by orderly concatenating the nonzero elements. The CC (RC) operation makes the not-all-zero column (row) vectors to be orderly tied up together. Figure 1 shows how a sparse column vector B is condensed into a 'dense' vector $\underline{B}$ by VEC operation. Figure 2 shows how a sparse matrix A is condensed in two steps, HEC then RC, into a 'dense' matrix $\underline{A}$.

$$B = \begin{bmatrix} b_{11} \\ 0 \\ b_{31} \\ 0 \\ 0 \\ b_{61} \end{bmatrix} \quad \xrightarrow{\text{VEC}} \quad \underline{B} = \begin{bmatrix} b_{11} \\ b_{31} \\ b_{61} \end{bmatrix}$$

Figure 1　A sparse column vector B is condensed into a 'dense' vector $\underline{B}$ by VEC operation

$$A = \begin{bmatrix} a_{11} & 0 & a_{13} & 0 & 0 & 0 \\ 0 & a_{22} & a_{23} & a_{24} & 0 & 0 \\ 0 & 0 & a_{33} & 0 & 0 & 0 \\ a_{41} & 0 & 0 & a_{44} & a_{45} & 0 \\ 0 & 0 & 0 & 0 & 0 & 0 \\ 0 & 0 & 0 & 0 & 0 & a_{66} \end{bmatrix} \quad \xrightarrow{\text{HEC}} \quad \begin{bmatrix} a_{11} & a_{13} \\ a_{22} & a_{23} & a_{24} \\ a_{33} \\ a_{41} & a_{44} & a_{45} \\ 0 \\ a_{66} \end{bmatrix}$$

$$\xrightarrow{\text{RC}} \quad \underline{A} = \begin{bmatrix} a_{11} & a_{13} \\ a_{22} & a_{23} & a_{24} \\ a_{33} \\ a_{41} & a_{44} & a_{45} \\ a_{66} \end{bmatrix}$$

Figure 2　A sparse matrix A is condensed in two steps, HEC then RC, into a 'dense' matrix $\underline{A}$

3. LINEAR CONTENT-ADDRESSABLE SYSTOLIC ARRAY

3.1 Matrix-Vector Multiplication

The recurrences in the matrix-matrix multiplication of C= A·B can be expressed as:

$$c_{ij}^{(0)} := 0$$

$$c_{ij}^{(t+1)} := c_{ij}^{(t)} + a(i,u)*b(u,j)$$

$$c_{ij} := c_{ij}^{(n)}$$

For matrix-vector multiplication (k=1), the index j in the recurrences should be equal to one. We shall discuss this case first.

The linear systolic array of N_A cells is depicted in Figure 3. Each cell is connected to its neighboring cells and is associated with a CAM which stores a 'dense' row of $\underline{A}$. Each memory word of the CAM consists of three fields: the value of nonzero elements, its column number, and its row number. The 'dense' vector $\underline{B}$ is serially input into the leftmost cell and travels through the array horizontally. The data words for $\underline{B}$ has four fields: the element value, its row number, its column number, and an end-of-vector bit.

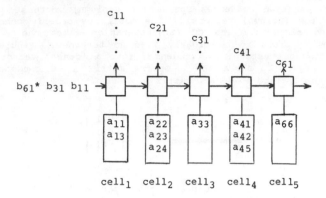

Figure 3   Content-addressable systolic array for the matrix-
            vector multiplication of $A \cdot B$

Each cell uses the row number of the received element as the key to read an element of the 'dense' row vector which is stored in the associated CAM. If the row number of the received element equals to the column number of a stored element, then the stored element will be read out to the cell for multiplication. Let's respectively use I_l, O_r, and O_t to represent the left, the right, and the top ports of a cell, use $\underline{b}$, RN, CN, and EOV to denote the four fields of received words, and use $\underline{a}$ and $\underline{c}$ to indicate the read element and the element of resulting matrix. The cell function can be described as follows:

Receive data word ($\underline{b}$, RN, CN, EOV) from I_l port;
Read $\underline{a}$ from CAM by key RN;
If memory word is found then $\underline{c} := \underline{c} + \underline{a}*\underline{b}$;
Send the received data word ($\underline{b}$, RN, CN, EOV) to the right neighbor via O_r port;
If EOV = '*' then send $\underline{c}$ and its associated row and column numbers via O_t port

Notice that the end element b_{N_b} marked by '*' is used to trigger out the desired results along the cells. In Figure 4, the snapshots of the

time	cell$_1$	cell$_2$	cell$_3$	cell$_4$	cell$_5$
1	$c_{11}=a_{11}b_{11}$				
2	$c_{11}=c_{11}+a_{13}b_{31}$	–			
3	$-/(c_{11})$	$c_{21}=a_{23}b_{31}$	–		
4		$-/(c_{21})$	$c_{31}=a_{33}b_{31}$	$c_{41}=a_{41}b_{11}$	
5			$-/(c_{31})$	–	–
6				$-/(c_{41})$	–
7					$c_{61}=a_{66}b_{61}/(c_{61})$

Figure 4   The snapshots of the process shown in Figure 3

process on the examples of Figure 1 and Figure 2 are recorded. All the output elements are especially enclosed with parentheses. It is clear that the system takes N_A+N_b-1 time steps to complete the process.

3.2 Matrix-Matrix Multiplication

We can extend the use of the above design to matrix-matrix multiplication of $A \cdot B$. The input 'dense' matrix $\underline{B}$ is obtained from B by **VEC** then **CC** operations and the 'dense' column vectors of $\underline{B}$ are serially input in an element-by-element fashion. An example of compressed matrix $\underline{B}$ is shown in Figure 5. The input/output data streams through the array is illustrated in Figure 6, where $\underline{A}$ is the same matrix as in Figure 2. Figure 7 shows the corresponding snapshots of the process. No matter how many columns of B has, N_A+N_b-1 time steps are required to complete the multiplication.

$$B = \begin{bmatrix} b_{11} & 0 & b_{13} & 0 & 0 & 0 \\ 0 & 0 & b_{23} & 0 & 0 & 0 \\ 0 & 0 & 0 & 0 & 0 & 0 \\ b_{41} & 0 & 0 & b_{44} & 0 & 0 \\ b_{51} & 0 & 0 & 0 & 0 & 0 \\ 0 & 0 & 0 & 0 & 0 & 0 \end{bmatrix} \xrightarrow{\textbf{VEC \& CC}} \underline{B} = \begin{bmatrix} b_{11} & b_{13} & b_{44} \\ b_{41} & b_{23} & \\ b_{51} & & \end{bmatrix}$$

Figure 5 A sparse matrix B is condensed in two steps, **VEC** then **CC**, into a 'dense' matrix $\underline{B}$

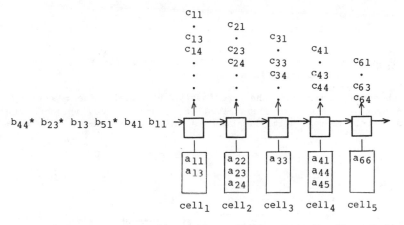

Figure 6 Content-addressable systolic array for the matrix-matrix multiplication of $A \cdot B$

time	cell$_1$	cell$_2$	cell$_3$	cell$_4$	cell$_5$
1	$c_{11}=a_{11}b_{11}$				
2	–	–			
3	$-/(c_{11})$	$c_{21}=a_{24}b_{41}$	–		
4	$c_{13}=a_{11}b_{13}$	$-/(c_{21})$	–	$c_{41}=a_{41}b_{11}$	
5	$-/(c_{13})$	–	$-/(c_{31})$	$c_{41}=c_{41}+a_{44}b_{41}$	–
6	$-/(c_{14})$	$c_{23}=a_{22}b_{23}/(c_{23})$	–	$-/(c_{41})$	–
7		$c_{24}=a_{24}b_{44}/(c_{24})$	$-/(c_{33})$	$c_{43}=a_{41}b_{13}$	$-/(c_{61})$
8			$-/(c_{34})$	$-/(c_{43})$	–
9				$c_{44}=a_{44}b_{44}/(c_{44})$	$-/(c_{63})$
10					$-/(c_{64})$

Figure 7 The snapshots of the process shown in Figure 6

4. PROBLEM PARTITION

The interesting and important issue of problem partition is raised when the matrix size exceeds the array size. In our case, this means that N_A, the number of not-all-zero rows of A, is greater than the actual array size p. Various techniques have been proposed to partition the matrix into pXp or nXp submatrices [4, 11, 13, 15, 17, 18]. Here, we suggest a method to rearrange large sparse matrices into 'narrow' matrices as follows. First, the sparse matrix A is compressed by the **RC** operation. Then we partition the intermediate matrix into $\lceil N_A/p \rceil$ pXn submatrices and orderly concatenate them to form a pX($\lceil N_A/p \rceil$ n) matrix. Finally, this matrix is further compressed by the **HEC** operation into a 'dense' matrix $\underline{A}$. Take A in Figure 2 as an example and let p = 3, by the above procedure, a 'dense' matrix $\underline{A}$ is obtained as traced in Figure 8.

$$A=\begin{bmatrix} a_{11} & 0 & a_{13} & 0 & 0 & 0 \\ 0 & a_{22} & a_{23} & a_{24} & 0 & 0 \\ 0 & 0 & a_{33} & 0 & 0 & 0 \\ a_{41} & 0 & 0 & a_{44} & a_{45} & 0 \\ 0 & 0 & 0 & 0 & 0 & 0 \\ 0 & 0 & 0 & 0 & 0 & a_{66} \end{bmatrix} \xRightarrow{\text{RC}} \left[\begin{array}{cccccc} a_{11} & 0 & a_{13} & 0 & 0 & 0 \\ 0 & a_{22} & a_{23} & a_{24} & 0 & 0 \\ \hline 0 & 0 & a_{33} & 0 & 0 & 0 \\ \hline a_{41} & 0 & 0 & a_{44} & a_{45} & 0 \\ 0 & 0 & 0 & 0 & 0 & a_{66} \end{array}\right]$$

$$\xRightarrow{\text{'narrowing'}} \left[\begin{array}{cccccc|cccccc} a_{11} & 0 & a_{13} & 0 & 0 & 0 & a_{41} & 0 & 0 & a_{44} & a_{45} & 0 \\ 0 & a_{22} & a_{23} & a_{24} & 0 & 0 & 0 & 0 & 0 & 0 & 0 & a_{66} \\ 0 & 0 & a_{33} & 0 & 0 & 0 \end{array}\right]$$

$$\xRightarrow{\text{HEC}} \underline{A}=\begin{bmatrix} a_{11} & a_{13} & a_{41} & a_{44} & a_{45} \\ a_{22} & a_{23} & a_{24} & a_{66} \\ a_{33} \end{bmatrix}$$

Figure 8   A large sparse matrix A rearranged into a
'dense' matrix $\underline{A}$

An enhanced version of the previously described linear content-addressable systolic array, as illustrated in Figure 9, can implement the partitioned matrix-vector multiplication of $A \cdot B$, where $\underline{B}$ is the same

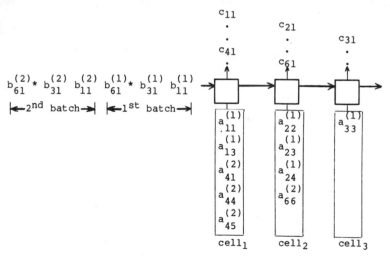

Figure 9   Content-addressable systolic array for the partitioned
matrix-vector multiplication of $A \cdot B$

vector as in Figure 1. 'Dense' row vectors of $\underline{A}$ are preloaded into the CAM's. Each memory word consists of four fields: the element value, its column number, its row number, and a block number. The additional block number, ranging from 1 to $\lceil N_A/p \rceil$, indicates in which the section of the row partition of the intermediate 'dense' matrix is located.

Because the size of the 'dense' matrix $\underline{A}$ is greater than the array size, $\lceil N_A/p \rceil$ duplicate sets of 'dense' column vectors of $\underline{B}$ must pass through the array. For this matter, the data words of $\underline{B}$ must contain an extra field to indicate the batch number (denoted as **BN**). Each cell uses the row number and the batch number of the received element as the keys to read an element stored in the associated CAM. If the row number and the batch number of the received element are respectively equal to the column number and the block number of a stored element, the stored element is read out for multiplication. The following is the function description of a cell:

Receive data word ($\underline{b}$, **RN**, **CN**, **EOV**, **BN**) from I_1 port;
Read $\underline{a}$ from CAM by keys **RN** & **BN**;
If memory word is found then $\underline{c} := \underline{c} + \underline{a}*\underline{b}$;
Send received data word ($\underline{b}$, **RN**, **CN**, **EOV BN**) to the right neighbor via
  O_r port;
If **EOV** = '*' then send $\underline{c}$ and its associated row and column numbers via
  O_t port

Using the above mentioned method, the linear systolic can also apply to matrix-matrix multiplication. Since the input 'dense' vector or matrix B are duplicated as many as $\lceil N_A/p \rceil$ times, $N_A + \lceil N_A/p \rceil N_b - 1$ time steps are required for the whole process.

5. CONCLUDING REMARKS

We have presented a linear systolic array which is equipped with content-addressable memories for fast computation of sparse matrix multiplication. A typical large sparse matrix contains about 2% to 5% nonzero elements [11], so the loading time of the condensed matrix $\underline{A}$ should be very short compared to the execution time. Once the 'dense' row vectors are stored in the CAM's, arbitrary many vectors or matrices $\underline{B}$ can be fed into the system in a pipelined fashion. When the matrix size is too big, our design can easily be enchanced without enlarging the actual array to do the problem partition. Because the cells are linearly connected and the data flow is in one direction, the cells can be implemented with pipelined function units so that the whole system become two-level pipelined achieving even higher data rate [19].

REFERENCES

[1] H. T. Kung and C. E. Leiserson, "Systolic arrays (for VLSI)," Sparse Matrix Symposium, SIAM, 1978, pp. 256-282.

[2] J. G. Nash, S. Hanson and G. R. Nudd, "VLSI processor arrays for matrix manipulation," in VLSI System and Computations, H. T. Kung, B. Sproull, and G. Steele, Eds, Computer Science Press, Oct. 1981, pp. 367-378.

[3] T. Y. Young and P. S. Liu, "VLSI arrays for pattern recognition and image processing: I/O bandwidth considerations," in VLSI for Pattern Recognition and Image Processing, K. S. Fu, Ed, Springer-Verlag, 1984, pp. 25-42.

[4] K. Hwang and Y. H. Cheng, "Partitioned matrix algorithms for VLSI arithmetic systems," IEEE Transactions on Computers, Vol. C-31, No. 12, Dec. 1982, pp. 1215-1224.

[5] H. Y. Chuang and G. He, "A versatile systolic array for matrix computation," Proc. 12th Annual Symposium on Computer Architecture, 1985, pp. 351-322.

[6] S. Y. Kung, "VLSI signal processing: from transversal filtering to concurrent array processing," in VLSI and Modern Signal Processing, S. Y. Kung, H. J. Whitehouse, and T. Kailath, Eds, Prentice-Hall Inc., Englewood Cliffs, New Jersey, 1985, pp. 127-152.

[7] H. J. Whitehouse, J. M. Speiser, and K. Bromley, "Signal processing applications of concurrent array processor Technology," in VLSI and Modern Signal Processing, S. Y. Kung, H. J. Whitehouse, and T. Kailath, Eds, Prentice-Hall Inc., Englewood Cliffs, New Jersey, 1985, pp. 25-41.

[8] W. E. Snyder and C. D. Savage, "Content-addressable read/write memories for image analysis," IEEE Transactions on Computers, Oct. 1982, Vol. C-31, No. 10, pp. 963-968.

[9] J. S. Hall, "A general-purpose CAM based system," in VLSI System and Computations, H. T. Kung, B Sproull, and G. Steele, Eds, Computer Science Press, Oct. 1981, pp. 379-388.

[10] K. Hwang and F. A. Briggs, "Computer Architecture and Parallel Processing," McGraw-Hill, 1984.

[11] O. Wing, "A content-addressable systolic array for sparse matrix computation," J. of Parallel and Distributed Processing, Vol. 2, No. 2, 1985, pp. 170-181.

[12] K. H. Chu and K. S. Fu, "VLSI architecture for high speed recognition of context free languages and finite state languages," Proc. 9th Conf. on Computer Architecture, April 1982, pp. 43-49.

[13] R. P. Tewarson, Sparse Matrices, Academic Press, New York, 1973.

[14] I. S. Duff, "A survey of sparse matrix research," Proc. IEEE, Vol. 65, No. 4, April 1977, pp. 500-535.

[15] S. Pissanetsky, Sparse Matrix Technology, Academic Press, New York, 1984.

[16] M. Veldhorst, "Approximation of the consecutive ones matrix augmentation problem," SIAM J. Computing, Vol. 14, No. 3, Aug. 1985, pp. 709-729.

[17] P. Kuekes and R. Schreiber, "Fast Sparse matrix computation," Digest COMPCON-81, 1981, pp. 251-253.

[18] D. Heller, "Partitioning big matrices for small systolic arrays," in VLSI and Modern Signal Processing, S. Y. Kung, H. J. Whitehouse, and T. Kailath, Eds, Prentice-Hall Inc., Englewood Cliffs, New Jersey, 1985, pp. 185-199.

[19] H. T. Kung and M. S. Lam, "Fault-tolerance and two-level pipelining in VLSI systolic arrays," Proc. MIT Conference Advanced Research in VLSI, Jan. 1984, pp. 74-83.

Part III

IMPLEMENTATION EXAMPLES

SIGNAL PROCESSING VLSI DEVELOPMENTS:
PROSPECTS   THROUGH   EXPERIENCE

Rikio Maruta   and   Takao Nishitani

C&C Systems Research Laboratories, NEC Corporation
4-1-1, Miyazaki, Miyamae-Ku, Kawasaki, 213 Japan

Abstract Trends of signal processor configurations in near future are clarified through development experience on various signal processing VLSI chips in a varaety of fields. The developed chips include general purpose single-chip signal processors in both fixed-point and floating-point formats, a general purpose DSP chip set with reconfigurable hardware structure, and application specific DSP chips in both dedicated hardwired design and software programmable structures. Some new trials are also reported for improving software development environment, and for extending software controllable application fields toward real time motion video signals.

1. GENERAL PURPOSE SIGNAL PROCESSORS

Of various signal processing chips, general purpose programmable signal processors are most useful for expanding DSP applications. Algorithm researchers can easily evaluate and improve their new ideas through real time simulation. Reduction in the turn-around-time speeds up the technological progress. Equipment manufacturers can have a compact and cost-effective means for their product development. Modification in system function can easily be done by software without any hardware modification.

Since these merits were foreseen at the planning stage of DSP VLSI product development in late 1970's, general purpose signal processors have been placed in the core of the DSP product line at NEC. Thus, uPD 7720 signal processor family was planned and developed[1]. The family includes an EPROM and CMOS versions. The usefulness of the EPROM version is recognized not only for the algorithm evaluation but also for equipment production, especially when product features need to be changed according to customer's requirements. The CMOS version cuts its power consumption amount by one seventh. Low power consumption signal processors expands digital signal processing

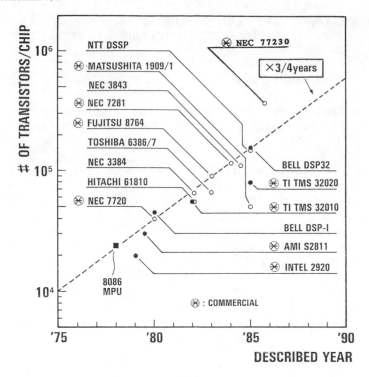

Fig. 1. Programmable Signal Processor Complexity.
Complexity is measured by numbers of transisters.

application fields to mobile radio communication systems.

The uPD 7720 application history, based on investigations on papers found in international conferences, is summarized in appendix 1. Application fields cover voiceband data modems, low bit rate speech coders, vocoders and speech recognizers. As appendix 1 shows only reported works, number of applications would be much more increased in actual.

The second generation has come with much more enhanced arithmetic operation capability. The newly developed 32-bit floating-point signal processor uPD 77230, having operation capability of 13.4 MFLOPS (Mega Floating-point Operations Per Second), is the first product in the uPD 77200 series[2]. An EPROM, a 24-bit fixed-point, and other versions are planned to follow. Field experiences of the predecessor are projected to the new architecture.

Figure 1 shows relationship between a number of

transisters in a chip and the chip development year. Since uPD 7720 and Bell's DSP-1 got great success in many application fields, trials on general purpose programmable signal processor development have been reported in every year. Every time when new chips come out, they have more enhanced functions than existing ones. Complexity in the number of transisters increases 3 times during every 4 years.

As the processing hardware resorces of general purpose signal processors increase, programming becomes increasingly difficult. One solution for this problem may be to develop a "DSP Compiler". Employment of a floating point arithmetic has rejected one barrier for using compiler systems. Troublesome "scaling" problems, required for fixed-point arithmetic digital signal processing applications, are eliminated in floating-point processing. However, the efficiency achievable with compilers for horizontal microprogram control signal processor, reported so far, is still very low. Exploiting AI technologies would be needed for having a drastic improvement in efficiency.

An alternative for reducing the load of programmers, which seems to be much easier than a DSP compiler at present, is to fully utilize DSP program libray. Since most of DSP functions are very regular and systematic, it can be expected that required jobs are efficiently programmable with some limited number of subroutine sets. If some provisions are incorporated in a signal processor architecture so that subroutines stored in the program ROM can be optionally called by giving simple commands, programming becomes very simple. This feature has been taken into uPD 77230 architecture design, which can be seen in Fig. 2. Being provided with two separate RAM's and a program STACK loadable from the data bus, uPD 77230 can be used in such a way that one RAM for "command" store, the other RAM for normal data store, and the program ROM for storing subroutines. What the user should do are finding an appropriate processing flow from the functional block diagram, in which each block corresponds to one subroutine stored in the program ROM, and generating the sequence of commands according to the processing flow to store it in the command memory.

The above method is very primitive. However, it indicates one direction of further efforts for reducing the load of programming general purpose signal processors. In addition to such command system approach, efficiency improvement of DSP compilers, as described before, will surely be pursued. There is also a possibility that a new machine architecture suitable for high level DSP languages is proposed.

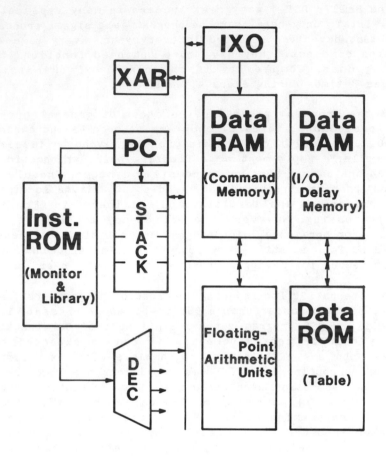

Fig. 2. Command System for uPD 77230. Floating-point signal processor has two Data RAM's, one Data ROM and one Instruction ROM. Instruction ROM contains general purpose digital signal processing subroutines and a monitor. A command sequence is stored in one Data RAM, while I/O data as well as inner state data in a digital signal processing. Data ROM is used for look-up tables. The monitor program fetches command, whose content is the beginning address of the corresponding subroutine in the Instruction ROM, from the Data RAM to PC (Program Counter) through Stack. Then, the corresponding subroutine is activated. After the subroutine processing, processor control returns to the monitor. Then, the monitor increments the content of IXO, acting as command sequence counter, and fetches the succeeding command to PC.

2. APPLICATION ORIENTED SIGNAL PROCESSORS

Although general purpose signal processors are very useful for algorithm development and equipment manufacturing, some applications which can assume a large production volume may justify the development of specific DSP VLSI chips. Application specific design can maximize efficiency, thereby enabling reducing cost, size, and power consumption. Equipment manufacturers also may necessitate proprietary devices in order to maintain their speciality over their competitors.

One example in shift from the use of general purpose signal processors to the use of a dedicated design chip is the interporation filter for CD (Compact Disc) players to reduce group delay distortion due to the analog reconstruction filter. Use of uPD 7720 signal processors allowed quick development and to avoid any initial risk. After receiving good reaction from customers, a dedicated chip, which enabled a drastic reduction in power consumption and cost, was developed[3]. The chip is now used by other CD player manufacturers, too. A similar example can be seen in telecom business. Echo cancellers were first built with multiple uPD 7720's, but soon a dedicated chip with enhanced capability and with a less power consumption was developed[4]. The chip is available for only in-house use, and is employed in proprietary telecom equipment.

Application specific DSP chips need not rely on a dedicated hardwired design. For applications in which modifications are foreseen, the software programmable signal processor structure is desirable. The architecture can be optimized for a specific category of applications. The most part of chip design can be based on the earlier design of general purpose signal processors.

Programmable signal processors for MODEM and the CCITT standard 32kbps ADPCM algorithm implementation are good examples. The MODEM signal processor was developed to provide more enhanced capabilities suitable for adaptive equalizer implementations, required for many kinds of high speed MODEM equipment[5]. Although data modem specifications are strictly defined as CCITT recommendations, I/O interfaces are different in every application systems. Software modification capability gives flexible interface for such occasions, as well as capabilities for implementing many different data modem chips.

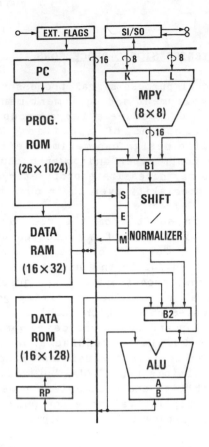

Fig. 3. CCITT Standard ADPCM Codec Processor.
The architecture of the chip has a programmable
signal processor, having Program ROM, Data RAM,
Data ROM and arithmetic units, such as
MPY(Multiplier), Shift/Normalizer and ALU.

The ADPCM processor[6], shown in Fig. 3, was developed to
efficiently perform the special floating-point format
operations with different bit-precision specifications which is
peculiar to the CCITT 32kbps algorithm. Employment of the
software approach enabled early start of chip development well
before finalization of the algorithm, and also enable to allow
algorithm modifications which are quite possible in earlier
stages of standardization. With architectural optimization and
CMOS technology employment, the low power consumption
conforming to the requirement of ordinary telecommunication
equipment has been successfully achieved.

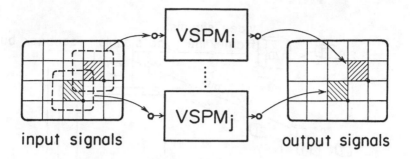

Fig. 4. Multi-processor approach by dividing one frame picture into sub-frame pictures. Every sub-frame picture is processed by every VSPM processor.

One of new application areas of software programmable signal processors would be real time video processing. If one picture frame is divided into some two-dimensional sub-regions as shown in Fig. 4, software programmable signal processors (VSPMs in Fig. 4) can process them at real time speeds. A processor called VSP (Video Signal Processor)[7], having multi-processor configuration shown in Fig. 5, has been constructed using discrete SSI/MSI/LSI components, demonstrating its usefulness for video processing algorithm evaluations. VLSI chips with a similar concept may be realized.

Another important progress in signal processors might be the inclusion of very high accuracy A-to-D and D-to-A conversion capabilities. Oversampling techniques which can lead high resolution with simple analog circuitry have been actively investigated. Computer simulation results[8] and hardware evaluation results using test chips indicate that a second-order predictive coder with a two-level quantizer is a possible candidate.

3. CONCLUSION

High level language developmet is shown to become an important key factor for general purpose signal processors in order to simplify complex microprogramming procedure, and to minimize program development period. On the contrary,

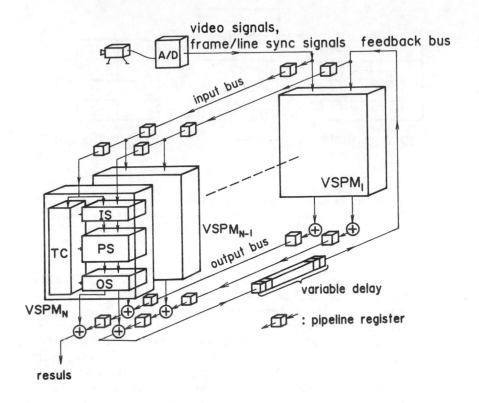

Fig. 5. Video Signal Processor. Every VSPM has the same structure, composed of IS(Input Storage), PS(Processing System), OS(Output Storage, and TC(Timing Controller). Every VSPM are combined with input video bus, output video bus and feed-back video bus. Employment of feed-back video bus enables interframe processing as well as two dimensional processing.

application specific processors have a varaety of approaches. Although hardwired implementation seemd to be suitable for such purposes, programable signal processor approaches are shown to be useful for some application specific fields. Next generation programmable signal processors are possibly concerning on motion picture processing and/or on built-in high accuracy analog interface circuits.

REFERERENCES

(1) T. Nishitani, R. Maruta, Y. Kawakami and H. Goto,"A Single-Chip DigitalSignal Processor for Telecommunication applications", IEEE J. Solid-state Circuits, vol.SC-16, No.4, Aug. 1981.

(2) T. Nishitani, I. Kuroda, Y. Kawakami, H. Tanaka and T. Nukiyama, "Advanced Single-Chip Signal Processor", Proc. IEEE ICASSP'86, vol.1, April, 1986.

(3) B. Hirosaki, Y. Toshimitsu, S. Ishihara, H. Nakada, K. Akiyama and K. Nosaka,"A CMOS-VLSI Rate Conversion Digital Filter for Digital Audio Signal Processing", Proc. of IEEE ICASSP'84, vol. 3, 1984.

(4) N. Furuya, Y. Fukushi, Y. Itoh, J. Tanabe and T. Araseki,"High Performance Custom VLSI Echo Canceller", IEEE Conference Record of ICC-85, 1985.

(5) M. Yano, K. Inoue, and T. Senba, "An LSI Digital Signal Processor", Proc. of IEEE ICASSP'82, vol.2, 1982.

(6) T. Nishitani, I. Kuroda, M. Satoh, T. Katoh, R. Fukuda and Y. Aoki,"A CCITT Standard 32 Kbps ADPCM LSI Codec", Proc. of IEEE ICASSP'85, vol.4, 1985.

(7) T. Nishitani, I. Tamitani, H. Harasaki, M. Yamashina, T. Enomoto, "Video Signal Processor Configuration by Multiprocessor Configuration", Proc. of IEEE ICASSP'86, vol.2, 1986.

(8) A. Yukawa, R. Maruta and K. Nakayama, "An Oversampling A-to-D Converter Structure for VLSI Digital Codec's", Proc. of IEEE ICASSP'85, 1985.

Appendix 1.   NEC 7720 Application History

1980   Push-button Signaling Receiver (NEC)

1981   Speech Recognizer (INFOVOX)

1982   LPC Vocoder (MIT)
        32kbps ADPCM (NEC)

1983   Echo Canceller (NEC)
        32kbps Delta-Modulation (A/S Electrik Bureau)
        TDHS/ADPCM (Univ. of Notre Dame)
        RELP Vocoder (Chalmers univ. of Technology)
        4800bps Modem (Siemens AG)
        High Level Language (Twente Univ.)
        PCM Test Equipment (CNET Lannion)
        Radiotelephony (Institut fur Ubertrangungstechnik)
        Low Coefficient Sensitivity Filter
                        (Tokyo Institute of Technology)

1984   32kbps ADPCM (BNR)
        64kbps ADPCM (Tokyo Univ.)
        16kbps Subband coder (British Telecom)
        16kbps APC codec (NEC)
        Speech Recognizer (Philips)
        Text to Speech (ETSI)
        HF Modem (NEC)
        PSK Carrier Processing (NEC)
        Analog Enchription (NEC)
        High Level Language (Helsinki Univ. of Technology)
        Wave Digital Filter (Linkoping Univ.)

1985   9600bps Modem (UDS Inc.)
        Multi-pulse Vocoder (NEC)
        Speech recognizer (Waseda Univ.)
        19.2kbps Modem (NEC)
        Music synthesizer (Osaka Univ.)
        MF Receiver (Univ. of Ottawa)
        LPC Vocoder (GTE)
        Subband Coder for Radio Telephone (AEG Telefunken)
        Mobile Radio Terminal (Philips)

ARCHITECTURAL FEATURES OF THE MOTOROLA DSP56000

DIGITAL SIGNAL PROCESSOR

Kevin L. Kloker
Motorola, Inc.
Corporate Research Laboratories
1301 E. Algonquin Road
Schaumburg, IL 60196

ABSTRACT

The Motorola DSP56000 is a high performance, user-programmable digital signal processor (DSP) implemented in 1.5 micron, double level metal, CMOS technology. The processor architecture is highly parallel, executing 10.25 million instructions per second with a 20.5 MHz clock. Precision arithmetic is performed by a 24 x 24 bit parallel hardware multiply-accumulator with 56 bit product accumulation. The integrated circuit also contains five on-chip memories - two 256 x 24 Data RAMs, two 256 x 24 Data ROMs and one 2048 x 24 Program ROM - organized into three separate memory spaces. Input/output capability is provided by a parallel host microprocessor interface, an asynchronous serial communications interface, a time division multiplexed, synchronous serial interface and a full-speed memory expansion port. A flexible instruction set with unique parallel move operations and addressing modes provides extremely high performance for a wide range of DSP and microprocessor applications. The architectural features contributing to the high performance of Motorola's DSP56000 are described in this paper.

INTRODUCTION

The Motorola DSP56000 is the first member of a new family of special-purpose microprocessors designed specifically for digital signal processing applications.[1,2] It is a fourth generation DSP IC and contains many new features to overcome the limitations of previous generation DSP devices. It represents three major achievements for advanced digital signal processing applications.

Speed

The DSP56000 achieves a 97.5 nsec instruction cycle time using a 20.5 MHz on-chip crystal oscillator or external clock. The processor delivers a peak throughput rate of 10.25 million instructions per second (10.25 MIPS), consisting of 10.25 million multiply and accumulate operations, 20.5 million addressing calculations, 20.5 million data transfers, 10.25 million instruction prefetches and 10.25 million program flow control operations per second. It executes many DSP benchmarks 2-4 times faster than recently announced, third generation DSP devices.[3,4,5]

Precision

The DSP56000 achieves 24 bit, fixed point data precision with 144 dB of external dynamic range. This dynamic range is sufficient for all real world applications, especially since the majority of data converters are 16 bits or less. Internal to the Data ALU, 56 bit accumulators provide 336 dB of internal dynamic range. With 56 bit accumulation, no precision is lost during intermediate calculations. The larger data size maximizes speed by allowing all arithmetic to be performed in single precision. With 8 extra bits of margin against overflow, roundoff and truncation errors, 24 bit data offers improved precision for advanced applications and ease of use by eliminating the intermediate scaling needed with 16 bit signal processors.

System On A Chip

The core processor consists of three separate execution units with a large set of local registers and arithmetic units. On-chip communication is supported by a multiple bus architecture having four internal data buses and three internal address buses. The DSP56000 also contains five on-chip memories, three sophisticated on-chip peripherals and a full-speed memory expansion port. Achieving complete DSP system integration on a single CMOS chip, the 88 pin integrated circuit interfaces easily into both single-chip applications and systems needing expanded memory and multiple processors. Putting these memory and peripheral resources on-chip maximizes execution speed and minimizes system chip count.

DSP56000 ARCHITECTURE

A block diagram of the DSP56000 architecture is shown in Figure 1. The IC architecture consists of a highly-parallel core processor which executes the DSP56000 instruction set and other on-chip resources such as memory and peripherals which provide storage and input/output (I/O) capability for the core processor. On-chip memory and peripherals are not considered part of the core and may vary from one family member to another. The chip pinout interfaces both the core processor and on-chip peripherals to external devices.

The instruction set defines three separate memory spaces - X Data, Y Data and P Program - which are each 65,536 locations by 24 bits wide. The total addressing capability is 196,608 twenty four bit words (or 589,824 bytes). All three memory spaces can be accessed in parallel in the same instruction cycle. Non-core resources (memory and peripherals) are memory-mapped into these memory spaces to provide a clean hardware and software interface with the core processor.

Core Processor

The core processor consists of three separate execution units - the Data ALU, the Address ALU and the Program Controller - connected by a multiple bus architecture. These execution units operate in a parallel rather than a pipelined fashion; i.e., each execution unit works on the same instruction at the same time. Working in parallel, the three execution units provide all of the resources to execute most DSP56000 instructions in a single instruction cycle. The instruction set is designed to allow flexible control of these parallel processing resources. Many instructions allow the user to keep each execution unit busy, thus enhancing program execution speed.

The architecture of each execution unit is optimized to support its role in instruction execution. Each execution unit contains a set of registers, arithmetic units and executable operations. These operations are register-oriented rather than memory-oriented. Each execution unit operates on its own local registers - source operands are read from registers within the execution unit, modified by arithmetic unit operations and the results are stored in registers within the same execution unit. The execution units are not pipelined and all operations execute in a single Instruction cycle. Lack of data pipelining contributes to the ease of programming and avoids architectural bias toward one algorithm at the expense of another.

The internal multiple bus architecture consists of four data buses and three address buses connecting the various resources on the chip. One address bus and one data bus is associated with each of the three memory spaces (X Data, Y Data and P Program) while the fourth data bus, called the Global Data bus, is shared by all three memory spaces. This bus structure supports general register-to-register,

register-to-memory and memory-to-register data movement and executes up to three data transfers (one for each memory space) in parallel. Referring to Figure 1, the resources connected to each bus indicate the primary data transfer functions. The Global Data bus physically extends the local X, Y and P Data buses to communicate with remote chip resources. The Global Data Bus can be connected to another data bus via the internal data bus switch. The internal data bus switch can transfer data between two data buses without any pipeline delays. Since the internal data bus switch can access each memory space, the bit manipulation unit is built into the switch.

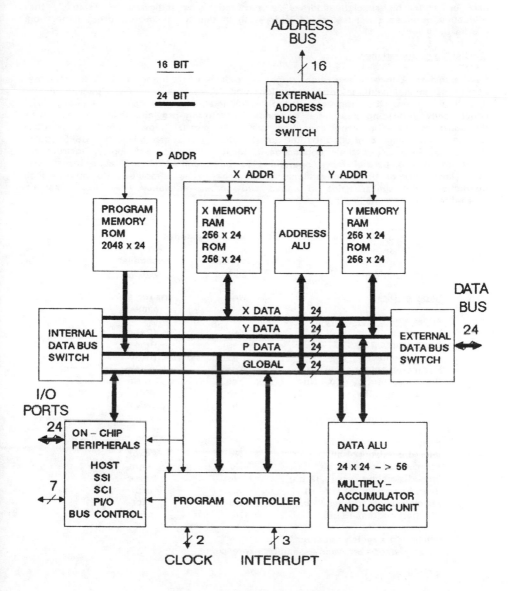

Figure 1. DSP56000 IC Architecture

DSP56000 INSTRUCTION SET

The DSP56000 has an easy to learn, microprocessor-style instruction set that is efficient for many different algorithms. The instruction set has a "reduced instruction set computer (RISC)" simplicity with its register-oriented (load-store) operations and single-cycle execution of most instructions. Enough functionality is built into each instruction to achieve high performance for small code loops. Sixty-two instructions are provided, grouped into arithmetic, logical, bit manipulation, program control, loop and move catagories. All instructions are one or two 24 bit words in length. The second instruction word is only used when 24 bit immediate data or a 16 bit absolute address is required. As indicated in Table 1, the DSP56000 provides a rich set of 14 addressing modes to minimize address generation overhead.

Parallel Move Operations

Data transfers between execution units or between execution units and memory occur in parallel with internal execution unit operations. These "parallel move operations" allow the register-oriented execution units to be kept busy by concurrently preloading new input operands and storing previous results. Used with the local registers in each execution unit, the parallel move operations provide "software pipelining" controlled by the user. Thus, users can adapt the DSP56000's parallel architecture to their application. Most arithmetic and logical instructions allow up to two parallel move operations to be specified in the same instruction. This allows two or three conventional instructions to be combined into one parallel instruction, with similar gains in speed and coding efficiency. A flexible set of parallel move operations is provided.

Table 1. DSP56000 Addressing Modes

Addressing Mode	Address Modifier	Assembler Syntax
Register Direct	No	any register name
Address Register Indirect		
No update	Yes	(Rn)
Postincrement by 1	Yes	(Rn)+
Postdecrement by 1	Yes	(Rn)-
Postincrement by offset Nn	Yes	(Rn)+Nn
Postdecrement by offset Nn	Yes	(Rn)-Nn
Predecrement by 1	Yes	-(Rn)
Indexed by offset Nn (Rn and Nn are unchanged)	Yes	(Rn+Nn)
Special		
Immediate data (24 bit)	No	#expr
Absolute address (16 bit)	No	expr
Immediate short data (8,12 bit)	No	#expr
Short jump address (12 bit)	No	expr
Absolute short address (6 bit)	No	expr
I/O short address (6 bit)	No	expr

where   n = register number 0-7
        expr = any valid assembler expression

DATA ALU

The Data ALU execution unit performs arithmetic and logical operations on data operands. As shown in Figure 2, it consists of four 24 bit general purpose input registers, six special output registers organized as two 56 bit accumulators, two 56 bit data shifter/limiters, a 56 bit accumulator shifter and a multi-function multiply-accumulator (MAC) ALU unit, which has two 56 bit inputs and one 56 bit output. The two 56 bit MAC ALU inputs are used for addition, subtraction, comparison, logical operations, etc. During multiply and multiply-accumulate operations, one of the 56 bit inputs serves as an accumulator input while the other 56 bit input is reconfigured as a 24 bit multiplicand and a 24 bit multiplier input. Multiplication is performed using a signed, two's complement fractional data representation common to digital signal processing algorithms. A powerful set of multiply (mpy) and multiply-accumulate (mac) instructions are provided with optional "Round to Nearest Even" rounding and positive or negative product accumulation.

The MAC ALU contains a 24 by 24 bit, parallel hardware multiply-accumulator circuit with 56 bit accumulation. It performs all operations in a single 97.5 nsec instruction cycle. This is in contrast to common two stage, pipelined architectures employing a multiplier separated from an adder by a product pipeline register. Since the MAC ALU is not pipelined, results are immediately available and many algorithms execute faster. The 56 bit accumulators store a complete 48 bit fractional multiply-accumulator result plus 8 bits of integer data called an accumulator extension. The extension portion of the 56 bit accumulator provides 8

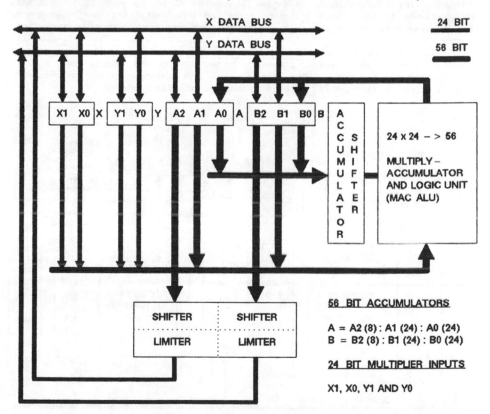

Figure 2. Data ALU Block Diagram

bits of overflow protection during intermediate calculations and allows at least 256 repetitive multiply-accumulate operations before an overflow is possible. The extension bits eliminate the need to scale down the input data to avoid accumulator overflows caused by word growth in repetitive calculations.

Two data shifter/limiter circuits provide block floating point and overflow protection features. In each shifter/limiter, a copy of the accumulator data is optionally shifted left or right one bit to perform dynamic scaling. If the output of the shifter cannot be stored in the destination without overflow (because the extension portion of the shifter output is in use), the limiter substitutes the nearest data value (-1 or +1) having the same sign as the accumulator data.

A 56 bit accumulator shifter is included on one of the MAC ALU inputs for one bit left or right shifts. The Data ALU does not contain a barrel shifter, but the MAC ALU can be used for multi-bit shifting operations. By multiplying the 24 bit data by a constant or variable, left or right shifts from 1 to 23 bits may be performed in a single instruction. This method minimizes chip area and opcode space. The instruction set can efficiently use the MAC ALU for shifting and bit field operations.

ADDRESS ALU

The Address ALU execution unit calculates addresses to locate data operands in memory. The block diagram of Figure 3 consists of twenty four 16 bit address registers, two 16 bit address arithmetic units and three address output multiplexers. The Address ALU provides up to two memory addresses each instruction cycle and updates them independently with two address arithmetic units. The Address ALU provides a very flexible addressing capability with 11 effective addressing modes using three types of address arithmetic - linear, modulo and reverse carry. The type of address arithmetic defines the type of data structure (array, sample shift register, queue, bit-reversed FFT buffer, etc.) being accessed in memory.

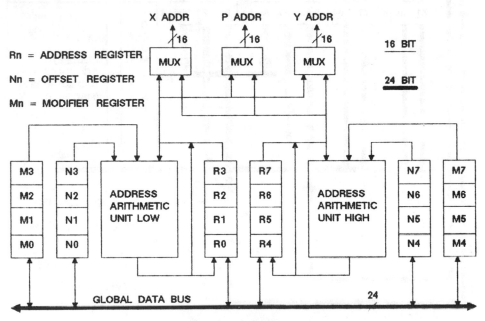

Figure 3. Address ALU Block Diagram

The 8 address registers Rn (n=0-7) are used as address pointers to locate data operands in memory. The 8 offset registers Nn are used as optional offset values to update address registers. The 8 modifier registers Mn select the type of address arithmetic performed by each address arithmetic unit. Each address register Rn is assigned an offset register Nn and a modifier register Mn having the same register number, n, for use in address calculations. For example, the address calculation (R0)+N0 postincrements the contents of address register R0 by the contents of offset register N0 using the type of address arithmetic specified by the contents of modifier register M0.

The Address ALU output multiplexers allow any address register Rn to be used as a pointer to any memory space. Complex data pairs are efficiently organized with the real part in X memory and the imaginary part in Y memory at the same address. Being able to access both the real and imaginary parts of a complex number with one address register avoids duplicating address pointers.

Address Arithmetic

During an address calculation on address register Rn, the contents of the assigned modifier register Mn are decoded by the address arithmetic unit to control the address arithmetic used. Linear address arithmetic (Mn=$FFFF) is identical to conventional microprocessor address calculations, where the address modification is performed by a binary adder. Linear addressing is used for addressing data arrays and tables. Reverse carry address arithmetic (Mn=0) is performed by propagating the adder carry in the reverse direction; i.e., from the most significant bit to the least significant bit of the adder. Reverse carry addressing is equivalent to bit-reversed addressing with simpler hardware. For sequential access of N point bit-reversed FFT data or coefficients, a postincrement by the value N/2 using reverse carry address arithmetic will generate the correct address sequence. Modulo address arithmetic keeps an address register pointing within a modulo region defined by a lower boundary address and an upper boundary address. Automatic wraparound occurs if the address pointer increments above the upper boundary or decrements below the lower boundary. Each address arithmetic unit actually calculates two addresses - with and without wraparound - and selects the address pointing within the modulo region. The modulo size is specified by the contents of the modifier register plus one (Mn + 1 = upper boundary - lower boundary). The lower boundary address must have as many least significant zeros as the modifier value has significant bits. For example, a modulo region of size 21 (Mn=20) can have a lower boundary address at any integer multiple of 32 (0, 32, 64, 96, 128, 160, etc). The upper boundary address is the lower boundary address plus Mn. Modulo address arithmetic is used to simulate a shift register in memory for digital filters, delay lines, queues (FIFO's) and waveform generators. Modulo addressing eliminates wasted cycles required to move data by simply updating an address pointer.

PROGRAM CONTROLLER

The Program Controller execution unit performs instruction flow control, instruction decoding and exception processing. It consists of a program address generator (PAG), an instruction decoder, an interrupt controller and a bus controller, shown functionally in Figure 4. During normal instruction execution, the PAG calculates the address of the instruction word three locations ahead of the currently executing instruction word. The instruction fetch and decode cycles are overlapped (pipelined) with the instruction execution cycle and are normally transparent to the user. The fetch and decode cycles become visible during change-of-flow instructions, external program memory wait states, and when the PAG must wait for access to internal program memory or the external address and data bus.

The DSP56000 provides hardware DO loop control to replace the software "decrement counter and branch" instruction normally associated with DO loops. The usual change-of-flow overhead is eliminated by fetching the loop instructions as if they were straight-line code. Straight-line coding is not needed to maximize speed since compact, looped code runs at the same speed. The loop count (LC), loop starting address and loop ending address (LA) are stored in PAG registers and processed in parallel with the executing program. Inside the DO loop, the instruction fetch address is compared to the loop ending address. When the end of the loop is detected, the loop count is tested for one. If the loop count is not one, it is decremented and the instruction word at the loop starting address is fetched. If the loop count is one, the DO loop execution is complete and normal sequential instruction fetches resume. Nested hardware DO loops are provided by saving loop parameters on a separate hardware stack. The system stack is 32 bits wide and 15 locations deep. The double width allows two registers to be transferred to/from the system stack every instruction cycle. DO loops and interrupts share the system stack, which can be extended to any depth by moving the stack data to/from memory using software operations.

A sophisticated interrupt controller reduces the timing overhead associated with servicing the 18 DSP56000 interrupts. Each peripheral device and external interrupt pin may be programmed via the interrupt priority register (IPR) to one of three interrupt priority levels (IPL) so that time-critical interrupts are always serviced

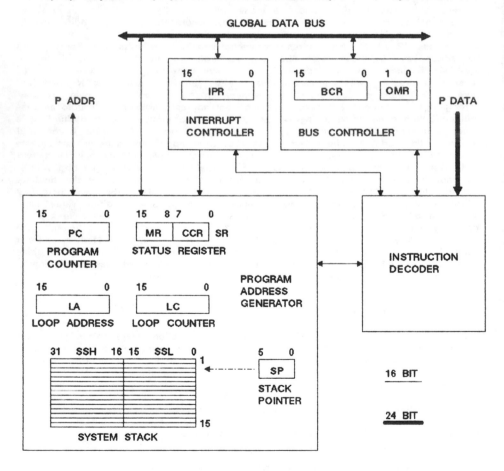

Figure 4. Program Controller Block Diagram

first. The two external interrupt pins IRQA and IRQB may be programmed as level sensitive or negative edge triggered. On-chip peripherals have separate interrupt vectors for each interrupting condition so the cause of the interrupt is already known when the interrupt service routine is entered. Some interrupt vectors are considered "error-free", meaning that no error conditions are associated with the interrupt request. In this case, a data transfer can service the request immediately without checking error flags. A separate interrupt vector is provided for the same interrupting condition with errors so that a longer interrupt routine can check the error flags.

In operation, the interrupt controller synchronizes and prioritizes pending interrupts to determine the highest priority, unmasked interrupt request. Second, the instruction fetch stream is temporarily redirected to fetch two interrupt instruction words at the interrupt vector addresses. The two interrupt instruction words at the interrupt vector addresses are fetched into the instruction pipeline without waiting for the current instruction to finish execution. The program counter (PC) is held constant since the interrupt controller provides the two interrupt vector addresses. Finally, normal instruction fetches resume based on the PC contents immediately after the two interrupt instruction words have been fetched. The usual interrupt vector change-of-flow is avoided so wasted cycles to empty the instruction pipeline and refill it are eliminated. The two interrupt instruction words are decoded and executed. They may be two single-word instructions or one two-word instruction.

If execution of the two interrupt instruction words does not cause a change-of-flow, the interrupt routine is called a "fast interrupt". Fast interrupts do not save any machine status so instructions which modify the machine state should not be used. One or two MOVE instructions are typically used to service an error-free interrupt request. Fast interrupts do not require a return from interrupt (RTI) instruction since no context switch is performed. Fast interrupts can process up to 1.7 million interrupts (transferring 10.25 million bytes) per second, yet consume only 33% of the total execution time. This performance level rivals dedicated direct memory access (DMA) hardware. Fast interrupts are a software alternative to DMA with several advantages. Unlike DMA, fast interrupts can service peripheral devices with the total flexibility of software. Fast interrupts can use any addressing mode with linear, modulo or reverse carry address arithmetic. They can service on-chip or off-chip peripherals with little or no extra hardware.

If either of the two interrupt instruction words is a jump to subroutine (JSR) instruction, the interrupt routine is called a "long interrupt". Programming a JSR instruction at the interrupt vector addresses causes a long interrupt routine. When the JSR instruction is decoded, a context switch is performed by saving the current PC and status register on the system stack and updating the interrupt mask in the status register. The PC is loaded with the JSR destination address and the long interrupt routine begins execution. If desired, additional registers may be saved by software operations. The long interrupt routine is terminated by the usual RTI instruction. Long interrupts allow true vectoring via the JSR destination address and are a natural extension of the fast interrupt mechanism.

The bus controller allows external memory, peripherals and slave processors (microprocessors or DSP IC's) to be accessed through a memory expansion port. Separate 16 bit address and 24 bit data buses are used to multiplex the 3 internal address buses and 4 internal data buses off-chip, respectively. The bus controller determines the number of external memory requests based on the value of each address and the memory maps defined by the operating mode register (OMR). If only one external memory access is requested per instruction cycle, the request is granted immediately and no extra cycles are required. If two or three external memory accesses are requested in a given instruction cycle, a minimum of one or two extra instruction cycles, respectively, will be required to complete the instruction. The bus controller can schedule 10.25 million external transfers per second using a synchronous bus defined by seven bus control signals. External transfers with no wait states require a memory access time of 55 nsec. The bus control register (BCR) may be programmed to insert 0-15 wait states for four catagories of external bus requests. Each wait state is one clock cycle, or 48.75 nsec with a 20.5 MHz clock. This allows fast and slow access devices to be mixed on the external bus. The Bus Request and Bus Grant control signals allow DMA with external memory concurrently with internal DSP56000 instruction execution.

ON-CHIP RESOURCES

The DSP56000 provides a large set of on-chip memory and I/O peripherals to support the core processor. The on-chip memory resources include a 256 x 24 bit X Data RAM, a 256 x 24 bit Y Data RAM, a 256 x 24 bit X Data ROM, a 256 x 24 Y Data ROM and a 2048 x 24 bit Program ROM. Microcomputer-style I/O capability is provided by three on-chip peripherals sharing 24 programmable, general purpose port pins. Non-core resources are supported by standard move, bit manipulation and jump on bit condition instructions. In addition, a special move peripheral (movep) instruction supports memory-to-memory data transfers between memory-mapped peripheral devices and any memory space.

General Purpose I/O Pins

Twenty four DSP56000 pins may be programmed as general purpose I/O pins similar to those found on Motorola's single-chip microcomputers. This allows up to 24 pins to be used as I/O flags for synchronization and control purposes. Individual pin control, data direction and data transfer functions are provided by six internal memory mapped registers. Each port pin may also be programmed to serve as a dedicated on-chip peripheral pin.

Host Processor Interface

The Host interface is a byte-wide, full duplex parallel port which may be connected directly to the data bus of a host processor. The host processor may be any of a number of industry standard microcomputers or microprocessors, a DSP IC or DMA hardware. The Host interface appears as a memory-mapped peripheral occupying 8 bytes in the host processor address space. An 8 bit bidirectional data bus and 7 control lines are used by the host processor to control data transfers. Fourteen internal registers provide double-buffered data transfers between the host processor and the DSP56000 using asynchronous polling or interrupts. In DMA mode, an external DMA controller can perform DMA transfers with the Host interface using the Host Request and Host Acknowledge handshake lines. DMA initialization commands are provided to setup the Host interface DMA channel. A special host command feature allows the host processor to issue a vectored interrupt request to the DSP56000 program. The host processor may select one of 32 DSP56000 host command interrupts to be executed by writing a vector address register. Host commands are useful for debugging and on-line diagnostics, implementing control protocols and DMA setup.

Serial Communications Interface

The Serial Communications Interface (SCI) provides full duplex, serial communications with a wide variety of serial devices, including microprocessors, DSP IC's, terminals and modems, either directly or via RS232C-style lines. The SCI interface supports industry standard asynchronous character modes with parity and multidrop options. The multidrop option Includes a wake-up on idle line and wake-up on address bit capability. A synchronous shift register mode allows I/O expansion and high speed, synchronous data transmission. The SCI consists of separate transmit and receive sections and a programmable baud rate generator. Seven internal registers provide doubled-buffered data transfer and control functions. Three I/O pins are used for transmit data, receive data and baud rate clock functions.   The baud rate can be generated internally or externally with asynchronous rates up to 320K bits per second and synchronous rates up to 2.5M bits per second.   The internal baud rate generator can also function as a general purpose timer when not used by the SCI interface.

Synchronous Serial Interface

The Synchronous Serial Interface (SSI) provides a full duplex, double-buffered serial port which allows the DSP56000 to communicate with a wide variety of serial devices.   These include one or more industry standard CODEC A/D-D/A's, DSP IC's, microprocessors and serial peripheral devices.   The SSI interface consists of separate transmit and receive sections and a common SSI clock generator.   The clock generator defines the serial bit rate, serial word size and the number of words per serial frame.   The data in each serial frame is controlled by software, allowing any user protocol to be implemented.   Several clock and frame sync timing options provide flexible, synchronous serial communications at rates up to 5M bits per second.   Three to six I/O pins are used, depending on the operating mode.   The SSI interface has eight internal registers.

Three SSI operating modes support the requirements of different serial devices.   The normal operating mode is used for periodic devices which transmit or receive one data word every serial frame.   One time slot is defined for data transmission at the start of each serial frame.   A CODEC A/D-D/A converter is an example of such a periodic device.   The on-demand operating mode is for non-periodic communications such as from one DSP56000 to another.   No time slots are defined for transmission and data is transmitted as soon as it is available.   The network operating mode defines from 2-32 time slots per serial frame which can be used for creating a network of communicating serial devices.   Each device can transmit or receive during one or more assigned time slots.   The time slot assignments for each serial device are determined by the user's software.   In network mode, multiple DSP56000's can communicate with no "glue chips".   Network mode can also interface the DSP56000 directly to time division multiplexed, serial I/O channels used in telecommunications applications.

DSP56000 PROGRAMMING EXAMPLE

The assembler source program for an Nth order real, finite impulse response (FIR) digital filter is shown in Figure 5.   The N data samples are stored in X Data memory and the N filter coefficients are stored in Y Data memory.   This natural partitioning is desirable since both X and Y memory spaces can be accessed in the same instruction cycle.   The four instructions at the label "start" initialize modulo N address pointers for the data samples and filter coefficients.   The data sample shift register is time-shifted each sample time by incrementing r0 by 1 modulo N.   Modulo N addressing is also used to automatically wraparound the coefficient address pointer r4 to the first filter coefficient each sample time.   The seven instructions at the label "fir" form a simple loop to get an "input" sample, perform the FIR filtering operation and store the filter "output".   The actual filtering operation is performed by repeating (rep) the multiply-accumulate (mac) instruction.   The Data ALU registers "x0" and "y0" are multiplied, the product is added to accumulator register "a" and the result is stored back in accumulator "a".   In the same instruction, two parallel move operations are used to preload the same Data ALU registers with the next data and coefficient from "x:" Data memory and "y:" Data memory, respectively.   Each parallel move operation also specifies a "+" postincrement or "-" postdecrement by 1 (modulo N) address register update in each Address ALU arithmetic unit.   Although not visible in the programming example, the Program Controller fetches new instruction words based on the current instruction execution, hardware DO loop conditions and any pending interrupts.   It is easy to see that the instruction set can keep all three execution units, three data buses and three address buses busy executing in parallel.

Label	Opcode	Operand	X Move	Y Move	Comment
start	move	#data_ptr,r0			;load address pointers
	move	#coef_ptr,r4			
	move	#N-1,m0			;load modulo N modifiers
	move	m0,m4			
fir	movep		x:input,x0		;get input sample
	clr	a	x0,x:(r0)-	y:(r4)+,y0	;preload first tap
	rep	m0			;do N-1 taps
	mac	x0,y0,a	x:(r0)-,x0	y:(r4)+,y0	;fir filter kernel
	macr	x0,y0,a	(r0)+		;round last tap
	movep		a,x:output		;store filter output
	jmp	fir			

Figure 5. Finite Impulse Response Digital Filter Program

CONCLUSIONS

The DSP56000 is a fourth generation, high performance digital signal processor. A wide range of hardware and software features provide unmatched speed, precision and systems integration in an easy to program, microprocessor-style device. Containing a complete DSP system on a single CMOS chip, it represents the state of the art for advanced signal processing applications. A summary of DSP56000 benchmarks is given in Table 2.

Table 2. DSP56000 Benchmark Summary

Benchmark	Performance	
N Tap Real FIR Filter with Data Shift	97.5	nsec per tap
N Tap Real, LMS Adaptive FIR Filter with Data Shift	292.5	nsec per tap
N Real, Cascaded IIR Biquad Filters (4 coefficients)	390	nsec per filter
N Tap Complex FIR Filter with Data Shift	390	nsec per tap
256 Point Complex FFT (Radix 2, looped code)	0.706	msec
1024 Point Complex FFT (Radix 2, looped code)	4.994	msec
Two Dimensional Convolution (3x3 coefficient mask)	975	nsec per output
Find Maximum Absolute Value and Index in Array	195	nsec per point

REFERENCES

[1] "Motorola's Sizzling New Signal Processor", Electronics, pp. 30-33, March 10, 1986.

[2] J. Bates, "Motorola's DSP56000: A Fourth Generation Digital Signal Processor", presented in Session 18 at the 1986 MINI/MICRO NORTHEAST Convention, Boston, Massachusetts, May 13, 1986.

[3] S. Abiko et al., "Architecture And Applications Of A 100-NS CMOS VLSI Digital Signal Processor", in Proceedings of the IEEE-IECEJ-ASJ International Conference on Acoustics, Speech, and Signal Processing, 1986, paper 8.3, pp. 393-396.

[4] T. Nishitani et al., "Advanced Single-Chip Signal Processor", in Proceedings of the IEEE-IECEJ-ASJ International Conference on Acoustics, Speech, and Signal Processing, 1986, paper 8.7, pp. 409-412.

[5] J. R. Boddie et al., "The Architecture, Instruction Set And Development Support For The WE DSP32 Digital Signal Processor", in Proceedings of the IEEE-IECEJ-ASJ International Conference on Acoustics, Speech, and Signal Processing, 1986, paper 8.10, pp. 421-424.

FLOATING-POINT DIGITAL SIGNAL PROCESSOR MSM6992

AND ITS DEVELOPMENT SUPPORT SYSTEM

Yoshikazu Mori, Toshio Jufuku, Hisaki Ishida,[*]
and Noboru Ichiura
Research Laboratory and VLSI R&D Center[*]
Oki Electric Industry Co., Ltd.

10-16, Shibaura 4-chome, Minato-ku, Tokyo 108, Japan

Abstract

We have successfully developed a new signal processor – the MSM6992 – as a second-generation DSP. This DSP has various features – including floating-point arithmetic, high-speed operation, a large-capacity memory space, etc. – that provide a much greater processing capability than is possible with first-generation DSPs. In addition, a highly efficient development support system has also been developed to facilitate program development for this DSP. Furthermore, high-level languages are under study for the purpose of enhancing the development environment for the MSM6992. This paper describes the architecture and development support system for this DSP as well as the high-level language research that is in progress.

1. Introduction

The upgrading of the functions and performance of digital signal processors is being aggressively pursued against the background of the progress of VLSI and signal-processing technologies in recent years. This new signal processor has been developed for the purpose of applying it to advanced signal-processing systems.

In order to prepare for applications to advanced signal-processing systems, the following design principles were adopted when developing the MSM6992:[1]

(1) The employment of floating-point arithmetic to allow the simplification of program development and the expansion of the dynamic range.

(2) The realization of an insruction cycle of 100ns, which is more than double the conventional DSP operation speed, thereby making possible the realization of complex algorithms in real time.

(3) The expansion of program- and data-memory space to allow applications that require a large-capacity memory.

(4) The provision of a very efficient program development support system to improve the efficiency of program development.

The MSM6992 signal processor, which was developed on the basis of the above design principles, is illustrated in Figure 1.

This DSP is equipped with high-speed arithmetic units that execute arithmetic for 22-bit floating-point data that consists of
16 bits for the mantissa and 6 bits for the exponent, a 256-word
internal data RAM and a 1024-word instruction ROM. Furthermore,
the program memory and the data memory can be expanded to a maximum of 64K words each by the use of external devices. This DSP
integrates 125K transistors within a single chip through the employment of the 2-micron CMOS technology.

2. Architecture

Figure 2 shows the block diagram of the MSM6992. It consists
of four major blocks: The Arithmetic Block that executes
floating-point and fixed-point arithmetic, the Internal Data
Memory Block, the Instruction Control Block, and the Interface
Control Block.

Arithmetic Block
The Arithmetic Block consists of a floating-point multiplier,
a floating-point ALU, etc. These arithmetic units execute high-
speed arithmetic for floating-point data and can also execute arithmetic for fixed-point data.

Internal Data Memory Block
The Internal Data Memory Block consists of two planes of 22-
bit x 128-word RAM and an address control unit - the ADU - that
generates addresses. The memory can be handled as two planes of
128 words or one plane of 256 words to provide high flexibility
for the handling of the internal data memory. Furthermore, this
data memory allows read/write operations in one machine cycle,
thereby increasing the throughput of the memory.

Instruction Control Block
The Instruction Control Block consists of an instruction ROM
of 32 bits x 1024 words and a sequencer. The MJR which is connected to the internal data bus can be selected to be the source
of program addresses, thereby enabling indirect jumps and indirect subroutine calls. In addition, two loop counters are provided as support for the dual loop operation.

Interface Control Block
This DSP is equipped with a 22-bit parallel interface to interface with the external devices. By using this parallel interface, an external data memory, an 8-bit or 16-bit micro processor
and a DMA controller can be connected, thereby allowing a flexible system construction.

Table 1 tabulates the instruction set of this DSP. The instructions for this processor consist of instructions for
floating-point arithmetic, fixed-point arithmetic, logical operations, shift operations, transfers, input/output, jumps, etc.
This DSP can simultaneously execute a maximum of eight different
instructions to provide a high processing capability.

The following paragraphs describe the data format and the
structure of the arithmetic block for floating-point arithmetic,
which are the greatest features of the architecture for the
MSM6992.

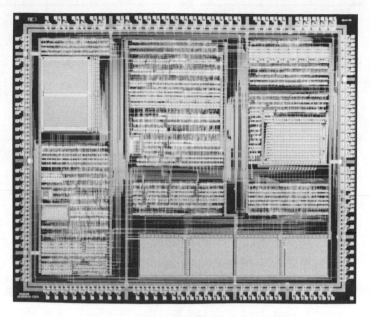

Figure 1. MSM6992 photomicrograph

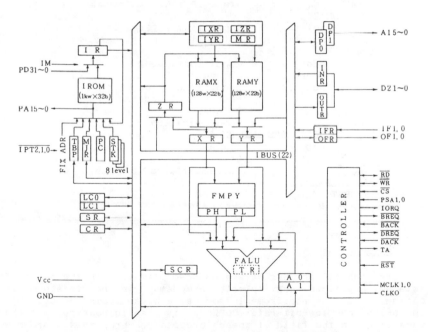

Figure 2. Block diagram of MSM6992

Table 1.   Instruction set of MSM6992

FLOATING POINT ARITHMETIC		
FADD	reg1, reg2	Floating-point Add
FSUB	reg1, reg2	Floating-point Subtract
FABS	reg	Take absolute value
FCHS	reg	Change Sign bit
FLD	reg	ALU through
TRNSXL		Data type conversion fix to float
TRNSLX		Data type conversion float to fix
FIXED POINT ARITHMETIC		
ADD	reg1, reg2	Add
SUB	reg1, reg2	Subtract
ADDC	reg1, reg2	Add with carry
SUBC	reg1, reg2	Subtract with carry(borrow)
LOGICAL OPERATIONS		
AND	reg1, reg2	Logical AND
OR	reg1, reg2	Logical OR
XOR	reg1, reg2	Logical exclusive OR
NOT	reg	Complement all bits
SHIFT OPERATIONS		
SAR	reg, sc	Arithmetic shift right
SAL	reg, sc	Arithmetic shift left
FSCALR	reg, sc	Floating-point scal down
FSCALL	reg, sc	Floating-point scal up
MOVES		
MOV	dst, src	Move a value
LXY	idm1, idm2	Load to XR & YR
LX	idm	Load to XR
STA	acc	Store in accumulator
STM	idm	Store in intenal_data_memory
INPUT/OUTPUT		
IN	pointer	Input from external_data_memory
OUT	pointer	Output to external_data_memory
JUMPS		
Jcc	address	Conditional Jump
CALL	address	Subroutine Call
JSRcc	address	Conditional subroutine call
PUSH		Push a next address
RET		Return
LOOP		Loop operation
ADDRESS REGISTERS MANIPULATION		
SET	reg, imm	Set a value
MODFY	reg, imm	Modify a value
INC	TBP	Increment a value
MISCELLANEOUS		
FLGRST	flag	Flag reset
NOP		No operation

3. Data Format

First-generation DSPs have been used primarily in the field of telecommunications, particularly in modems. In the telecommunications area, the requirement for speech analysis processing, centering on low-bit-rate CODECs, is also increasing. At the same time, in the field of speech processing, proposals for new algorithms, centering on speech recognition, are being aggressively made in recent years. For speech analysis processing, the linear predictive coding method is being applied as an effective method in various fields. However, the realization of this analysis method by the use of fixed-point DSPs has been difficult to accomplish due to their limited dynamic ranges. In the field of low-bit-rate CODECs, bit-rate reduction and quality-upgrading for speech are being pursued. As a result, CODEC algorithms are becoming more sophisticated year by year. To operate these complex algorithms, an acceleration of DSP processing speed to a great extent - when compared with conventional DSPs - has been necessitated.

In order to improve the limited dynamic range and the slow operating speed that have been the problems with first-generation DSPs, a 22-bit floating-point data format has been adopted for this DSP for the two purposes of the realization of (1) the required dynamic range and (2) a high-speed arithmetic operation by the use of the 2-micron CMOS technology. The data format supported by this DSP is illustrated in Figure 3. This floating-point data format consists of 22 bits per word - 6 bits for the exponent and 16 bits for the mantissa. Both exponent and mantissa are expressed by the 2's complement. By the use of this data format, complex speech analysis processing, which has been difficult to accomplish with conventional DSPs, can now easily be realized. As a result of the development of this DSP, the application range of DSPs has been greatly broadened, and also includes speech processing.

4. Design of Arithmetic Block

When designing this DSP, the 22-bit floating-point data format was adopted and, at the same time, the realization of a high-speed operation of 100ns for the machine cycle was targeted. Furthermore, the 2-micron CMOS technology was used as the device technology to realize this DSP. In addition, to shorten the design period for the MSM6992, the building-block method was adopted as the design method. In view of these factors, the following approach was adopted for the design of the arithmetic units.[2] First of all, the ALU, which requires the longest processing time, has been divided into two stages and has been designed to have a pipeline structure for parallel operations in each of the two stages. As to the multiplier, in order to realize a high-speed operation and a compact size, the multiplication array

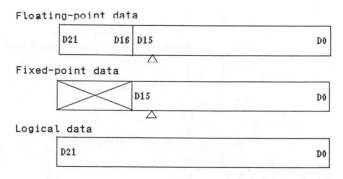

Figure 3. Data Format

for the mantissa has been manually designed. As a result, a multiplier that executes 16 bits x 16 bits of multiplication in 55ns has been realized. Furthermore, the ALU has been designed by the use of automatic routing system. In order to prevent an increase in the delay time caused by the automatic routing, adequate verification has been performed by a timing verification system, thereby resulting in the realization high-speed operations. In addition, by utilizing the arithmetic unit that has been provided for floating-point arithmetic, the ALU makes possible the arithmetic shift for fixed-point data and high-speed division for floating-point data. Figure 4 illustrates the block diagram of the arithmetic unit.

5. Development Support System

For general-purpose digital signal processors, it is important to provide program development support systems. Figure 5 shows the development support system for this DSP.

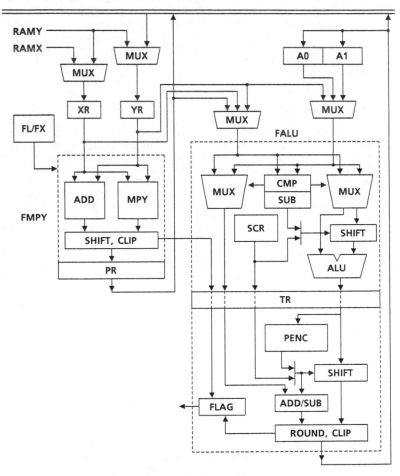

Figure 4. Diagram of Arithmetic Block

Language Processor

A high-function macro assembler has been developed as the language processor for this DSP. As will be described latter in this paper, it is difficult to develop high-level languages for DSPs. For this reason, an upgrading of the functions of the assembler has been accomplished for the MSM6992. The major features of this assembler include powerful macro function, automatic selection of instruction type, etc. The automatic selection of instruction type is a function that automatically determines the instruction type on the basis of the combination of instruction mnemonics. Furthermore, either a relocatable format or absolute format can be selected as the object format for this assembler.

Debugger/Emulator

A debugger and emulator have been provided as debugging tools for programs. By the use of these tools, the program can be operated in real time in a target system, thereby performing the debugging function. Figure 6 shows the structure of the emulator and the debugger commands.

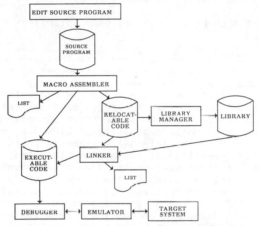

Figure 5. Development Support System

Debugger commands
Memory Dump, Enter,
Fill, Compare, Verify,
Display registers,
Go, Trace, Untrace,
Read File, Write File,
Assemble, Disassemble, etc

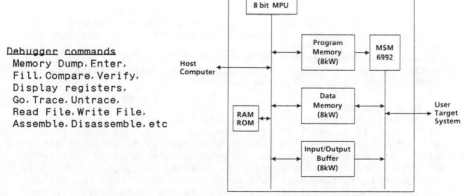

Figure 6. Block diagram of Emulator

6. Study of High-Level Language

As described previously, a development support system has been developed for this DSP. In order to improve the efficiency of program development, however, the language processor must be further upgraded. In an effort to develop high-level languages, the development of cross-compilers for DSPs has been attempted, but such attempts has not always been successful. The development of high-level languages for DSPs has thus been extremely difficult.[3]

The major reason for this is the existence of a great semantic gap between a high-level language and the structure of a DSP. To be more specific, conventional compilers are structured for computers that employ SISD-type architecture whereas DSPs employs SIMD-type architecture. Because of the difference in the architectures, the efficiency of the execution of the codes generated by the compilers is extremely low in DSPs. This is because DSPs greatly differ from ordinary computers in the following aspects:

1. Pipeline architecture
2. Horizontal-type micro-instruction
3. Plural numbers of memory areas

The functions of a compiler that has been optimized for DSPs are discussed below.

DSPs feature a high arithmetic capability that is made possible by the pipeline structure for a plural number of high-speed arithmetic units and an improvemnt in the data supply capability to the arithmetic units. For this rerason, in order to maximize the arithmetic capability of DSPs, a careful data allocation that is based on the addressing method to the internal data memory is required. While the arithmetic capability of DSPs is dependent on the arithmetic speed of the internal arithmetic units, it also requires the pipeline operation of the arithmetic units for the repetition of the identical processing of a large amount of data as well as the efficient execution of the processing of the sum of product. In order to realize this pipeline operation in a processor, the parallel evaluation and execution of arithmetic processing that is independent of the input data are necessary. In other words, in order to improve the efficiency of the arithmetic processing of the DSPs, the optimization of the following is essential:

1. Optimization of allocation of data
2. Optimization of arithmetic processing

The allocation of data is an extremely difficult problem and will be studied further.

Our basic study this time has been conducted by limiting the subject to the optimization for arithmetic processing and the optimizing method is summarized below.

Optimizing Method

As mentioned above, two major problems exist in connection with the development of high-level languages. In order to simplify the handling of the problem when the optimization of arithmetic processing is studied, an intermediate language that is to be positioned between the existing assembly language and the future high-level languages have been provided so that the optimization can be accomplished when translating into the assembly language from the intermediate language.

The intermediate language is the instruction set that is the target of the future high-level language and is the virtualization of the MSM6992 architecture. Figure 7 shows the virtualized model of the MSM6992 to which the intermediate language is applied. For the virtualization, a non-pipeline structure and general-purpose registers for the internal data memory have been employed. The internal memory space consists of 256-word general-purpose registers, two accumulators, etc. As a result, the instruction set has been made to be at the same level as that of micro-processor. By virtualizing the architecture to this level, a programmer can perform programming without involving himself with the details of the hardware of the processor, unlike the case with conventional DSPs.

The following section describes the optimizing pre-processor that generates the source program for the assembly language from the description of the intermediate language.

Optimizing Pre-processor

The optimizing pre-processor that is presently being developed is intended for the development of an assembly language at the micro-instruction level from the above mentioned intermediate langugage and provides two stages of optimizing processing, as mentioned below.

The optimizing process as accomplished by this type of pre-processor usually develops micro-instructions from the intermediate language and is made to execute the compaction processing between micro-instructions. Through such optimizing processing, waste processing that is otherwise caused between micro-instructions after their development is eliminated and the optimization effect can be achieved to some extent. It is still inadequate, however, when compared with the optimized coding by manual designing. This is caused by the fact that the order of execution for the instructions when optimization is performed is determined at the intermediate language stage, and the pipeline's

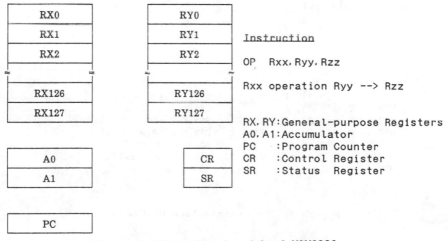

Figure 7. Virtualized model of MSM6992

IIR Filter

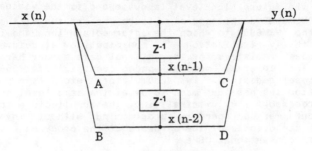

```
LD    x[n], A0              :  X(n)-- > A0  ...........  ①
MAC   A, x[n-1], A0, A0     :  A*x(n-1)+A0-- > A0  ....  ②
MAC   B, x[n-2], A0, A0     :  B*x(n-2)+A0-- > A0  ....  ③
MPY   C, x[n-1], A1         :  C*x(n-1)-- > A1  .......  ④
MAC   D, x[n-2], A1, A1     :  D*x(n-2)+A1-- > A1  ....  ⑤
ADD   A0, A1, y[n]          :  A0+A1-- > Y(n)  ........  ⑥
MOV   x[n-1], x[n-2]        :  x(n-1)--> x(n-2)  ......  ⑦
MOV   A0, x[n-1]            :  A0--> x(n-1)  ..........  ⑧
```

Flow analysis

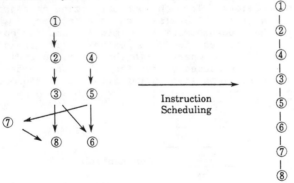

```
LD    x[n], A0              :  X(n)-- > A0
MAC   A, x[n-1], A0, A0     :  A*x(n-1)+A0-- > A0
MPY   C, x[n-1], A1         :  C*x(n-1)-- > A1
MAC   B, x[n-2], A0, A0     :  B*x(n-2)+A0-- > A0
MAC   D, x[n-2], A1, A1     :  D*x(n-2)+A1-- > A1
ADD   A0, A1, y[n]          :  A0+A1-- > Y(n)
MOV   x[n-1], x[n-2]        :  x(n-1)--> x(n-2)
MOV   A0, x[n-1]            :  A0--> x(n-1)
```

Figure 8. The optimization process

Developed micro-instructions
(LXY XI n l, dummy)
                   (FLD   XR)
                              (STA A0)
(LXY XI n−1 l, A)
(LXY XI n−1 l, C)
                (FADD PH. A0)
                (FLD   PH)(STA A0)
(LXY XI n−2 l, B)   (STA A1)
(LXY XI n−2 l, D)
          (FADD PH. A0)
          (FADD PH. A1)(STA A0)
                        (STA A1)
(LXY XI n−1 l, dummy)(FLD   A0)   (STM XI n−2 l)
                         (MOV YR. ALU)
          (FADD YR. A1)
          (FLD   A0)   (MOV ZR. ALU)(STM YI n l)
                         (MOV ZR. ALU)(STM XI n−1 l)

After compaction
(LXY XI n l, dummy)
(LXY XI n−1 l, A)   (FLD   XR)
(LXY XI n−1 l, C)   (STA A0)
(LXY XI n−2 l, B) (FADD PH. A0)
(LXY XI n−2 l, D) (FLD   PH)(STA A0)
                  (FADD PH. A0)(STA A1)
                  (FADD PH. A1)(STA A0)
(LXY XI n−1 l, dummy)(FLD   A0)(STA A1)   (STM XI n−2 l)
                         (MOV YR. ALU)
          (FADD YR. A1)
          (FLD   A0)   (MOV ZR. ALU)(STM YI n l)
                       (MOV ZR. ALU)(STM XI n−1 l)

Figure 8. The optimization process(continue)

interlocks - which exist between the instructions at the intermediate language stage - therefore remain unreleased.

In order to solve the problem and further enhance the optimization effect, it is desirable to simulataneously execute a
number of instructions by utulizing the processor's pipeline. To
do this, the optimizing pre-processor has been provided with a
function that identifies the pipeline's interlocks that can be
identified beforehand when developing the program and which have
been described by the programmer by the use of the intermediate
language, executes the scheduling of the instructions at the intermediate language level on the basis of the results of the in-
terlock identification, and maximize the parallelism of the instructions.[4] In this way, the optimum coding can be obtained by
developing the micro-instructions and then executing the compaction processing at the micro-instruction level after first having
performed the optimizing processing at the intermediate language
level and have determined the order of execution for the instructions.[5] Figure 8 illustrates the optimization process.

Conclusion

A new DSP - the MSM6992 - has been successfully developed as a
second-generation digital signal processor. This DSP executes a

high-speed floating-point arithmetic and features an excellent expandability. Furthermore, as a program development support system for this DSP, an assembler, linker, library-manager, and emulator have been developed. In addition, to complement the programming environment of the DSP, the development of high-level languages is under study. Finaly, as part of the study of the high-level language, the development of an optimizing pre-processor is being pursued.

Acknowledgments

The authors would like to thank S. Nakaya, T. Higashi and H. Ichiki for their direction and encouragement throughout this project. They would also like to thank I. kawanaka, M. kanou, T. Nakamura, M. Iida and A. Nomura.

References

[1] Y. Mori, et al, "Architecture of High-speed 22-Bit Floating-point Digital Signal Processor", Proceedings of the International Conference on Acoustics, Speech, and Signal Processing 1986, pp. 405-408.
[2] T. Jufuku, et al, "Design of New High-speed Digital Signal Processor", Proceedings of the IEEE Custom Integrated Circuits Conference 1986, pp. 82-85.
[3] S. Ono, et al, "Evaluation of a High Level Language Oriented Program Development System for High Performance DSP DSSP1", Proceedings of the International Conference on Acoustics, Speech and Signal Processing 1986, pp2915-2918.
[4] J. Hennessy, et al, "Postpass Code Optimization of Pipeline Constraints", ACM Transactions on Programming Languages and Systems, vol. 5, no. 3 (July 1983), pp. 422-448.
[5] D. Landskov, et al, "Local Microcode Compaction Techniques", Computing Surveys, vol. 12, no. 3 (Sep. 1980), pp. 261-294.

CHAPTER 23

SINGLE CHIP DIGITAL SIGNAL PROCESSORS

by

Charles L. Saxe and Arif Kareem
Tektronix Inc.
P.O. Box 500
Beaverton Oregon 97077

INTRODUCTION:

Since 1984 the Portables Advanced Development group of Tektronix has been providing "high end" digital scopes with high performance signal processing capability. This capability is provided via programmable array processors, the first of which is called the M275. In the 2430 digital scope the M275 handles all waveform data manipulation including: interpolation, waveform scale and offset, CCD error corrections for transfer inefficiency and fix pattern errors, adaptive pre-filtering, waveform transfer, as well as many other functions. This component has been a major success in that it allows the 2430 to have a screen update rate far beyond that of any other digital scope on the market, a feature much appreciated by 2430 customers.

A number of digital storage scopes in design will use either the M275, or for those needing the highest level of performance, a new generation is currently in development: TriStar.

M275 ARCHITECTURE:

The M275 block diagram is shown in Figure 1. Here it can be seen that there are four main units: an "Arithmetic Unit", an "Address Computation Unit", a "Memory Control Unit", and an "Instruction Fetch Unit with built-in loop-control". All four of these subsystems simultaneously process data in a single clock cycle. Consequently one instruction can do an arithmetic computation, fetch (or store) a point from data memory, update the the pointer to this data, fetch an instruction (and process a branch condition), and finally decrement and test a loop counter. The standard instruction format allows the programmer to independently control all of the above units for each clock cycle of the program execution. For example the following instruction will perform the functions listed below:

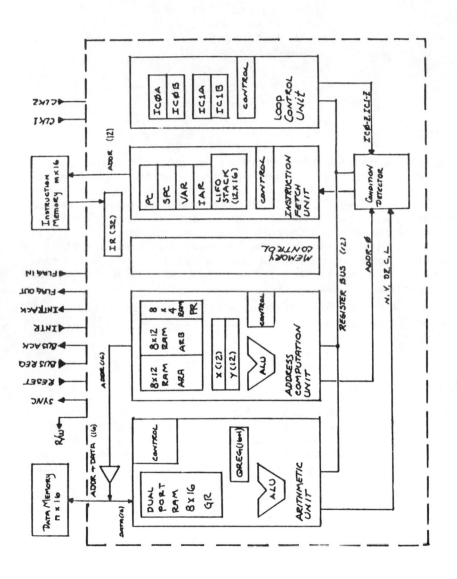

Figure 1 : M275 Block Diagram

Example: ADD 3,4 ; ADX 5 ;   READ 7   ;   RTN,IOZ,F

- The "ADD 3,4" says "add general-registers 4 and 3 and store in 3".

- The "ADX 5" says "use the contents of address register 5b as an address to data memory. Then add the contents of address-offset-register "X" to 5b and write the sum to 5b".

- The "READ 7" says "read to contents of the addressed memory location and write the results in general-register 7".

- The "RTN,IOZ,F" says "return from subroutine conditionally, if iteration-counter- 0 is not zero". If the counter isn't zero it will be decremented automatically.

It is important to note that ALL of the above operations are occurring in a single clock cycle. At the cycle "end" all results are stored to their respective locations. Since this is a single cycle machine there are no stored states other than the condition code flags, and the various registers. Also note that the above operations are only a few of the many possible opcodes that have been provided. The instruction set is rich with operations optimized to the algorithms used in the digital scope environment. Although the Arithmetic Unit data path is 16 bits, double precision is provided on all arithmetic functions.

ARCHITECTURAL TRADEOFFS

Balancing the many tradeoffs between power, speed, silicon area, functionality, memory speed, cost, etc. remains a key aspect of VLSI signal processor design. In design of the M275 many of the above tradeoffs led to some rather unusual conclusions for a DSP system. Since the M275 was designed on a single metal 3u HMOS process and die cost was of great concern, minimizing total silicon area was very important. A careful study of the algorithms led us to the conclusion that a 2 bit "Booths" multiply in which the multiplier could be of varying length would result in only a 10% slowdown of the total execution time in the critical program loops. This is a result of several factors. For example, much algorithm time was spent in symmetrical FIR filtering. In such a filter it is possible to do an add first, followed by a multiply - reducing the number of multiplies by a factor of two. In addition the coefficients were normally 8 bits in length. With such a filter the average number of cycles per multiply is only two per data point. Since operands and results come from external data memory, and are fetched in parallel with arithmetic operations, the multiply steps can be overlapped with both the data fetch and operand address calculations. Even in algorithms using full 16 bit multiplications, the M275's 4 parallel units result in a very respectable execution speed. For example a 512 point FFT with 16 bit multiplies can be done in 24 milliseconds, using a 300 nSec. cycle, with a program only 40 lines long. This is in comparison to about 35 mSec for the TMS32010 which uses a 200 nSec parallel multiplier, a 200 nSec clock and requires thousands of in-line program locations. (The TMS32010 became available about

Address Computation Unit
Simplified Block Diagram

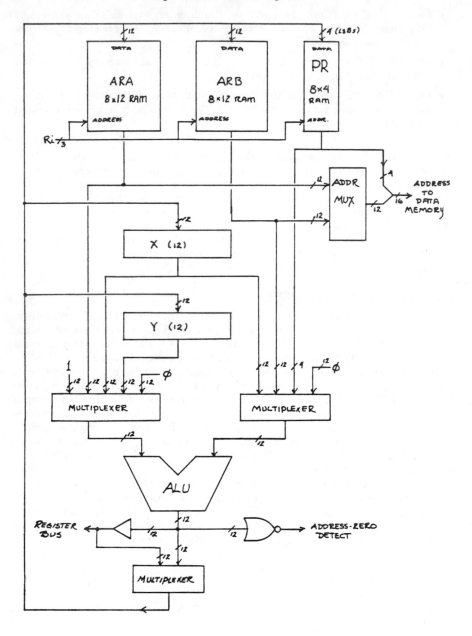

Figure 2

the same time as M275.) The M275's speed advantage, even in the case of the FFT with 16 bit multiplies were a result of the system being very well balanced. Each of the units; Arithmetic, Address Calculation, and Instruction Fetch, perform equally well on the class of programs for which the machine was designed.

ADDRESS COMPUTATION

Operand address computation required careful attention in design of M275 because the required algorithms needed to operate on large, complex arrays of data. What was needed was methods to both initialize pointers, and modify them in tight program loops. Many DSP algorithms consist of several levels of tightly nested loops that are themselves contained in several levels of less speed critical loops. This fact had great impact in the design of both the Instruction Fetch/Loop Control unit and the Address Computation Unit (ACU). The ACU design (refer to Figure 2) is based on a concept we will call "dual rank" registers. The application of a dual rank ACU register will be described with reference to a very simple FIR filter. In an FIR filter (convolution) the program must keep track of a number of addresses. For example in Figure 3 it can be seen that in a convolution of the top waveform with the lower coefficients there are at least four pointers being modified.

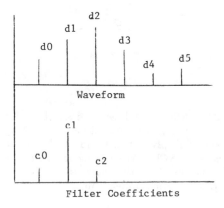

Figure 3

In the first calculation frame the value: $r0 = (d0*c0 + d1*c1 + d2*c2)$ is computed. One pointer, which we shall call a "reference pointer" must keep track of the start of the waveform frame that is being used (d0 on frame 0, d1 on frame

1, etc.) Another pointer, which will be called the "working pointer" must move up through the waveform during each pass of the inner program loop (do:d1:d2 on pass 1; d1:d2:d3 on pass 2, etc.). A similar set of pointers are required for the coefficients. Note that the above two pointers are related. The working pointer starts from the value of the reference pointer and moves up. When it reaches its highest value it is moved back to the NEXT value of the reference pointer. It is this fundamental concept that resulted in the use of dual rank registers.

Figure 4 shows the organization of the 8 dual rank address registers of M275.

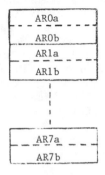

Figure 4

In the above convolution AR0a could be used as the reference pointer to the beginning of a waveform frame; AR0b would be the working pointer. AR0b can be updated during each multiply. At the end of the convolution frame, AR0a and AR0b are updated at ONCE: AR0a is moved to the beginning of the next frame, and AR0b is moved back to the same value. This sets up both pointers for the next iteration of the critical inner loop.

LOOP CONTROL

In the above convolution example it was shown how the pointers can be updated at the same time that the arithmetic operations are performed. It was not mentioned how the count of waveform points is performed. A convolution has two fundamental counts to maintain, the the number of multiplies in the inner loop, as well as the number of frames to be computed. On the M275, since the multiplies are done 2 bits at a time, another count can be used to keep track of the number of multiply steps performed. Two Iteration Counters are provided to help in these tasks. Again these counters are dual rank in nature, with pairs of "reference" counters and "working" counters. The reference and working counters are loaded together in the same instruction cycle. If one of the Iteration Counter zero flags is referenced as a condition for a branch instruction, the working counter is automatically decremented if it is not already zero. If the value was zero the starting value contained in the reference section is auto-loaded into the working section, preparing the counter for the next pass through the loop. It is important to again note that this all happens by simply referencing an IC zero flag as a branch test condition, allowing all testing, decrementing, and re-initializing to be done in parallel with the arithmetic operation. In addition to being fast, this structure is very easy for a programmer to keep track of. All he or she must do is initialize the pair together with a simple one step operation, and then simply reference an IC zero flag as a branch condition.

If more than two loop counters are needed Address registers can be used. Since these are also dual rank the reference count can be kept in the "a" portion of the register and the "b" portion is used as the working counter. The decrement can be done in a single overlapped clock cycle but this of course prohibits an address calculation from being performed on this cycle. For most M275 algorithms currently in use the two iteration counters suffice for tight inner loops, and the less speed critical outer loops use address counters for counting loop iterations.

OTHER M275 FEATURES:

To allow for communications with another processor the M275 has a bus-request system that allows another machine to take control of it's data memory. During such a time the other processor can write into the M275's data memory, a list of messages. Following the bus request, the other processor can interrupt the M275 and cause it to process the message. Also included in this chip is an overflow trap that allows testing for overflows during a high speed algorithm, without wasting any cycles.

TRISTAR ARCHITECTURE:

TriStar is a second generation 32 bit, single chip, programmable digital signal processor now being developed on a 1.5u double metal CMOS process. TriStar, like the M275, is a single cycle machine with various units operating in parallel. TriStar's application differs in several respects from M275. One important difference will be the word length of data. M275's data was often 8 bits whereas TriStar's data will often be 12 to 16 bits. The goal in development of TriStar was to provide about 10 times the throughput of M275. This was achieved in several ways, the first being a shorter clock cycle of 160 nSec. Secondly, each of the major units of the machine was expanded. In the Arithmetic Unit a single cycle will allow a three port RAM access, a full 16 x 16 multiply and barrel shift, and a 32 bit accumulate. In addition much experience has been gained in actually building instruments with M275. Program bottlenecks that have occurred in M275 programs have helped suggest other improvements that increase the machines throughput. For example many M275 programs were plagued with the need to perform saturation arithmetic at various word lengths. Hardware in the TriStar now helps with this function. For most applications of M275 and TriStar fixed point arithmetic is sufficient. For those cases where greater dynamic range in desired TriStar includes a block floating point capability. Since the TriStar datapath is a full 32 bits, single precision arithmetic will nearly always be sufficient, but double precision is supplied for future needs.

Speeding up the Arithmetic Unit of TriStar required an appropriate improvement to the machines ability to fetch operands. One major change is the inclusion of two ACU's with two full ports to two data memories. With this structure two data points can be fetched (or stored) each clock cycle. In addition specialized instructions have been included to allow specialized addressing modes such as modulo queues and FFT bit reversal.

The block diagram of this processor is shown in Fig. 5.

APPLICATION OF TRISTAR

Presence of two directly accessible external data memories is useful for applications like FFT computation, filtering and data transfer. Most DSP algorithms are based around the multiply-accumulate operation. An on chip multiplier/accumulator in the TriStar allows for a single cycle implementation of such an operation. Following the multiplier is a barrel shifter which is useful for scaling of data, and which also supports the block floating point arithmetic mode of the processor. The possibility of overflow during a chain of successive calculations on a block of data is a concern in all DSP applications. Provision for overflow trap as well as saturation arithmetic in the TriStar addresses this concern. In addition, the block floating point mode automatically provides the type of scaling typically introduced by an algorithm designer to expand the dynamic range but applies it only when the size of the data requires scaling. Like the M275, the standard instruction format allows the programmer to independently control all of the parallel operating units.

IMPLEMENTATION

The first phase of implementation, following specification of the architecture and instruction set, is development of a register transfer level, system simulator. In the M275 this was done in APL; in the TriStar, C. The system block diagram is then converted to MOS logic as the first step to physical design. Both of the above IC's have been implemented with a combination of full custom and auto-routed standard cells. Data paths, RAM's, PLA's, and other

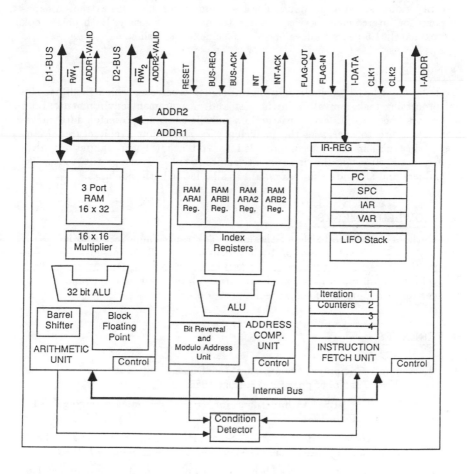

TRISTAR SIGNAL PROCESSOR BLOCK DIAGRAM

Figure 5

regular structures are implemented on CALMA systems. When possible, cell based design is used - most often on random control logic. The final assembly of the IC is done on a CALMA system, after which a transistor level netlist is extracted from the artwork (using MASKAP). This netlist is then simulated with rnl, a switch level simulator. The chip is also simulated at the logic level to verify global speed and timing. The logic simulation includes extracted capacitances for interconnect between blocks to increase accuracy. Both chips took approximately 1.5 years to develop from specification to silicon.

CONCLUSION:

We have described two single chip digital signal processors which facilitate signal processing tasks prevalent in the test and measurement environment. Both processors use a single cycle instruction, parallelism and powerful data movement structure to overcome the inefficiency of single bus architectures. Many special instructions have been provided to enhance the throughput for algorithms used often in the instrument environment. Development tools, like assemblers and software simulators, are available for both machines.

ACKNOWLEDGEMENTS

The authors would like to thank the many Tektronix employees who have contributed to the development of both processor IC's. This includes Eric Etheridge, Vic Hansen, Greg Kogan, Dave Mckinney, Al Meyer, Dan Milliron, Thuy Nguyen. In additions the authors are grateful to Fred Azinger and George Walker for development of software support tools.

REFERENCES

[1] J. Allen, " Computer Architecture for Digital Signal Processing," Proc. IEEE, Vol. 73, No. 5, pp. 852-873, May 1985.

[2] R.E. Owen, " VLSI Architectures for Digital Signal Processing," VLSI design 5, pp. 20-28, 1984.

[3] S. Magar, R. Hester, R. Simpson, " Signal Processing uC builds FFT-based Spectrum Analyzer," Electronic Design, pp. 149-154, August 19, 1982.

[4] L.R. Rabiner and B. Gold, Theory and Application of Digital Signal Processing, Englewood Cliffs, New Jersey: Prentice-Hall Inc. 1975, ch. 11, pp. 627-657.

CHAPTER 24

HOUGH TRANSFORM SYSTEM

Joseph J. Dituri, F. Matthew Rhodes, Dan E. Dudgeon

Lincoln Laboratory, Massachusetts Institute of Technology
Lexington, MA 02173-0073

ABSTRACT

The design of a Hough Transform System utilizing a Wafer Scale
Integrated Circuit (IC) is described. This system is used in the
extraction of line parameters from data points containing co-linear data.
An application of the Hough Transform involving the determination of the
rotation rate of an object from Doppler measurements is also described.
The Wafer Scale IC is built on a single 3-inch wafer using 5 µm two-level
CMOS technology enhanced by a unique laser restructuring capability.

INTRODUCTION

This paper describes the design of a Hough Transform System utilizing
Wafer Scale technology and the application of this system in determining
the rotational rate of an object from Doppler measurements. The simplified
Doppler image of a rotating object is shown in Fig. 1. Across the object a
velocity gradient proportional to rotation speed is detected. The object's
rotational speed can be determined by estimating the slope of this velocity
gradient using the Hough Transform method.

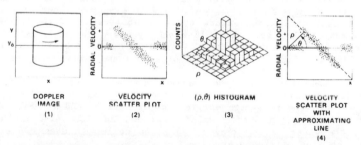

| DOPPLER IMAGE (1) | VELOCITY SCATTER PLOT (2) | (ρ,θ) HISTOGRAM (3) | VELOCITY SCATTER PLOT WITH APPROXIMATING LINE (4) |

A. FOR A GIVEN VALUE y_0 IN (1), PLOT RADIAL VELOCITY AS A FUNCTION
   OF x (2).

B. FOR EACH DETECTED POINT IN (2) AND EACH CANDIDATE θ, INCREMENT
   THE BIN IN (3) THAT CORRESPONDS TO $\rho = x \cos\theta + \text{vel} \sin\theta$.

C. THE RESULTING PEAK IN (3) GIVES THE ρ AND θ VALUES
   FOR THE APPROXIMATING LINE SHOWN IN (4).

Fig. 1. Application of Hough Transform to Velocity
Gradient Estimation.

A block diagram of this line extraction system is shown in Fig. 2. It consists of a Restructurable VLSI (RVLSI) wafer configured for the Hough Transform, a histogram memory and a controller. This line fitting technique is very insensitive to the inclusion of extraneous data points and can be used as a functional box for the extraction of line parameters in any two-dimensional space.

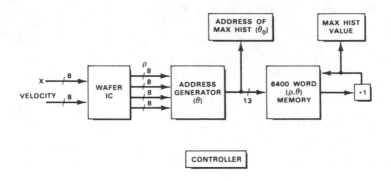

Fig. 2. Hough Transform System.

RVLSI Wafer

Previously, a single wafer processor [1] was designed and fabricated utilizing the principle of RVLSI. The computational resources on that wafer are adequate to implement a Hough Transform, provided that a different set of discretionary interconnections are used. There are two cell types on that wafer, a serial-pipeline multiply-accumulate cell (M2AC) and a parallel-to-serial converter cell (P/S). A functional schematic of the M2AC cell is shown in Fig. 3. All data are input and output serially. The overall cell delay is 38 clock cycles, but because of the pipelining, new data words can be input every 16 clock cycles. The M2AC cells can be customized in various ways by using a laser to make or break connections. The multiplier coefficients (W) are 16 bits long and are preprogrammed in.

The second cell type, the parallel-serial converter cell, is a double-buffered corner turning memory that is used to convert 16-word frames of 16-bit data from bit-parallel to bit-serial, and vice-versa. The floor plan of a single 3-inch wafer is shown in Fig. 4. The central portion of the wafer is an array of 352 M2AC cells. Eleven PS cells are placed on each side of the wafer.

Hough Transform

The Hough Transform is a method by which a line in the Cartesian coordinate system can be described by a point in polar coordinates. The line in Fig. 5a can be parameterized by a line orthogonal to it that passes through the origin. This orthogonal line can be described by an angle (θ) and a length from the origin to the intersection point (ρ). This pair (ρ, θ) is the polar coordinate of the intersection point and can be formulated by the following equation:

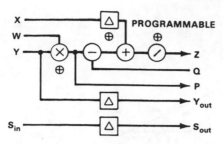

Fig. 3.   M2AC Cell.

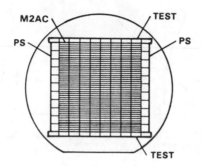

Fig. 4.   Wafer Floor Plan.

$$\rho = x \cos \theta + y \sin \theta \tag{1}$$

Equation (1) defines the Hough Transform between Cartesian coordinates and polar coordinates.   By fixing the values of ρ and θ in Eq. (1), the equation for a line in Cartesian coordinates is formed.   Conversely by fixing the values of x and y in Eq. (1) a relationship is formed between ρ and θ that forms a curve in polar coordinates (Figs. 5b,c).   By combining these two statements we can say that a line in Cartesian coordinates is represented by a point in polar coordinates and certain curves in polar coordinates map into a point in Cartesian coordinates.   This then implies that transforming a set of co-linear points in Cartesian coordinates into a set of curves in polar coordinates will produce a single common intersection point in the polar plane which represents the line (Figs. 5d,e).   This feature of the transform can be used to do line fitting of data.   In particular, the slope of a line can be determined by looking for peaks in the (ρ, θ)-plane.

Application of Hough Transformation

We use the Hough Transformation in our Doppler measurement system to determine rotation speed.   Figure 1 is a simplified velocity image of our scenario.   A velocity gradient, which is proportional to the rotation rate, exists across the object.   The estimation of this velocity gradient is achieved utilizing the Hough Transform to map points in the (x,velocity) space into polar coordinates.   We substitute velocity for y in Eq. (1) and perform the required calculations.   The result is a set of curves in polar coordinates where many of these share a common intersection point (ρ_0, θ_0).   The gradient which is represented by this point equals $(-\cos \theta_0/\sin \theta_0)$.

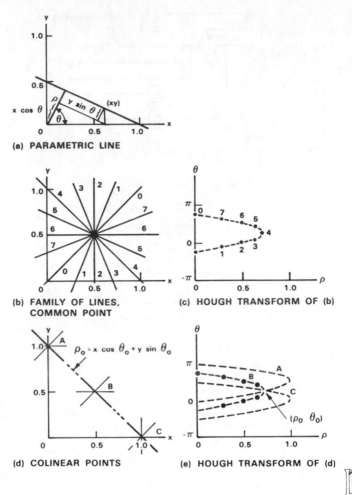

Fig. 5. Hough Transform.

Our original application of the Hough Transform utilized an array processor for the calculations. For each (x,velocity) pair of data points we calculated a curve in polar coordinates according to Eq (1). We obtain acceptable accuracies in the gradient calculation by utilizing a θ step increment of 2° through each pass of the equation. Since we can ignore negative values of ρ in our polar coordinate plots we only have to perform the calculations for 3 of the 4 quadrants of θ. In addition, since we have certain physical bounds on the object rotation rate (and therefore bounds on θ_0) we don't have to pass θ through all possible values. We were able to confine θ to an area of 40° in each quadrant, thus giving us a total of 60 θ values to use in our calculations of Eq. (1).

The method of extraction of θ_0 is as follows: Consider a two-dimensional memory array of 60 columns wide (θ) by 100 columns high (ρ). In Eq. (1) we scale the sine and cosine functions by an appropriate location in this memory can be thought of as a point in the polar coordinate plane. For each input pair (x,velocity) and each of 60 possible value for θ, a value of ρ is calculated according to Eq. (1). The (ρ, θ) pairs are then histogrammed in the memory array by adding a one in the appropriate memory location. After this operation is repeated for all (x,velocity) points in our image, the memory location with the largest count (ρ_0, θ_0) corresponds to the line in the Cartesian coordinate plane. The gradient of that line, which is proportional to the rotation rate, equals $\cos\theta_0/\sin\theta_0$

This process is very time consuming on our array processor. For each (x,velocity) pair we had to perform 60 calculations of the complexity of Eq. (1). After each (ρ, θ) calculation the corresponding memory word had to be read, updated, and returned to memory. After this process was done the memory was then searched for the largest value and the corresponding memory address was converted to θ_0. For a 1000 pixel image the amount of processing time was approximately 120 msec.

Hough Transform On a Wafer

It was observed that the multitude of M2AC cells on our wafer could be efficiently utilized for this Hough Transform process. The wafer can easily implement the calculation of 60 values of (ρ, θ) for each input pair. The calculation of Eq. (1) can be performed by a pair of M2AC cells configured as shown in Fig. 6a. We extended the processing to handle 64 values of θ and this requires only 128 out of the total of 352 M2AC cells to be functional. The wafer architecture is shown in Fig. 6b. Two PS cells are utilized, one each for the x and velocity values with 16 bits of accuracy. In order to minimize the fanout requirements of the P/S cells we repeated the x and velocity values for a full frame (16 cycles). With this configuration each of the 16 serial outputs produces identical data and drives only 4 M2AC cell pairs. The M2AC cells have their coefficients preprogrammed to reflect the multiplication coefficients (scaled cosine and sine θ). These M2AC cells produce 64 serial outputs which must be converted to bit parallel format by 4 S/P cells. Since the cosine and sine tables were scaled to limit the value of ρ to be less than 100, only the 8 LSBs of each S/P cell are brought to the chip's output pins.

Hough Transform System Operation

Figure 2 is a block diagram of our line extraction system. Utilizing a clock cycle of 400 nsec (2.5 MHz) we input a new (x,velocity) pair every 16 clock cycles (6.4 μsec). During this processing time 64 (ρ, θ) values must be read out of the S/P cells and updated in the polar coordinate memory. In order to accomplish this we created a memory utilizing static RAMS that can do a read, update, write cycle in 100 nsec. After all points are inputted into the system the (ρ, θ) memory is scanned for the largest value. In this system a 1000 pixel image can be analyzed in approximately 6.4 msec. The element limiting the clock cycle rate is not the wafer chip but the speed of the (ρ, θ) memory. In order to double the clock rate we could split the (ρ, θ) memory into two parts and have each part service two of the four S/P cells. Since each S/P cell corresponds to an independent section of θ space, the two (ρ, θ) memories can be easily separated. In this mode 1000 pixels could be processed in 3.2 msec.

In a more general system where the resulting values of θ_0 are not as well defined or limited we could use two wafer chips with a resulting range

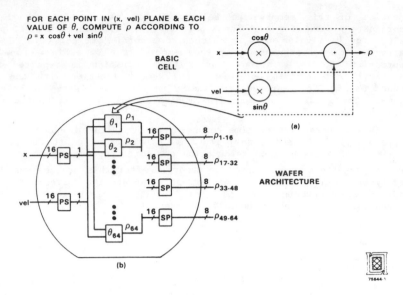

Fig. 6. Hough Transform.

of 256° of θ. The sine and cosine tables could also be modified to increase or decrease the θ step according to gradient accuracy requirements.

CONCLUSIONS

The design of a Hough Transform System used for gradient estimation has been described. This system utilizes a wafer-scale IC to calculate the rotation rate of objects in significantly less time than that required using a standard array processor. A Hough Transform wafer is being constructed from an existing wafer populated with M2AC cells and S/P cells. Laser restructuring will be used to interconnect the required 128 M2AC cells and six S/P cells to implement the Hough Transform computation.

ACKNOWLEDGMENT

The authors gratefully acknowledge Kevin Damour for his contributions to the software development and benchmarking tasks of this project.

REFERENCE

[1] S.L. Garverick and E.A. Pierce, "A Single Wafer 16-Point 16-MHz FFT Processor," Proceedings of the 1983 IEEE Custom Integrated Circuits Conference, pp. 244-248.

A CMOS Video FIR Filter and Correlator

K. Ramachandran
D. F. Daly
R. R. Cordell

Bell Communications Research
Red Bank, NJ 07701-7020

ABSTRACT

A very large scale integrated circuit (VLSI) chip has been designed to perform digital signal processing functions that are important for conditioning video image data. This chip performs four stages of a finite impulse response (FIR) filtering operation with 8 bit coefficients and 9 bit data samples. If additional stages are required, N chips can be connected together to perform 4N operations. In another mode, the chip also performs four stages of the correlation operation. This sixteen thousand transistor chip is fabricated in low power silicon complementary metal oxide semiconductor (CMOS), 2 micron technology with two level metal interconnections. It operates at clock rates up to 21 million samples per second.

I. Introduction

A CMOS VLSI digital signal processing chip has been designed and fabricated that will perform two functions of importance in processing video images - a programmable finite impulse response filter [FIR] and a correlator. The first of the functions implemented on this chip is the finite impulse response filter equation expressed as

$$y_n = \sum_{m=1}^{m=M} a_m \, x_{n-M+m}$$

(1)

where M is the number of taps in the filter. The filter is realized on the chip in the form described by the system diagram on Fig. 1. In cascaded form, si in Fig. 1 gets its input from y(n) of the previous stage. Each tap of the filter requires a multiplier, an adder, and one register for delay. The summation of four products is achieved by a series of four pipelined adders, each of which accumulates one additional product. Furthermore, the same elements - multiplier, adder and register - can be used to compute the terms of the correlation function which is described as

$$R\left(xy\right)_n = \sum_{m=0}^{m=M-1} x_m\, y_{n+m}$$

(2)

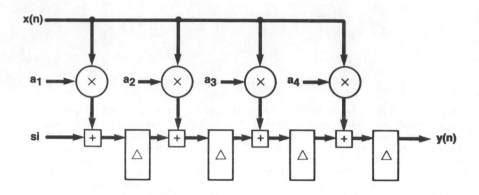

Fig. 1 - System Diagram of FIR Filter Algorithm

II. Chip Architecture

These algorithms are implemented with the chip architecture shown in Fig. 2 Each tap in this figure contains the multiplier, adder and delay register (now called the Sum Register) that was contained in Fig. 1. Each tap also contains three new registers - the Data Register, the Multiplier Pipeline Register and the Product Register. Since these registers provide identical delay in each tap, they have no effect on the algorithm. They merely delay by three data clock cycles (DCK), the time required to get from the input data to SUMOUT. This "pipelining" of the data through the multiplier, the most complex part of the data path, reduces the number of gates between registers, thereby enabling the circuit to run at a higher clock rate.

The main difference between the FIR and the correlator is in the way coefficients are fed into the multiplier. For the FIR function, the coefficients are stored on the chip and are applied to the data stream continuously. (These coefficients do not change with time.) For the correlator, the "coefficients" are really a second data stream that is applied simultaneously to all taps on the chip.

These two methods of handling the "coefficient" data stream are illustrated in the chip block diagram, Fig. 2. When the MODE input is 0, the chip is configured for the FIR operation. The Load Registers are interconnected to form an eight bit wide shift register between the taps. Coefficients are shifted from tap to tap by the coefficient load clock, CLCK. In this configuration, each Coefficient Register receives its input from the associated Load Register. After the coefficients have been shifted through the Load Registers to the appropriate tap, a positive-going edge of CUCK, the coefficient update clock, moves them

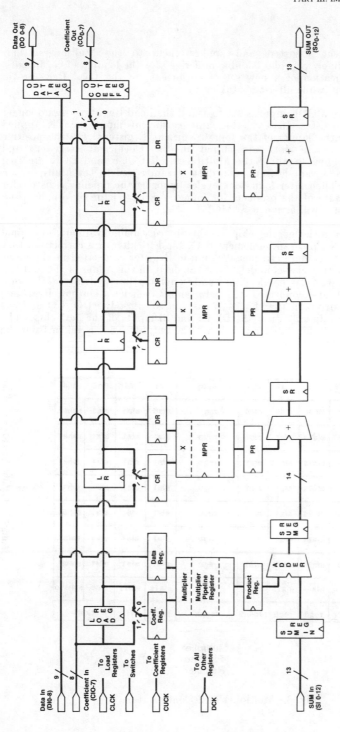

Fig. 2 - Chip Architecture

into the Coefficient Registers and applies them to the Multipliers. This arrangement effectively provides "double buffering" for the loading of coefficients, allowing the application of a new set of coefficients to the Load Registers in background fashion, while filtering is taking place.

For correlation, the MODE input is set to 1 and the Load Registers are no longer used as a shift register. Instead, the same coefficient input data is applied simultaneously to the inputs of the four Coefficient Registers. On the positive edge of CUCK, these data are stored in the Coefficient Registers and simultaneously applied to all of the Multipliers. If CUCK and DCK are tied together, then a new sample from each data stream is fed to the Multiplier on each clock cycle. Then after four clock cycles (ignoring the pipeline registers for the moment) the sum of the products for four consecutive pairs of samples from the two data streams will arrive at SUMOUT.

For all of these algorithms the chip is directly cascadable without additional hardware. Hence, N chips can implement a 4N tap FIR filter or a correlation that spans 4N points of the data streams. A Sum-in Register is inserted at the input of the chip so that the inter-chip delay and the delay in the Adder of the first tap do not add together. To balance this additional pipelining delay in the sum path, DI and CI are connected to Data and Coefficient Output Registers, respectively, that drive off the chip. To cascade N chips, the DO, CO and SO outputs of one chip directly drive the DI, CI and SI inputs of the next chip. The SI inputs of the first chip are grounded and the output signal is taken from the SO outputs of the last chip.

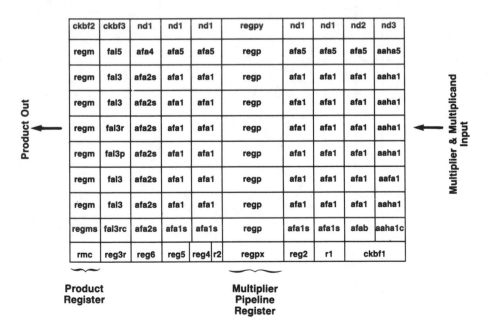

Product Out (left label)

Multiplier & Multiplicand Input (right label)

ckbf2	ckbf3	nd1	nd1	nd1	regpy	nd1	nd1	nd2	nd3
regm	fal5	afa4	afa5	afa5	regp	afa5	afa5	afa5	aaha5
regm	fal3	afa2s	afa1	afa1	regp	afa1	afa1	afa1	aaha1
regm	fal3	afa2s	afa1	afa1	regp	afa1	afa1	afa1	aaha1
regm	fal3r	afa2s	afa1	afa1	regp	afa1	afa1	afa1	aaha1
regm	fal3p	afa2s	afa1	afa1	regp	afa1	afa1	afa1	aaha1
regm	fal3	afa2s	afa1	afa1	regp	afa1	afa1	afa1	aafa1
regm	fal3	afa2s	afa1	afa1	regp	afa1	afa1	afa1	aaha1
regms	fal3rc	afa2s	afa1s	afa1s	regp	afa1s	afa1s	afab	aaha1c
rmc	reg3r	reg6	reg5	reg4 r2	regpx	reg2	r1	ckbf1	

Product Register

Multiplier Pipeline Register

Fig. 3 - Multiplier Cell Configuration

III. Design

The most critical module on the chip is the multiplier. This multiplier generates the 17 bit product of a 9 bit data word and an 8 bit coefficient. The multiplier is a two's-complement Pezaris array, [1],[2], equipped with carry-save adders. The cell array for the Multiplier is shown in Fig. 3. Two design improvements were made to enable the multiplier to generate a new product every 70 ns. First, one stage of pipelining was introduced in the middle of the Pezaris array. Second, the bottom row of the array utilizes a Manchester carry chain with carry bypass to reduce the effect of carry ripple on the delay time of the multiplier. The multiplier contains 2600 transistors and occupies an area of 1.5 sq. mm. The output of the multiplier is rounded to 13 bits, summed with the output of the previous tap and stored in a 14 bit accumulator. This accumulator also utilizes a Manchester carry chain with carry bypass to meet the speed requirements of the chip. The most significant bit provides one bit of overload protection. The least four significant bits provide for control of roundoff error accumulation. Under the assumption of mean square noise accumulation, one will still get 9 bits of good data after 256 taps. By the same reasoning, only the top 13 bits of the accumulator need to be sent to the next chip.

The flip-flops in this circuit are static and require only complementary clock signals that need not be non-overlapping. The clock signal, DCK, is converted from TTL to CMOS and buffered at the chip input. This single-phase buffered clock signal drives a single clock distribution net that is routed throughout the chip. At each group of registers there is a local clock driver circuit which produces the clock and clock-bar signals for the group of flip-flops. A clock driver consists of a series of two inverters to generate clock and a series of three inverters to generate clock-bar. The inverter strings in the local driver are scaled so that there is equal delay between the clock distribution net and the clock or clock-bar inputs to the individual flip-flops. Considerable attention was paid to equalizing relative clock loading so as to maintain fast clock edges and to minimize clock skew. The smaller clock nets for CUCK and CLCK are handled in a similar fashion.

The simultaneous switching of several output buffers can create large current spikes which generate power supply noise because of wiring inductance in the chip's package. This noise could seriously degrade the operation of the main "core" circuitry of the chip. It was therefore decided to provide separate Vdd and Vss pins and wiring for the core circuitry, so that there would be no sharing of common package inductance by the I/O buffers and the core circuits. Thus, one pair of power pins supplies Vss and Vdd to the core circuitry, while another pair supplies Vss and Vdd to the output buffers and input protection diodes.

IV. Design Methodology

The complete architecture of the chip at the gate level was designed and simulated first on a Mentor workstation. The structured custom layout of the circuit was then carried out using Bellcore's version of the MULGA symbolic design system [3],[4]. This system enables the designer to construct the transistor level topology of individual cells on a virtual grid and then connect the individual cells by abutment. An entire array of these abutted cells ("tiles") can then be converted from the symbolic realization to a physical realization and compacted according to a specific set of design rules. An example of a leaf cell layout and its conversion to mask is shown in Fig. 4. Symbolic design makes it possible to quickly and efficiently develop the topology of the circuit without reference to a specific set of design rules. Then by updating the design rule table, one can automatically convert a single symbolic data base into any number of different CMOS technology design files.

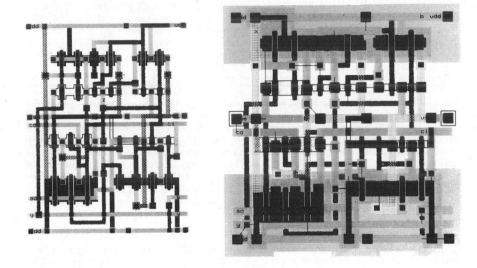

Fig. 4 - And Full Adder (AFA1) Symbolic and Mask Layout

Each of the four modules in a tap was created by symbolic design of individual cells which were then compacted and pitchmatched to optimize silicon area and performance. The modules were then interconnected by an automated river router. Next, the taps were interconnected by the same river router. Finally, a utility called Makeframe was used to create a frame of I/O pads and output buffers and to route connections from the core of the chip to the pads. At each stage in the design process, the cells or the modules (which are composed of cells or other modules) were verified by simulation. The MULGA system can extract simulation files for either a circuit simulator (SPICE), a timing simulator (EMU) or a logic simulator (SOISIM).

V. Performance and Results

This chip was fabricated using 2 micron drawn gate n-well CMOS with two level metal interconnections. First silicon was fully operational and met all performance objectives. Figure 5 is a picture of the chip. The CMOS process used for this chip comes from VLSI Technology Inc. Characterization data on the wafer and chips indicate that the process used for these chips was between "typical" and "slow" for both the n-channel and p-channel transistors. The overall chip characteristics are summarized in Table I.

TABLE I

Technology:	2 um n-well CMOS 2 level metal
Die Size:	246 x 199 mils (6250 x 5050 microns)
Power:	70 mW at 10 MHz
Device Count:	16,000 transistors
Maximum Freq.:	21.0 MHz
Pin Count:	68
Package:	Leadless Ceramic Chip Carrier

Packaged chips were tested using a sample set of functional vectors in the FIR configuration. These vectors stress the multiplier, the adder and all the register paths. At room temperature and 5 volts, all vectors pass up to a frequency of 21.0 MHz. If the voltage is reduced to 4.5 volts, the vectors still pass up to 19.0 MHz. Conversely, if the voltage is increased to 5.5 volts, the vectors pass up to 23.0 MHz.

Fig. 5 - Die photograph of FIR filter chip

VI. Acknowledgements

The design of this chip was possible, in large measure, because of the CAD environment developed at Bellcore by Dave Boyer, Anna Kalaschnikow and Steve Okun. Special thanks are due to Gino Cheng who contributed to both the CAD development and the layout of the multiplier.

[1] Pezaris, S.D. (1971) A 40ns 17bit x 17 bit Array Multiplier, IEEE Trans. Computers, vol C-20 pp.442-447, Apr. 1971

[2] Chin, S.K. and Henlin, D.A. (1983) Two's Complement Array Multiplier Techniques, General Electric Technical Information Series, R83ELS007, Apr. 1983

[3] Weste, N. (1981) MULGA- An Interactive Symbolic Layout System for the Design of Integrated Circuits, Bell System Technical Journal, Vol. 60, No. 6, July-Aug. 1981, pp. 823-857.

[4] Boyer, D.G. and Cordell, R.R. (1985) Structured Custom VLSI Design Methodology, Proc. IEEE International Conference on Computer Design, pp. 324-325, October 1985.

A RASTERIZATION OF TWO-DIMENSIONAL FAST FOURIER TRANSFORM

Wentai Liu & Thomas Hughes
Department of Electrical & Computer Engineering
North Carolina State University
Raleigh, NC. 27695-7911

William T. Krakow
Microelectronics Center of North Carolina
Research Triangle Park, NC. 27709

ABSTRACT

A scheme for rasterizing the two-dimensional Fast Fourier Transform (2DFFT) with size NxN is presented. Through the reorganization of the flow graph of the 2DFFT based on the radix-2 1DFFT algorithm, the scheme is devised to emulate the window operations associated with the flow graph. Accordingly, this paper presents a novel 2DFFT pipelined architecture with $2logN$ processors. VLSI implementation of the proposed architecture is reported in this paper. The proposed architecture avoids using the *corner-tuning* memory, which is required in most literatures based on the radix-2 algorithm, for the matrix transposition associated with the 2DFFT. The data can be continuously driven into the arithmetic processors in a pipelined fashion. It is specially tailored for real-time applications if both the input and output devices are raster-scan devices. The architecture consists of only two kinds of components, namely a data commutator, and a butterfly unit. It is modular and is suitable for VLSI implementation. The scheme can be generalized for three-dimensional FFT.

1. INTRODUCTION

The two-dimensional **Discrete Fourier Transform** is defined as follows:

$$X(k_1,k_2) = \sum_{n_1=0}^{N-1}\sum_{n_2=0}^{N-1} x(n_1,n_2)W_N^{n_1k_1}W_N^{n_2k_2} \qquad (1)$$

This transform is generally evaluated either by the classical row-column algorithm (radix-2 algorithm) [1] or by the vector radix algorithm [2,3].

Many applications of digital signal processing require the evaluation of discrete Fourier transforms (DFT's) of multidimensional sequences. The case of two-dimensional discrete Fourier transform has been widely applied in the area of filtering, image enhancement, image coding, image compression and restoration, radar detection, computerized tomography, nuclear magnetic resonance (NMR) tomography and seismic analysis [4].

Since the appearance of the original Cooley-Tukey algorithm in 1965 [5], the standard methods of computing the two-dimensional discrete Fourier transform have been governed by the **separability** of two-dimensional DFT. Two-dimensional Fourier transform has been decomposed into two one-dimensional DFT by using the one-dimensional fast Fourier Transform (FFT) to execute the transform either in row-column-wise or in column-row-wise format. The case of one-dimensional FFT has been carefully studied and implemented in both software and hardware since 1965 [6].

Matrix transposition and high I/O bandwidth are two major problems associated with the use of a row-column (or column-row) algorithm for the two-dimensional DFT. Therefore techniques to efficiently store the data in the secondary storage device such that it can avoid the matrix transposition or minimize the traffic between the main memory and secondary memory are very crucial in the execution of the FFT.

In many applications the matrix is stored on a mass storage device, *e.g.* a disk or a tape, where the smallest record that can be easily accessed is either an entire row or column. The way in which the data is stored facilitates data access for one-dimension but impedes data access of another dimension. The decomposition approach severely degrades the performance of the implementation of a two-dimensional FFT either in conventional machines (especially virtual memory machines) or SIMD machines. The efficient matrix transposition algorithm as shown in [7, 8], matrix transposition is still a serious bottleneck even for most super vector machines in performing vector gathering operations. One way to ease the matrix transposition problem is to employ a *staging memory* proposed by Batcher [9] between the memory and the processor. The staging memory can reformat or corner-turn the array of data.

The naive or *direct* algorithm for computing the one-dimensional DFT in Equation (1) required N^2 multiplications. However the 1-D FFT is an algorithm to reduce the requirement to (N log N) multiplications. Algorithms for the 1-D FFT can be classified into full parallel, iterative parallel, cascade, and scalar structure. Many different architectures for the 1-D FFT have been proposed or implemented. For example, discrete implementations for the 1-D FFT have been mentioned in [10, 6, 11]. Despain et al. use the CORDIC technique and VLSI technology to implement a 1-D FFT processor in a cascade structure [12]. Computational tasks involving two-dimensional discrete Fourier transforms are computationally intensive and need very high I/O bandwidth. SIMD machines have been proposed to map the row-column-wise decomposition (radix-2 algorithm) of a two-dimensional FFT [13, 14]. Array processors with radix-2 algorithm for 2DFFT using the *corner-tuning mechanism* have been proposed [15, 16]. Pipelined processor with vector radix-2x2 for 2DFFT has been proposed by Liu [17]. There is a need to have a highly efficient architecture for 2DFFT especially in the environment of *raster-scan input/output devices*. This has motivated us to search for new algorithms and to propose highly parallel and pipelined architectures for the two-dimensional FFT. The architectures should fully utilize the ideas of *pipelining* and *parallelism* which are important characteristics of VLSI design.

This paper reports a novel pipelined structure for the 2DFFT in the applications with the raster-scan input/output devices. We are able to derive this new architecture by using only the radix-2 algorithm even without using the *corner-tuning* memory or matrix transpose mechanism. The architecture promises to meet the real-time application requirements with wide range of resolution in the image. The structure allows sufficient parallel operations such that the processing time is limited solely by the time it takes to collect the data.

2. THEORY OF OPERATION

In this section, the operation of the pipelined 2DFFT is explained in terms of both the Sande-Tukey algorithm (Decimation-in-Frequency) and the Cooley-Tukey algorithm (Decimation-in-Time) for one-dimensional FFT. Flow graphs for a two-dimensional transform will be generated. Based on the flow graph, rasterization of the two-dimensional transform becomes possible.

Without loss of generality, we assume that the decomposition order for applying one-dimensional FFT algorithm on a two-dimensional input data with size NxN is column first then followed by row. A flow graph, which consists of *logN* one-dimensional stages, can be defined for every column or row. The key idea is to make a flow graph for a two-dimensional transform by merging all the one-dimensional column transformational flow graphs together at each stage (there are *logN* one-dimensional stages for every column transform) then followed by the merge of all the one-dimensional row transformational flow graphs. This is equivalent to concatenating the two-dimensional input data into a one-dimensional data according to the raster-scan order such that the butterfly operations (every column then row) can be executed in parallel. As we've mentioned previously, for every column of data, it consists of *logN* stages in the conventional flow graph. At the first stage, the transformation examines the data points displaced by half the data length $\frac{N}{2}$, the second by one quarter of the data length $\frac{N}{4}$, etc, until the last one (*logN's*) by 1. The displacement at stage i is $\frac{N}{2^i}$. The transformation performs the butterfly operations as defined in Figure 1. In the case of a two-dimensional transformation flow graph (column first then row second), the first $logN$ stages are devised to perform one-dimensional transformations for all N columns of data. At the first stage, the transformation examines the data points displaced by $\frac{N^2}{2}$, the second by $\frac{N^2}{4}$, ... , until the $logN's$ stage by N. The displacement at any stage between 1 and $logN$ is $\frac{N^2}{2^i}$. Then the one-dimensional transformations for all N rows of data starts at the stage $logN + 1$ and ends at the stage $2logN$. At stage $logN + 1$, the transformation examines the data points displaced by $\frac{N}{2}$, ... , until at stage $2logN$ by 1. The displacement at any stage between $logN + 1$ and $2logN$ is $\frac{N}{2^{(i-logN)}}$.

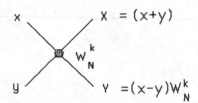

Figure 1. A DIF Butterfly Operation

In fact, at the same stage, all the operations associated with rows/columns can be operated in parallel. It takes $logN$ stages for column transformations. It is then followed by another $logN$ stages for row transformations. In total, it takes $2logN$ stages for a 2DFFT transformations. It is also clear that each stage needs only the data generated from the preceding stage. The output appears in the two-dimensional bit-reversal order. This makes the flow graph for a two-dimensional FFT appear as the one shown in the Figure 2. A validation of the flow graph can be provided by the following:

$$X_{01} = x_{01}^{(4)}$$

$$= x_{01}^{(3)} + x_{03}^{(3)}$$

$$= (x_{00}^{(2)} - x_{02}^{(2)}) + (x_{01}^{(2)} - x_{03}^{(2)})\omega$$

$$= \left\{[(x_{00}^{(1)} + x_{20}^{(1)}) - (x_{02}^{(1)} + x_{22}^{(1)})] + [a(x_{01}^{(1)} + x_{21}^{(1)}) \quad (x_{03}^{(1)} + x_{23}^{(1)})]\omega\right\}$$

$$= \left\{[(x_{00}^{(0)} + x_{20}^{(0)}) + (x_{10}^{(0)} + x_{30}^{(0)})] - [(x_{02}^{(0)} + x_{22}^{(0)}) + (x_{12}^{(0)} + x_{32}^{(0)})]\right\}$$

$$+ \left\{[(x_{01}^{(0)} + x_{21}^{(0)}) + (x_{11}^{(0)} + x_{31}^{(0)})] - [(x_{03}^{(0)} + x_{23}^{(0)}) + (x_{13}^{(0)} + x_{33}^{(0)})]\right\}\omega$$

It is easy to show that each of the stages may be realized with a basic component whose general form is shown in Figure 3. The figure is similar to the one in Groginsky and Works' paper [18]. Any module i alternatively transfers block of 2^i data samples into the delay line and into the arithmetic unit. When the data just fills the delay line, the arithmetic unit obtains a twiddle factor and begins its operation. The next block of 2^i input data samples are sent to the arithmetic unit that now produces two complex outputs in response to the two complex inputs it receives. One of the outputs is immediately transferred to the next stage while the other output is sent to the delay line. Thus during the interim period when the delay line is filled with fresh input data, the contents of the line containing the results of processing the earlier blocks are transferred to the next stage. The arithmetic unit computes the complex two-point transform as shown in Figure 1.

The module may be assembled into a system for computing 2DFFT with the size of NxN samples. The system consisting of $2logN$ modules is shown in Figure 3. Figure 1 shows that at most $\dfrac{N}{2^i}$ twiddle factors needed to be supplied in the pipelined processor.

Similarly, the flow graph based on the Cooley-Tukey algorithm can be derived. The output is also in the two-dimensional bit-reversal order. This is shown in Figure 4.

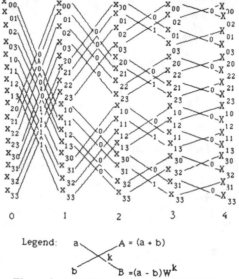

Legend:

$$a \diagdown \diagup A = (a + b)$$
$$ \times k$$
$$b \diagup \diagdown B = (a - b)W^k$$

Figure 2. A DIF Flow Graph for 2DFFT

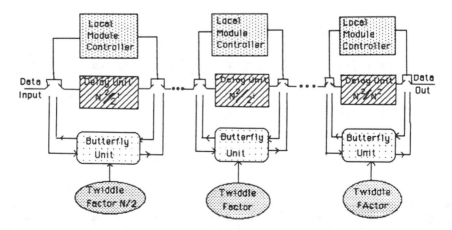

Figure 3. A Pipelined 2DFFT Processor

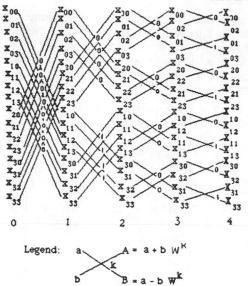

Legend: $A = a + b \ W^K$

$B = a - b \ \overline{W}^k$

Figure 4. A DIT Flow Graph for 2DFFT

3. COMPUTATION STRUCTURES AND IMPLEMENTATIONS

3.1. Computation Requirements

In this section, we compute the minimal computational requirement for the real-time applications with respect to the proposed architecture. The requirements are expressed in terms of maximal computation time in each pipelined stage, number of pipelined stages, and number of multipliers per pipelined stage. We assume that the real-time applications is operated in the environment of *TV frame rate with 30 frames per second*. Table 1 shows the requirements.

Image Size	Maximal Computation Time/Pipelined Stage	Number of Pipelined Stage	Number of Multipliers/Stage
128x128	2.2μs	14	1
256x256	540.0ns	16	1
512x512	132.0ns	18	1
1024x1024	33.0ns	20	1
2048x2048	8.0ns	22	1
4096x4096	2.0ns	24	1
8192x8192	0.5ns	26	1

Table 1. Computational Requirements for Real-time Applications (Radix-2)

285

The requirement can be relaxed if a vector radix 2x2 algorithm is used in the pipelined processor as reported in Liu's paper [17]. This is shown in Table 2. Table 1 and Table 2 shows the trade-off between *space* and *time* for various algorithms to implement 2DFFT. More relaxed constraints can be obtained if an iteratively parallel algorithm had been adopted as reported in Liu's paper [17]. This is shown in Table 3.

Image Size	Maximal Computation Time/Pipelined Stage	Number of Pipelined Stage	Number of Multipliers/Stage
128x128	8.8µs	7	3
256x256	2.2µs	8	3
512x512	518.0ns	9	3
1024x1024	132.0ns	10	3
2048x2048	32.0ns	11	3
4096x4096	8.0ns	12	3
8192x8192	2.0ns	13	3

Table 2. Computational Requirements for Real-time Applications (Radix 2x2)

Image Size	Maximal Allowed Computation Time/ Pipelined Stage	Number of Pipelined Stage	Number of ALU per Iterative Stage	Number of Multiplier/Stage
128x128	563.2µs	7	64	3
256x256	28.2µs	8	128	3
512x512	13.6µs	9	256	3
1024x1024	660.0ns	10	512	3
2048x2048	330.0ns	11	1024	3
4096x4096	160.0ns	12	2048	3
8192x8192	81.0ns	13	4096	3

Table 3. Computational Requirements for Real-time Applications (Iterative Radix 2x2)

Word length of data is an additional consideration of computation requirements. Table 4 gives the word length necessary at different image sizes to maintain errors below 2 % in a fixed point system. We defined word length = length (real) = length (imaginary).

Image Size	Word length(bits)	Error (%)
128x128	9	1.9
256x256	10	1.3
512x512	11	1.9
1024x1024	12	1.3

3.2. Implementation

Based on the tables above we decided to implement the Radix-2 Algorithm of image size 256x256. This limit was established not so much because of speed limitations (540ns cycle time is rather leisurely for CMOS) but to keep chip count and size reasonable.

Figures 2 and 3 give an idealized high level architecture which needs a few modifications to be realized in hardware. The starting point is the pipelined 2D FFT module of Figure 5.

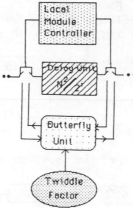

Figure 5. A Pipelined 2DFFT Module

3.2.1. Butterfly Unit

The *Butterfly Unit* (BU) has the task of performing the complex operation given by Figure 1. The total number of real-number (not complex-number) operations to be performed on this data in about 500ns are four multiplies and eight additions/subtractions. For the target technology of MOSIS's 3 micron scalable CMOS, some decomposition of the problem into several chips is necessary. Figure 6 is the partitioning we used. Two chips have some redundant functions but the advantage to a VLSI implementation is uniformity.

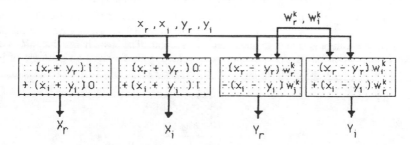

Figure 6. Chip Level Partition of the Butterfly Unit.

This chip design has been started with Berkley's Magic and locally developed standard cell tools [19]. The design uses 1.25 μm CMOS technology developed by the Microelectronics Center of North Carolina (MCNC). A 3 μm CMOS version (GE CMOS technology, designed by M. Bates [20]) of the design is shown in the Appendix.

An additional consideration while designing this system is to allow for future growth to designs of larger frame images ($\geq$ 256x256) and higher speed. These BU chips use pipelining mechanism, not so much needed to compute the function in 500ns, but so the system could handle pipelined BU needed in future designs.

3.2.2. Delay Unit and Switch

Implementing the *Delay Unit* (DU) and Switch of Figure 5 is not as simple as the BU. The data flow possible from Figure 5 are given below in Figure 7 If however the BU is to be piped then other data flows occur. For the cases of the pipeline length of BU less than or greater than the delay length of the DU, the four data flows of Figure 8 occur. If the pipe length of BU is equal to the DU's delay length, then only data flows in Figures 8.b and 8.d occur.

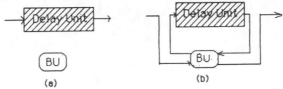

Figure 7. Data Flow for the Case Without Pipelining BU

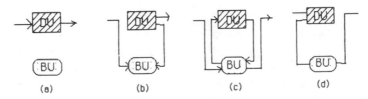

Figure 8. Data Flows for the Case with Pipelining BU

For our case of 256x256, the largest delay length is $\frac{N^2}{2}$ = 32K or half a frame buffer. However due to pipelining of the BU, the delay length needs to be ($\frac{N^2}{2}$ + BU pipe length). *The idealized DU of Figure 7 could be a simple shift register delay, but the data flows of Figure 8 and size of ($\frac{N}{2}$ + BU's pipe length) suggest the use of RAM.* 500 ns allows for 2 cycles of memory so that the required data flows can be accomplished by address arithmetic of RAM. Figure 9 gives the design of our RAM DU and switch. Data goes into the BU during the first memory cycle and outputs on the second. The switch on the left is in high impedance state when the RAM is in the *read cycle*. Figure 10 gives an example of how the RAM and switches can accomplish data flow of Figure 8.c.

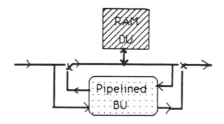

Figure 9 RAM Delay Unit and Switch

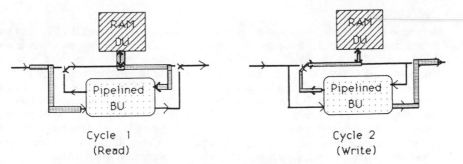

Cycle 1
(Read)

Cycle 2
(Write)

Figure 10 Data Flow for Figure 8.c Using Two Cycle RAM

3.2.3. Twiddle Factor (ω) Generator

The last major component of the Butterfly Unit to be specified is the ω table or Twiddle Factor Generator. The purpose of this component is to provide the correct ω during a valid AU cycle. This task could be met by ROM or RAM properly addressed but would require several chips. Our solution is to put the entire table ($2 * $ word length $* \frac{N}{2^i}$ bits) in a custom static variable length shift register chip. The primary features of this chip is possible delay lengths of 2^i and the ability to either shift data or hold it in place. Based on Figure 2, the chip should have the capability of holding data N cycles before it makes any shift. This is shown in Figure 11. For the case of Figure 12, the chip should have the capability of continuously shifting the data N cycles. The main difference between Figure 2 and 12 depends on the capability of the shift register chip. However Figure 2 certainly offers great advantage over Figure 12 because the design can be done without the matrix transposition.

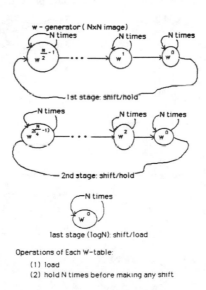

Figure 11 Control Schemes for the Delay Line

4. CONCLUSION

A scheme for rasterizing the two-dimensional Fast Fourier Transform (2DFFT) with size NxN is presented. Accordingly, this paper presents a novel 2DFFT pipelined architecture with $2logN$ processors. The proposed architecture avoids the *matrix transpose problem* which is required as reported in most literatures. The unique feature of the architecture is that the data can be continuously driven into the arithmetic processors in a pipelined fashion without using the *corner-tuning* mechanism. It is specially devised for the real-time applications if both the input and output devices are raster scan devices. The control of the proposed architecture is very simple and can be generated by a binary counter if Figure 7 is used. However if pipelining mechanism is used in the butterfly unit, it becomes much complicated as shown in Figure 8. It even becomes more complicated if the delay line is implemented using RAM.

Realtime computational requirements based on this architecture for varios image sizes are discussed. The chip implementation for the size 1024x1024 is definitely realizable using the state of art CMOS technology.

The scheme can be easily generalized to the case of three-dimensional FFT with $3logN$ stages. Currently we are investigating for the schemes for the rasterizations of the 2D Walsh Transform, 2D Cosine Transform, and 2D Hadamard Transform since we are able to show that these Transforms share the same communication pattern, namely *two-dimensional shuffle-exchange* in the paper [21].

Acknowledgements

This work has been partially funded by Microelectronics Center of North Carolina through the grant FAS# 5-30406 and a University of North Carolina System - Bell Northern Research Award. T. Hughes is also supported by a DuPont Graduate Fellowship. We are grateful for Mike Bates for providing the GE's 3μm CMOS design. Thanks to G. Mei, R. Salama and T. Hildebrandt for their helpful discussions. Finally special thank goes to Dr. R. Cavin for his continuous interest and encouragement for this project.

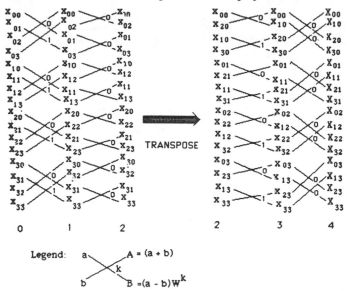

Figure 12 Another Flow Graph Based on Radix-2 DIF

5. REFERENCE

1. D. Dudgeon and R. Mersereau, *Multidimensional Digital Signal Processing,* Prentice-Hall Inc., New Jersey (1984).

2. G. Rivard, "Direct fast Fourier transform of bivariant functions," *IEEE Trans. Accoustic, Speech, Signal Processing* ASSP-25 pp. 250-252 (June 1977).

3. W. Liu, *Architecture for two dimensional Fast Fourier Transform,* US Patent pending No. 6-539540 (Oct. 5, 1983).

4. R. Gonzalez and P. Wintz, *Digital Image Processing,* Addison-Wesley, Reading, Mass. (1977).

5. J. Cooley and J. Tukey, "An algorithm for the machine computation of complex Fourier series," *Mathematical Computation* 19 pp. 297-301 (April 1965).

6. L. Rabiner and B. Gold, *Theory and Application of Digital Signal Processing,* Prentice-Hall Inc., New Jersey (1975).

7. J. Eklundh, "Efficient matrix transposition," pp. 9-35 in *Two-Dimensional Digital Signal Processing II,* ed. T. Huang,Springer-Verlag (1981).

8. W. Moorhead, *Private Communication*Nov. 1983.

9. K. Batcher, "Architecture of the MPP," *IEEE Computer Architecture for Pattern Analysis and Image Database Management,* pp. 170-174 (Oct. 1983).

10. G. Bergland, "FFT hardware implementations - an overview," *IEEE Trans. Audio Electroacoust.* AU-15 pp. 104-108 (June 1969).

11. H. Kung, "Special-purpose devices for signal and image processing: an opportunity in VLSI," CMU-CS-80-132, Computer Science Department, Carnegie-Mellon University (July 1980).

12. A. Despain, C. Sequin, C. Thompson, E. Wold, and D. Lioupis, "VLSI implementation of digital Fourier transform," UCB/CSD-82/111, EECS, University of California at Berkeley (Nov. 1982).

13. L. J. Siegel, "Image processing on a partitionable SIMD machine," pp. 293-300 in *Languages and Architectures for Image Processing,* ed. M. Duff and S. Levialdi,Academic Press (1981)

14. C. Joshi, J. McDonald, and R.Steinvorth, "A video rate two dimensional FFT processor," *Proceed. IEEE Int. Conference on Accoustics, Speech and Signal Processing,* pp. 774-777 (1980).

15. T. Chou, "An array architecture for parallel FFT processing," *Proceeding of ICCC82,* pp. 232-235 (1982).

16. F. Hsu, H. Kung, T. Nishizawa, and A. Sussman, "Architecture of the link and interconnection chip," *1985 Chapel Hill Conference on VLSI,* pp. 439-461 (1985).

17. W. Liu and D. Atkins, "VLSI pipelined architectures for two dimensional FFT with raster-scan input device," *Int'l Conference on Computer Design: VLSI in Computer (ICCD84),* pp. 370-375 (Oct. 1984).

18. H. Groginsky and G. Works, "A pipeline Fast Fourier Transform," *IEEE Trans. Computers* C-19(11) pp. 1015-1019 (Nov. 1970).

19. J. Rose, W. Krakow, and W. Liu et al, "VPNR: An improved placement-and-routing system," *Microelectronics Center of North Carolina (MCNC) Technical Bulletin*, (June 1986).

20. M. Bates, *VLSI design of a butterfly unit for a 2DFFT,* MS Thesis, North Carolina State University (May 1986).

21. W. Liu and G. Mei, "Parallel processing for a class of two dimensional image transformation," *Submitted for Publication*, (Feb 1986).

Appendix:

Design Summary (by M. Bates)	
I/O	132
Gates	4881 2-input Nand
Chip Size	260 mils x 272 mils
Clock Rate	8 MHZ
Technology	GE 3μm CMOS
Pipeline Stage	2

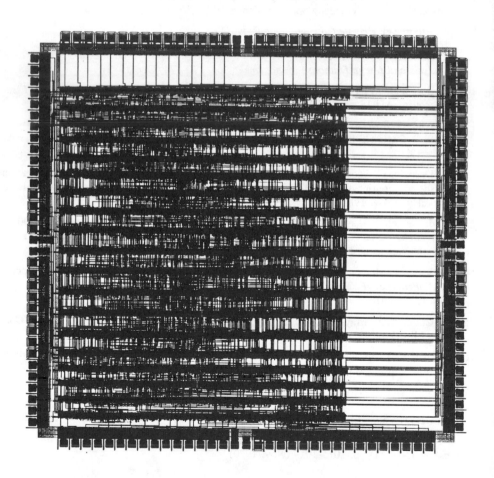

A VLSI FLOATING-POINT SIGNAL PROCESSOR

Paul M. Chau and Walter H. Ku
School of Electrical Engineering and
National Research and Resource Facility for Submicron Structures
Cornell University, Phillips Hall
Ithaca, New York 14853

ABSTRACT

Bit-serial architectures have been used effectively in the design of VLSI signal processors. They provide particular advantages in the areas of signal routing and communications as well as providing computational efficiency through the use of bit level pipelining. Utilizing dedicated systolic communication and control paths minimizes global interconnect and signal broadcasting problems, while maximizing extensibility and efficiency. Furthermore, bit level pipelining distributes both memory and processing elements in a modular and regular fashion, facilitating ease of design, layout, and amenability for silicon compilation. By employing *functional parallelism* in the form of parallel pipelines of bit-serial structures, wideband high performance signal processing capabilities can be achieved. A variety of fixed-point bit-serial VLSI signal processors has been designed and built. Although fixed-point computation with scaling (to accommodate possible word growth) has been satisfactory for a wide assortment of signal processing applications, floating-point computation will be necessary to address the growing needs of signal processing in wide dynamic range or low signal-to-noise environments. We describe a novel floating-point VLSI signal processor that embodies the above bit-serial VLSI signal processing concepts but has its capabilities extended to a floating-point arithmetic environment.

INTRODUCTION

The increasing needs of more sophisticated techniques to solve signal processing problems digitally at higher speeds and higher performance have led to increasing demands for greater computational capability. These requirements have led to the marriage of Very Large Scale Integration (VLSI) with Digital Signal Processing (DSP). Unlike general purpose computing machines, VLSI Digital Signal Processors are algorithm or application specific special purpose computers. They are designed to exploit the highly computationally intensive, low conditional branching, signal processing algorithms [1],[13]. Thus signal processing is greatly enhanced by the utilization of dedicated special purpose hardware [5]. VLSI design principles have generated many new architectures, such as systolic and wavefront processors, for the implementation of modern signal processing algorithms [14],[22].

Fourier Transform and Matrix Arithmetic Based Signal Processing

Fundamental to many signal processing systems is the need for the efficient computation of the Discrete Fourier Transform (DFT). This has led to considerable research into the well known group of algorithms of the Fast Fourier Transform (FFT). The Fourier transform has been one of the most important algorithms employed in signal processing, with applications in matched filtering, spectral estimation, and beamforming [15]. Recently, extensive research has been done on new techniques in the areas of adaptive digital processing and high resolution spectral estimation, which are eigenvalue/eigenvector-based methods. These techniques have demonstrated superior results compared to conventional methods using Fourier analysis and windowing. The more recent methods, when reduced to basic computations, involve linear algebra (matrix) type operations. Thus inner product (multiply/accumulate) calculations as well as $1/x$ and $SQRT(x)$ are needed.

The sophisticated adaptive and linear algebra based signal processing algorithms will require floating-point operations. Thus a VLSI floating-point signal processor capable of both the efficient computation of the Fourier Transform and the arithmetic for key matrix operations, such as inner product computation, would be highly desirable for precision intensive operations. There are also situations where the signal data to be processed dictate the need for floating-point. Fixed-point systems are adequate when signal processing applications have a high signal-to-noise ratio with narrow dynamic range. A common alternative to fixed-point, to avoid the cost of floating-point, has been the use of block floating-point. However, scaling is necessary to maintain dynamic range, and though some block floating-point processors calculate the scale factor, the problem of numeric range is pushed off to a host processor. If the data dynamic range is broad, this can create a computational bottleneck for the host processor, and hence dissipate speed advantages of having a special purpose peripheral. This is so because the overall system throughput is hampered by the slowest component, which in this case may be the host microprocessor. A floating-point solution will facilitate high performance needs in applications where signal-to-noise ratio is low and dynamic range is broad.

VLSI Bit-Serial DSP and Floating-Point Signal Processing

Currently there are chips available which implement floating-point arithmetic via a two or three chip system. There are also several fixed-point and block floating-point general purpose programmable signal processing chip/chip sets such as TI's TMS320 series and others [4]. All these chips tend to be conventional bus oriented bit-parallel designs. Very recently, VLSI floating-point digital signal processors were announced by AT&T Bell Labs, the WE-DSP32 [12], and by NEC, the uPD77230 [4]. The architectures of these machines also utilized a conventional bus oriented bit-parallel processor design but with floating-point capability. In these machines, the hardware complexity is greatly increased, since bit-parallel circuits are needed for exponent handling, barrel shifting, priority encoding for normalization, etc.

The design of arithmetic machines tend to be of two general classes: digit-serial and digit-parallel. There are advantages and disadvantages associated with both. Serial designs often will be more area efficient than a parallel design, but the speed of operation is limited by the system clocking and has greater latency. Parallel designs in general have faster throughput and shorter latency than a serial one, but at the cost of increased area requirements. However, the metric used to evaluate any design must consider the overall area-time-throughput efficiency. The tenets of VLSI bit-serial design for signal processors have been defined by various researchers [8],[18],[20], where the latter researchers advocate *"functional parallelism"* in the form of bit-serial word-parallel architectures. We have adopted these principles for our design effort and have extended these concepts to a floating-point environment.

COMPLEX FLOATING-POINT VLSI SIGNAL PROCESSOR ARCHITECTURE

An overview of our architectural approach is shown in Figure 1. We have utilized corner turning memory at the I/O interface so that data can enter or leave our chip in a conventional bit-parallel fashion. Bit-serial I/O ports are also provided as an alternative input/output means. Thus arrays of our complex floating-point signal processors can be configured into parallel pipelines for bit-parallel, bit-serial, or both, data transfers. This is shown in Figure 2. The block arrows indicate bit-parallel communication paths to accommodate current conventional bit-parallel host processors and peripheral chips, while the group of single arrows denotes bit-serial transfer of data. The architecture of the chip enables it to work as an autonomous, monolithic special purpose co-processor to a host microcomputer. Many of these chips can be configured for high performance modes of operation in pipelines, parallel pipelines, or arrays.

Inner Product Processors & Floating-Point Arithmetic for Matrix Operations

The core of our novel VLSI floating-point signal processor utilizes systolic bit-serial arithmetic units and shifters that are organized in an array of parallel pipelined bit-serial processors. It is optimized for the multiply-accumulate operation that is so prevalent in signal processing algorithms. The multiply/accumulate operation, commonly known as the inner product calculation, has been investigated by several researchers [2],[24]. These implementations were fixed-point structures. Our design incorporates an array of complex and fully floating-point (real and imaginary parts) bit-serial inner product processors. Figures 3a and 3b depict our normalized floating-point word format and the relative serial shifting of data in pre-arithmetic alignment.

At the heart of the machine is an array of parallel fully systolic bit-serial multipliers. A novel "linear half-barrel shifter" for denormalization/normalization that requires order (m) area rather than order (m^2) in conventional bit-parallel barrel shifters is utilized. These shift delay (SD) units are employed to prealign the operands before the bit-serial addition/subtraction to form the complex multiply/add results. Figures 4a and 4b illustrate the mantissa and exponent dataflow block diagram, respectively. We exploit the inherent latencies in bit-serial mantissa arithmetic to do the exponent arithmetic concurrently via bit-serial arithmetic. The denormalization (pre-arithmetic alignment) is done after the exponent differencing is complete. The exponent difference is deposited in a counter which is used by the shift control to select and inhibit the shifting of one mantissa relative to another. The post-arithmetic normalization employs simple logic and counters to tally the number of leading sign bits rather than bit-parallel priority encoders. This count is then passed on to the exponent handler and the mantissa handler to do the proper adjustments. The count is bit-serially added to the exponent while the count is used by the shift control to reverse and shift the results until normalization is achieved. Note that an advantage of our bit-serial approach is that the double precision result from the multiply is retained until the very last possible moment. Hence the result shift registers are double precision and their implementation require layout folding. Rounding of the results follow in a very natural way in a bit-serial LSB first type algorithm: carry-save full adders allow the injected round bit's carry-out to propagate towards the MSB. Thus in this kind of environment, an additional final carry propagating adder is avoided since the rounding is incorporated into the bit-serial addition/subtraction. Figures 6a and 6b shows the block diagrams for the normalization count and the parity/round logic.

The array of complex floating-point inner product processors enables our machine to be configured for the floating-point operations common to matrix based calculations. This includes primarily floating-point addition/subtraction and floating-point multiplication. With some minor modifications, floating-point division and floating-point square root can also be provided. These operations are performed in a bit sequential "on-line/semi-on-line" fashion. On-line algorithms [10] have been developed to enhance digit-serial arithmetic processing. The specifications of on-line operation require that the jth result bit be available after $j+d$ bits have been processed, where d is a small constant known as the on-line delay. Semi-on-line algorithms [21] are computationally less stringent in that they generate result bits only after the absorption of all operand bits and a small delay d.

On-line algorithms are of two classes: Right directed (ROL, process MSB first, LSB last) or Left directed (LOL, processed LSB first, MSB last), where MSB is Most Significant Bit (Digit) and LSB is Least Significant Bit (Digit). To provide for all of the floating-point operations using on-line techniques, it is necessary to have ROL ordering. Furthermore, there must be redundant re-coding of the operands. LOL on-line algorithms do not need redundant re-coding and hence simplify the processing, but then they are only applicable to addition/subtraction and multiplication. This would be sufficient for inner product and FFT type computations, and would preclude 1/x and SQRT(x). Bit-sequential LSB first adders and multipliers benefit from the natural directional flow of the carry propagation from low order to high order bits and have been a popular choice for bit-serial inner product type computations. However, LOL multiplication has the disadvantage of m time step latency before the most significant half of the $2m$ product is available, where m is assumed to be the number of operand bits.

Floating-point on-line computations have certain limitations [19] in which the ability to obtain normalized results is not always guaranteed unless one adheres to some restrictions. Since redundant re-coding is necessary, the additional complexity may exact not only an increased hardware cost, but in a bit-sequential environment, it would reduce the processing speed [21]. In addition to the less restrictive requirement of having result bits delayed by d after the last bit is absorbed, semi-on-line algorithms also require O(m) implementation complexity. Semi-on-line algorithms are classified as *forward*, when bits are processed as they are absorbed into a semi-on-line bit-serial processor, or as *backward* when processing starts after the last input bit is in the processor. Sips [21] defines bit-sequential inner product, division and square root algorithms, and proposes a full adder array structure for implementing semi-on-line floating-point arithmetic operations. Though we also employ similar bit-level type algorithms, we differ in our implementation approach. Clearly, in scenarios where all digits of the operands are not simultaneously available for processing, on-line algorithms offer intrinsic computational advantages. But if the operands are such that they are available, as in the case of dealing with conventional bit-parallel microprocessors and peripherals, then semi-on-line algorithms offer similar digit serial advantages as on-line algorithms with less stringent restrictions. A summary of each of the basic floating-point arithmetic operations are as follows.

Floating-Point Addition/Subtraction

Floating-point addition/subtraction can be summarized by the following steps:

    1. exponent difference (select larger exponent)
    2. pre-arithmetic mantissa alignment (right shift smaller exponent mantissa)
    3. mantissa arithmetic
    4. post-arithmetic rounding/truncation and normalization (inc/dec. exponent)
    5. check for overflow and underflow

When operating with normalized operands, in a "true add" of magnitudes (a + b, - a - b), there is the possibility of a 1 bit overflow, which requires, a 1 bit right shift to normalize. In a "true subtraction" of magnitudes (a - b, - a + b), if the pre-arithmetic alignment shift is greater than 1, then at most 1 bit left shift is needed for normalization. If the pre-arithmetic alignment shift is 0 or 1, then "catastrophic cancellation" may occur and normalization may require a left shift up to the number of mantissa bits [23]. To be compatible with the recent IEEE 754 Binary Floating-Point standard [7], the results must be able to be rounded to one of several rounding schemes. Thus in addition to the arithmetic guard bit (G) to the right of the LSB, a round bit (R) and sticky/spill bit (S) is needed to determine proper rounding. Due to the inherent sequential dependence of the above steps, to perform the steps in a bit-sequential manner embodies many inherent speed consequences since the time of steps 2 to 4 are each of $O(m)$ time. However, due to the fact that floating addition/subtraction follows the floating-point multiplication in an inner product, we can exploit the inherent $O(2m)$ latency of that operation to do both exponent differencing, one after the other, concurrent with the multiplication of the mantissa bits. It is necessary to place shift delay registers after the bit-serial multipliers and before each stage of bit-serial adders in order to align the operands properly. The exponent difference count is used by shift controllers to inhibit the generation and shifting of one operand's mantissa bits relative to another.

Floating-Point Multiplication

The basic steps in floating-point multiplication can be summarized as follows:

    1. add exponents (remove extra bias if exponents are biased)
    2. multiply mantissas (determine product sign)
    3. one bit left shift normalize if necessary (decrement exponent)
    4. check for overflow/underflow

The bit-serial multipliers are of $O(m)$ in complexity and require $O(2m)$ in time. Here steps 1 and 2 can be done concurrently. In the post-multiplication addition/subtraction stages exponent differencing is also needed. Since the exponent bit lengths are smaller than the mantissa bit lengths (typically about 1/3 or less, eg. IEEE 754 single is 8 bit exponent, 23 bit mantissa and double is 11 bit exponent and 53 bit mantissa), the addition/subtraction exponent differencing can follow after the multiplication exponent differencing. With LSB first bit-serial multipliers, there is an intrinsic $O(m)$ latency before the most significant half of the $2m$ product occurs. Additional shift delay registers can be inserted between the bit-serial arithmetic units to insure proper operand alignment. The shifting control of the result bits of one operand relative to another is dictated by the exponent difference. For LSB first bit-serial arithmetic, the rounding/truncation can be done bit-serially without an additional final carry propagating adder, since the carry will naturally flow towards the MSB using carry-save type full adders. Linear recurrences are accomplished by having the results of the inner product operation as a possible input after passing through delaying shift registers and feeding back to the multiplier's multiplexed input paths.

Floating-Point Division

Floating-point division can be summarized by the following steps:

    1. subtract exponents (restore bias if exponents are biased)
    2. pre-division alignment (to avoid possible overflow, right shift of the normalized
       dividend to insure dividend < divisor, & check operand for nonzero dividend)

3. divide mantissas (determine quotient sign)
4. check for overflow/underflow

Division is in general not a deterministic operation like addition, subtraction and multiplication due to the intrinsic iterative nature used to compute division. The approach that can be employed is the "pencil-and-paper" approach; ie, via shift and subtract. Either a restoring or non-restoring algorithm can be used [26]. The quotient bits are generated MSB first, in an iterative bit by bit manner until a negative remainder results. While multiplication can be significantly speeded up by various techniques over simple shift and add, division does not share this attribute. In a bit sequential arithmetic environment using shift and add/subtract methods, we now have multiplication and division sharing the same linear rate of solution convergence. Overall system throughput is increased by using the principle of *functional parallelism*, whereby many bit-serial processors are working concurrently.

Floating Point Square Root

Floating-point square root can be summarized as follows:

1. check if exponent is even; if not, increment exponent and right shift mantissa
2. divide exponent by 2 (shift right one)
3. divide mantissas
4. check for overflow/underflow

The square root function is similar in computation to division, though somewhat more complicated. It is also an indeterministic iterative calculation, with result bit generated MSB first. To perform division and square root requires modifications of the bit-serial hardware. Table A shows the semi-on-line recurrence for division and square root [21]. Note that the recurrence remainder R_j is shifted left relative to the subtracted divisor and requires a full addition and subtraction for the non-restoring algorithm and only subtraction for the restoring one. Using a modified version of the Gnanasekaran [11] fast bit-serial/parallel multiplier, the semi-on-line algorithms of Table A can be computed. Figure 5 shows a block diagram of the alterations needed for division.

An Inner Product Computer for Matrix Based Signal Processing

Much of the primary computational requirements of sophisticated real-time signal processing techniques can be reduced to a set of basic matrix operations [22], which would include matrix-vector multiplication, matrix-matrix multiplication and addition, matrix inversion, solution of linear equations, least squares solution of linear systems, eigensystem solution, and singular value decomposition. The first five operations are computed deterministically, and there exist a variety of systolic and wavefront cellular architectures [14]. The latter two operations and their generalized counterparts are iterative in nature and are currently under intensive research. Characteristics to the systolic and wavefront processors include the 5 presented arithmetic functions and the inner product calculation. Thus one of the goals of this VLSI floating-point digital signal processor was to provide the functionality to perform these 5 fundamental computations and the inner product calculation in particular, and to provide this functionality within the tenets of VLSI bit-serial systolic design principles. Then this processor would act as an inner product element and many of these elements would be arranged in linear or mesh arrays to perform the matrix based operations.

A VLSI Floating-point Bit-serial FFT Processor

In addition to performing basic arithmetic for matrix based operations, another goal was the efficient computation of the FFT. Thus in Figure 1, an FFT and IFFT adder/subtractor matrix for the special purpose computation of FFT/IFFT butterfly is included (the bypass bus is used when the FFT/IFFT operation is not desired). Our design approach is modular, thus either a standard radix-2 Cooley-Tukey Decimation-In-Time (DIT) butterfly, or our modified radix-4 Cooley-Tukey DIT butterfly which computes all power of 2 FFT/IFFT's sizes and not just those that are powers of 4 can be configured. This modified FFT algorithm is further explained in [16],[17] and shown in Figure 7. We selected these 2 algorithms for the Fourier Transform operation, since they are very well suited to the multiply/accumulate inner product processor operation, and hence amenable for our floating-point implementation approach. Other researchers have concurred with us in the assessment that the Cooley-Tukey FFT class of algorithms are in several aspects ideal for VLSI implementation, when compared with the Winograd and Prime Factor Fourier Transform algorithms [3],[8],[25]. This new architecture exploits the area, time, and throughput efficiencies of bit-serial signal processing, and the speed and I/O convenience of bit-parallel for interfacing with conventional commercial chip products. Our previous experience in custom VLSI chips for VLSI-DSP [16],[17] have demonstrated the efficacy of this approach. This novel VLSI floating-point digital signal processing architecture is an evolution of our previous work and is a solution to the high performance needs of sophisticated modern signal processing algorithms by providing the increased precision and dynamic range of floating-point arithmetic.

DESIGN AND PERFORMANCE CONSIDERATIONS

Numerical Performance

The numerical performance of our architecture, configured for the modified radix-4 FFT/IFFT operation, is shown in Figures 8 and 9. The former figure shows the dynamic range performance of single precision (32 bits) floating-point as compared to the 20 bit block floating point (24 bits) used in CUSP [16], and the 22 bit floating-point TRW format of Eldon [9]. Figure 9 shows our simulation results for the signal to noise numerical precision performance of our processor design when performing 32 bit floating-point FFT/IFFT on random white noise input with no DC offset, as compared with the performance of CUSP with its 20 bit block floating-point and with the performance of a commercial 16 bit fixed-point multichip system. Dynamic scaling is needed to accommodate bit growth in the fixed-point and block floating-point cases.

There are several advantages of floating-point arithmetic for signal processing. The primary benefit is clearly the increased dynamic range for the same number of bits. By grouping bits into a mantissa and an exponent field, a much wider dynamic range is achieved. Scaling is now handled automatically and no longer off loaded as a problem for a host computer. Furthermore, with a normalized mantissa field, the maximum number of significand bits is retained. Fixed-point and block floating-point arithmetic have been adequate in most past signal processing applications where there has been small to moderate computation-induced word growth. With short one dimensional Fourier transform lengths and input A/D conversions of 8 to 12 bits, 16 bits of precision have been adequate for many signal processing tasks. But with the advent of multi-dimensional signal processing and sophisticated matrix based signal processing, floating-point arithmetic is mandatory. In some high precision intensive calculations, double precision floating-point is required. Thus with VLSI technology, hardware intensive floating-point solutions to signal processing tasks are now cost effective.

Design and Layout Considerations

It can be seen from Figure 1 that a monolithic floating-point digital signal processor of this complexity requires a fully custom design to include all the arithmetic and control capability to function autonomously and to optimize the performance. Previous fully custom VLSI chip design efforts strongly demonstrate the need for a minimum of modular and regular cells that can be easily abutted together. By including data and control paths as part of the cells, the onerous task of global chip interconnection is minimized. It is possible to include bussing segments as part of individual cells so that when they are abutted, the interconnects are continuous, easing the connection problem. However, this is not a sufficient layout criterion since the global signal broadcasting problem is not solved. Long control and communication paths with multiple loads are susceptible to crosstalk noise and require hefty drivers or driving stages. Systolic control and data paths pipelined at the bit-level and included within the bit cell definition ameliorate both the global interconnect and signal broadcast problems. Having simple bit cells that are regular and modular greatly facilitates the ease of their design, placement and the expandability of internal word lengths. The need for the linear shift delay registers and the multiplier bit slices, both with bit lengths on the order of the operand bit sizes require "layout folding" into rectangular serpentine cell abutments reminiscent of *M. C. Escher tessellations*, as illustrated in Figure 10.

Figures of Merit

The successful implementation of high level signal processing algorithms and architectures in physical structures requires careful assessments of all the various trade-offs that go into a VLSI design. The minimization of computations such as the number of multiplies, which traditionally have been the criteria used to judge the merits of an algorithm, no longer is the pre-eminent standard to be used. For example, the preference of the FFT over the DFT algorithm was primarily due to the greater computational efficiency of the FFT. Subsequent to Cooley and Tukey's FFT, there have been several proposed Fourier algorithms that, based solely upon computational performance (minimized number of multiplications) are computationally more efficient. The Winograd and Prime Factor Fourier Transforms are two such algorithms. But VLSI design tenets dictate the importance of modularity, regularity, expandability and the systolic distribution of control signals and data flow. Extending these ideas to bit level computation with its simple processing structure and hence substantial capacity for massive parallelism, embodies the concepts of *functional parallelism* for VLSI bit-serial signal processing. With bit sequential processing, the computation is inherently pipelined with delay registers at a bit level. Thus memory is distributed as well as the arithmetic. In terms of computational efficiency, the Cooley-Tukey FFT is superior to the DFT, though not as efficient as the other aforementioned Fourier algorithms. However, the Cooley-Tukey FFT is very modular and regular as compared with the Winograd Fourier Transform Algorithm and is thus more suited to VLSI layout. It would appear that due to the global rather than local recursiveness of the FFT algorithm, it would not be amenable for systolic interconnection of FFT computational units (butterflies). Various schemes have been proposed for VLSI FFT implementation with some form of commutator network for the intercommunication of arrays of FFT butterflies. However, a systolic version of the FFT is possible when the concepts of bit-serial architectures are utilized in its construction [6].

Various "Figures of Merit" (FOM's) have been proposed for the comparative evaluation of VLSI designs. Two widely used FOM's include Thompson's *area-time* metrics and the VHSIC program's *gates-Hz/cm²*. The former is a theoretical device for lower and upper bounding the area and computation time of an algorithm implemented in VLSI, while the latter attempts to measure the per unit area logic density and speed of a VLSI chip (a high number here is desirable). Both FOM's allow trade-offs of one parameter versus another, as long as their composite is within a given threshold. The first FOM in particular is useful in establishing a theoretical optimal bound for the VLSI implementation and computation of the Fourier Transform. Thus alternative VLSI FFT processors can be related to this area-time bound to judge their optimality on a theoretical basis. Thompson has shown that the AT^2 lower bound for the VLSI computation of the Fourier Transform is area x time^2 = $\Omega(N^2 \log^2 N)$. Using systolic and bit sequential (radix-2) computational and data flow structures, a radix-2 Decimation-in-Frequency (DIF) version of our floating-point FFT approach, CUFLP, is within the area-time FFT bound as defined by Thompson as shown in [6]. Table B shows a comparison of some VLSI Fourier Transform implementations we have investigated. The second widely used FOM is more a physical rather than theoretical means of comparing alternative VLSI implementations. This metric is obtained from a gate count, operational clock speed and die size of a realized VLSI design.

Recently, Ward *et al* [25] have proposed a new FOM utilizing a cost function which characterizes an algorithm implemented in VLSI and its associated logic design, independent of the circuit technology employed. Configurations which have higher throughput by utilizing proportionally greater amounts of hardware or greater logic speeds are assessed as having the same cost. As an example, Ward found that when this costing FOM approach is applied to the FFT and its computationally more efficient counterpart, the Winograd Fourier Transform Algorithm (WFTA), though the WFTA has fewer arithmetic computations, it is less efficient than the FFT except in a fully parallel situation. These results take into account both control cost as well as arithmetic costs, and hence suggest that the irregular nature of the WFTA obtains arithmetic savings at the expense of greater control complexity costs. This recent finding, in addition to the area-time metric results, reaffirms our assessment that the FFT from an overall perspective has the characteristics that are optimal for VLSI implementation of the Fourier Transform.

CONCLUSIONS

We have presented the architecture of a novel VLSI bit-serial floating-point signal processor that embodies the systolic bit sequential concepts of *functional parallelism*. To meet the needs of signal processing applications which require broad dynamic range and maximum significand precision such as those in multi-dimensional and matrix based signal processing, floating-point computations will be necessary, and VLSI technology has made it cost effective. We are convinced that systolic bit sequential VLSI signal processing, which offers major advantages in terms of signal communication, modularity, regularity and expandability, is a natural candidate for extension to a floating-point environment. Undoubtedly, these floating-point concepts will eventually be incorporated into silicon compilers, which have been employed for the implementation of bit-serial VLSI signal processors [8], enabling the efficient and fast generation of special purpose floating-point signal processors.

REFERENCES

[1] J. Allen, "Computer Architecture for Digital Signal Processing", Proceedings of the IEEE, Vol. 73, No. 5, pp. 852-873 (May 1985).

[2] M. R. Buric and C. A. Mead, "Bit-Serial Inner Product Processors in VLSI", *Caltech Conference on VLSI Proceedings*, pp. 155-163 (January 1981).

[3] C. S. Burrus, "Computation of the Discrete Fourier Transform", *Trends and Perspectives in Signal Processing*, Vol. 2, No. 2, pp. 1-4 (1982).

[4] D. Bursky, "1986 ISSCC: Digital Chips", Electronic Design, pp. 101-153, (Feb. 20,1986).

[5] P. R. Cappello, VLSI Signal Processing, New York, NY: IEEE Press (1984).

[6] P. M. Chau and W. H. Ku, "VLSI Implementation of the Fast Fourier Transform", *SPIE Real-Time Signal Processing*, Vol. 698, (Aug. 18-22, 1986).

[7] J. T. Coonen, "An Implementation Guide to a Proposed Standard for Floating-point Arithmetic", Computer Magazine, pp. 68-79, (Jan. 1980).

[8] P. B. Denyer, and D. Renshaw, VLSI Signal Processing: A Bit-Serial Approach, Reading, MA: Addison-Wesley, (1985).

[9] J. A. Eldon, "A Floating-Point Format for Signal Processing", *ICASSP Proceedings 1982*, pp. 717-720 (1982).

[10] M. D. Ercegovac, "On-Line Arithmetic: An Overview", *SPIE Real Time Signal Processing*, Vol. 495, pp. 86-93, (1984).

[11] R. Gnanasekaran, "A Fast Serial-Parallel Binary Multiplier", IEEE Transactions on Computers, Vol. C-34, No. 8, pp. 741-744, August 1985.

[12] W. Hays, *et al*, "A 32-bit VLSI Digital Signal Processor", IEEE Journal of Solid State Circuits, pp. 998-1004 (October 1985).

[13] H. T. Kung, B. Sproull, and G. Steele, VLSI Systems and Computations, Rockville, MD: Computer Science Press, (1981).

[14] S. Y. Kung, "VLSI Array Processors", IEEE ASSP Magazine, v2, pp. 4-22 (July 1985).

[15] S. Y. Kung, H.J. Whitehouse, and T. Kailath, VLSI and Modern Signal Processing, Englewood Cliffs, NJ: Prentice Hall, (1982).

[16] R. W. Linderman, P. M. Chau, W. H. Ku, and P.P. Reusens, "CUSP: A 2-micron CMOS Digital Signal Processor", IEEE Journal of Solid State Circuits, pp. 761-769 (June 1985).

[17] R. W. Linderman, P. P. Reusens, P.M. Chau, and W.H. Ku, "Digital Signal Processing Capabilities of CUSP, A High Performance Bit-Serial VLSI Processor", *1984 ICASSP*, pp. 1611-1614 (Mar. 1984).

[18] R. F. Lyons, "A Bit-Serial VLSI Architecture Methodology for Signal Processing", VLSI 81, Academic Press, pp. 131-140, (1981).

[19] R. M. Owens, "Techniques to Reduce the Inherent Limitations of Fully Digit On-Line Arithmetic", IEEE Transactions on Computers, Vol. C-32, #4, pp. 406-410, April 1981.

[20] N. R. Powell, and J M. Irwin, "Signal Processing with Bit-Serial Word-Parallel Architectures", *SPIE Real-Time Signal Processing*, Vol. 154, pp. 98-104 (1978).

[21] H. J. Sips,"Bit-Sequential Arithmetic for Parallel Processors", IEEE Transactions on Computers, Vol. C-33, #1, pp. 7-20 (January 1984).

[22] J. M. Speiser and H. J. Whitehouse, "A Review of Signal Processing with Systolic Arrays", *SPIE Real-Time Signal Processing*, Vol. 431, pp. 2-6, (Aug. 23-25,1983).

[23] P. H. Sterbenz, Floating-Point Computations, Englewood Cliffs, NJ: Prentice Hall, (1974).

[24] E. E. Swartzlander, *et al*, "Inner Product Computers", IEEE Transactions on Computers, Vol. C-27, pp. 21-31 (January 1978).

[25] J. S. Ward, *et al*, "Figures of Merit for VLSI Implementation of Digital Signal Processing Algorithm", IEE Proceedings, Vol 131, part F, No. 1, pp. 64-70, (Feb. 1984).

[26] S. Waser and M. J. Flynn, Introduction to Arithmetic for Digital Systems Designers, New York: Holt, Rinehart, and Winston, (1982).

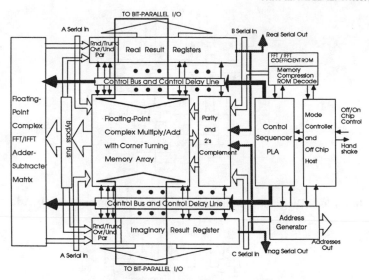

Figure 1: VLSI Floating-Point Digital Signal Processor

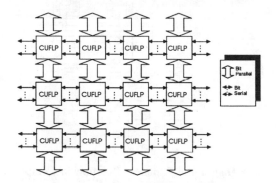

Figure 2: Bit-Parallel/Bit-Serial Systolic Array

Figure 3(a): Data Word Format

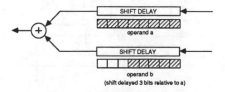

Figure 3(b): Relative Serial Shift

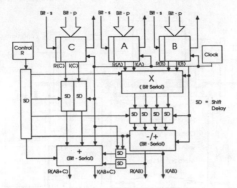

Figure 4(a): Block Diagram of Mantissa Dataflow in Complex Multiply/Add

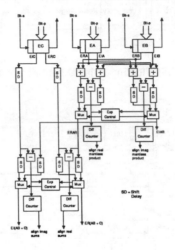

Figure 4(b): Block Diagram of Exponent Dataflow in Complex Multiply/Add

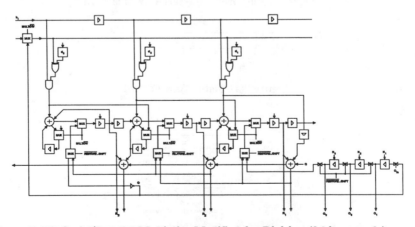

Figure 5: Bit-Serial/Parallel Multiplier Modified for Division (3-bit example)

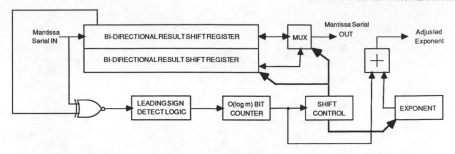

Figure 6(a): Block Diagram of Normalization Count Logic

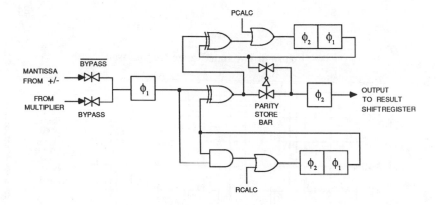

Figure 6(b): Parity/Rounding Logic Diagram

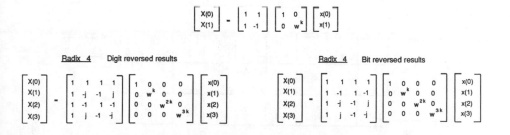

**Figure 7: Radix-2, Standard (Digit-Reversed) Radix-4 FFT
and Modified (Bit-Reversed) Radix-4 FFT**

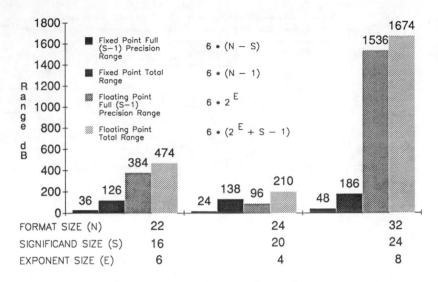

Figure 8: Dynamic Range Comparison

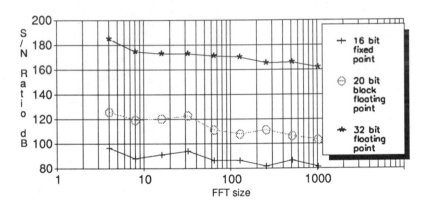

Figure 9: FFT/IFFT Numerical Performance Simulations

				Cell 3	Cell 2	Cell 1	Cell 0	
			Cell i-1	Cell i	Cell i+1			
Cell N-1	Cell N-2	Cell N-3	Cell N-4					

Figure 10: Rectangular Serpentine Tessallation Layout of Bit-Serial Cells

Square Root: $Q = \sqrt{A}$		Division: $Q = N/D$	
Step 1 [initialize]	$j = 0, R_0 = 2^{-1} \cdot A, q_0 = 1$	Step 1 [initialize]	$R_0 = 2^{-1} \cdot N, j = 0, q_0 = 1$
Step 2 [recursion] {non-restoring}	$D_j = \sum_{v=1}^{j} q_v \cdot 2^{-v} + 2^{-j-2} + \bar{q}_j \cdot 2^{-j-1}$ if $q_j = 1$ then $R_{j+1} = 2 \cdot R_j - D_j$ if $q_j = 0$ then $R_{j+1} = 2 \cdot R_j + D_j$	Step 2 [recursion] {non-restoring}	if $q_j = 0$ do $R_{j+1} = 2 \cdot R_j + D$ if $q_j = 1$ do $R_{j+1} = 2 \cdot R_j - D$
Step 3 [test]	if $R_{j+1} < 0$ then $q_{j+1} = 0$ if $R_{j+1} \geq 0$ then $q_{j+1} = 1$ if $j < m - 1$ then $j = j+1$ GOTO Step 2.	Step 3 [test]	if $R_{j+1} < 0$ then $q_{j+1} = 0$ if $R_{j+1} \geq 0$ then $q_{j+1} = 1$ if $j < m - 1$ then $j = j + 1$ GOTO Step 2
(Step2 [recursion]) {restoring}	$R_{j+1} = 2 \cdot R_j - q_{j+1} \cdot (D_j - \bar{q}_j \cdot 2^{-j-1})$	(Step 2 [recursion]) {restoring}	$R_{j+1} = 2 \cdot R_j - q_{j+1} \cdot D$

Table A: Bit-Sequential Semi-On-Line Algorithms

ELEMENT	AREA	TIME	AREA X TIME
AxT Optimal Mesh Processor	$N\log N$	$(N\log N)^{1/2}$	$(N\log N)^{3/2}$
Near AxT Optimal Mesh of Radix-4 CUSP units	$N\log^2 N$	$(N\log N)^{1/2}$	$N^{3/2}\log^{5/2} N$
PFA Array Element	$N\log N$	$N^{2/3}$	$N^{5/3}\log N$
CUSP Array Element	$N\log N$	$N\log N$	$(N\log N)^2$
CUFLP (Radix-2 DIF FFT Systolic Mesh Processor)	(Maximum Precision) $N\log N$	$(N\log N)^{1/2}$	$(N\log N)^{3/2}$
	(Constant Precision) $N\log(\log N)$	$[N\log(\log N)]^{1/2}$	$[N\log(\log N)]^{3/2}$

Table B: Area x Time Complexity of Alternative Architectures

CHAPTER 28

SYSTOLIC/CELLULAR PROCESSOR FOR LINEAR ALGEBRAIC OPERATIONS[1]

J. Greg Nash, K. Wojtek Przytula and S. Hansen

Hughes Research Labs
3011 Malibu Canyon Road
Malibu, California 90265

1.0 INTRODUCTION

One of the principal problems encountered in designing concurrent computing systems is that of providing a sufficiently general range of capabilities without undue addition of hardware and software complexity. Given any particular structured signal processing algorithm, it is usually possible to map it into some efficient concurrent processor implementation. However, in doing this one pays a price in the loss of generality. What is needed is a signal processor design which is sufficiently general that it can service signal processing needs across a wide variety of applications in present and future, unspecified systems. Our approach to doing this for image and signal processing is based on recognition of the fact that important operations often involve cellular and linear algebraic manipulations of data, as for example in the solution of various types of linear systems[1-3].

Our systolic/cellular linear algebraic processor (LAP) is intended for use in numeric signal and image processing applications. The LAP, which is based on a 16 x 16 array of PEs, provides all the linear algebraic capabilities shown in Table 1. This is efficiently achieved by using a regular, simple, easy to implement nearest neighbor interconnection scheme, with identical PEs.

Table 1.   Linear Algebraic Algorithms Capable of Being Executed on the Faddeev Based Systolic/Cellular Array

> Inner and Outer Products
> Triangular System Solution
> Matrix manipulation (multiplication, inverse, addition, transpose)
> Matrix factorization (LU and QR)
> Linear systems solution
> Least squares problems (e.g., under/over determined, constrained, generalized)
> Banded matrix system solution
> Eigenvalue Decomposition
> Singular Value Decomposition

---

1. Supported in part by NSF Grants ECS 8213358 and ECS 8016581 and Office of Naval Research Contract N00014-81-K-0191.

Our system design is based on just two algorithms, which greatly simplifies both hardware design and software coding. The Faddeev[4] algorithm offers a systematic means for achieving all operations, up to and including least squares solutions (full rank case) on the list in Table 1. The other is a singular value decomposition algorithm[5], which provides an approach for obtaining the last two operations on the list. Because both of these algorithms have a similar underlying structure, one unified architecture, meeting the criteria of the previous paragraph, can be used between them. We can also partition the Faddeev based problems in a numerically stable way for matrices of arbitrary size[6].

An example of an application we have been pursuing is that of robotics control. Kinematics of a robot arm is conveniently described in terms of translations and rotations in a three-dimensional Cartesian space. However, the arm is controlled by signals applied to its joint actuators. These signals are associated with joint translations and rotations. The Cartesian translation and rotation rates, which are used to express the task of control have to be converted into joint rates to determine the control signals. The relation between Cartesian and joint rates is expressed by means of a Jacobian matrix:

$$\begin{bmatrix} u \\ w \end{bmatrix} = J * \overset{\bullet}{q}$$

where v and w denote, respectively, translational and rotational Cartesian rates, $\overset{\bullet}{q}$ represents joint coordinate rates, and J is the Jacobian. Given the Cartesian rates and the Jacobian matrix, we want to compute in real-time the joint rates for the arm by obtaining the pseudo inverse of J, which is nx6, n being the number of joints. We have extensively simulated this problem on the LAP using the internal 32-bit fixed point number representation. (The 32-bit representation presently limits usage of our system to those computations where dynamic range requirements are not high.)

2.0 FADDEEV ALGORITHM

The basic algorithm is to find **CX+D** given **C, D** and equation **AX=B**. where **A** is of full rank. If the data is arranged as

$$\begin{array}{c|c} A & B \\ \hline -C & D \end{array}$$

and if a suitable linear combinations of the rows above the line (from **A** and **B**) are added to the row beneath the line (e.g., **-C+WA** and **D+WB**, where **W** specifies the appropriate linear combination), so that only zeroes appear in the lower left hand quadrant, then the desired result, **CX+D**, will appear in lower right hand quadrant. This follows because the annulment of the lower left hand quadrant requires that **W** = CA^{-1}, so that

$D+WB = D + CA^{-1}B$. Since $X = A^{-1}B$, we have the final result $D+WB = D + CX$, which appears in the lower right hand quadrant. Numerous matrix operations, such as multiplication, addition, and linear system solution are possible by selection of appropriate entries for A, B, C and D. The Faddeev algorithm is thus programmable by positioning the matrices in the appropriate quadrant before calculations begin. The simplicity of the algorithm is due to the absence of a necessity to actually identify multipliers of the rows of A and the elements of B (i.e., W doesn't have to be computed); it is only necessary to annul C. This can be done by ordinary Gaussian elimination.

One of the more important features of this algorithm is that it avoids the requirement for back substitution or solution to the triangular linear system. Instead, the values of the unknowns are obtained directly at the end of the forward course (elimination) of the computation.

For general matrix inputs Gaussian elimination doesn't guarantee numerical stability and fails for zero pivot elements. Pivoting techniques help, but they are difficult to implement on a mesh connected array because non-local communication is involved. "Growth" in element values is another negative by-product of this procedure. For these reasons we have modified the Faddeev algorithm to use a QR orthogonal factorization technique (Givens rotations) which avoids these difficulties. This technique also triangularizes matrices without changing column norms, a procedure necessary for solving over- and underdetermined systems of equations, i.e., least squares problems. It is necessary to divide the process of annulling the lower left hand quadrant into a two step procedure[7,8]. As shown in Figure 1, A is first triangularized by Givens rotations (simultaneously applied to B); after this is completed the remainder of the process can be accomplished by Gaussian elimination using the diagonal elements of B as pivot elements. It is important to

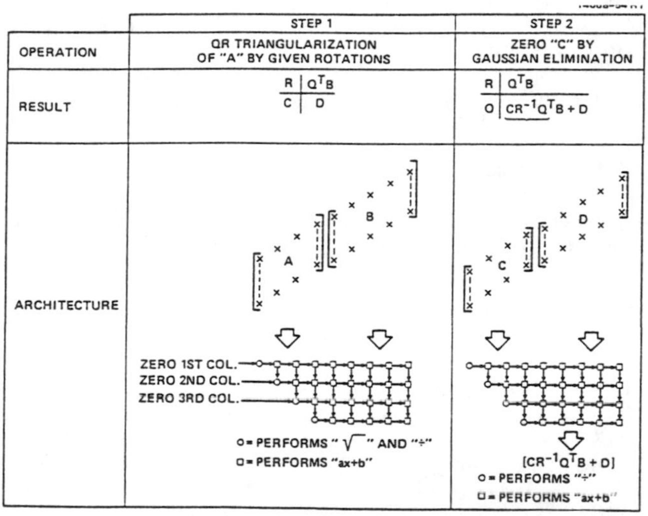

Figure 1. Modified Faddeev Algorithm.

note here that there are no restrictions on the coefficient matrix **A** other than it be full rank. This implies that the pivot elements along the diagonal must always be non-zero, insuring stability of the elimination procedure.

3.0 ALGORITHM IMPLEMENTATION.

In this section we describe a systolic implementation of the modified Faddeev algorithm. As detailed above, the first step required is a QR factorization of the **A** matrix, where **Q** is an orthogonal matrix, and **R** is an upper triangular matrix. This can be done using a triangular array of processing elements and passing the **A** matrix down through the array as shown in Figure 1. As can be seen the data is skewed in such a way that it is not necessary to broadcast the sin and cosine values along rows of processing elements (PEs). The purpose of the circular processors is to perform "rotations" on columns or vectors in such a way that zeroes are introduced (corresponding to alignment of components of the vector along a major axis) for all elements a_{ij}, $i>j$, in the jth column. In this way the resulting **R** matrix is upper triangular and is left stored as the "r" elements in the PEs of Figure 1. The rectangular extension of the triangular array will then contain Q^tB.

The second step in the modified Faddeev algorithm could be accomplished as shown in Figure 1. Here **C** and **D** are also passed down through the array of processing elements in a similar way. In this case the set of operations performed in each PE is slightly different. The PEs indicated by the circles each zero one column of **C** by pivoting on the diagonal elements of **R**. In this case after $O(n)$ time steps the result $CR^{-1}Q^tB+D$ will appear row by row coming out of the array at the bottom right.

The triangular structure of Figure 1 can be easily transformed into a square organization shown in Figure 2, which is the basis

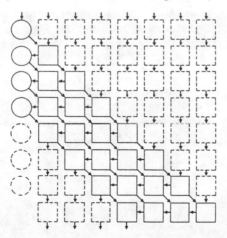

Figure 2.   Alternative square PE array.
Here the "dashed" boxes
correspond to delay elements.

for our systolic/cellular system design. The data flow will not
be entirely uniform for this case so that some additional
control capabilities will be required. The advantage of
arranging PEs in this fashion is two-fold. First, problems with
varying array sizes could be accommodated. Secondly, with a
square array of PEs other cellular processing options are
available. For example data in two matrices could be processed
in block format, as might be required in the simple addition of
two matrices.

4.0 PE DESIGN

The primary motivation for the designs discussed here is for an
area-time efficient arithmetic[9]. Presently, typical parallel
and serial/parallel multipliers have area-time products that are
$O(n^3)$, where n is the number of bits per operand. Our goal has
been to design circuits with $O(n^2)$ area-time products. This has
led to recursive algorithms which can be implemented in
serial/parallel structures based on carry-save full adder cells.
The carry-save approach eliminates the need for carry
propagation resulting in $O(n^2)$ area-time product.

The PE we have designed, shown in Figure 3, can be used to
implement 1-D and 2-D systolic arrays. This 32-bit fixed point
chip, based on an earlier version actually used in a linear
systolic array[10], was built in a bit slice fashion for ease of

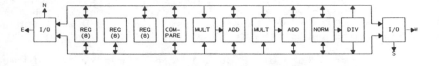

Figure 3. Organization and layout of processing element.

reconfiguration. The speed of this chip has been enhanced by a number of features. First, it uses dual buses so that two operands can be transferred per clock cycle. All arithmetic algorithms have the necessary control embedded in hardware and all arithmetic is done in a serial/parallel fashion for area-time efficiency. This adds considerable speed over a microcoded approach and eases the chip controller design complexity. In addition it has two multipliers, two adders, and a divider, which can all operate concurrently. The multiplier and divider run on their own set of high speed clocks that are separate, but in synchronism with, the slower system clocks.

The on-chip functional units, shown in Figure 3, also include 8 dual port registers, two sets of 8 single port registers, and a comparator. The normalizer circuit, which pre-scales divider operands, also has a parallel shifter associated with it.

Various communication options are available via the 64 bits of bi-directional I/O lines. The PE can be configured as having two 32-bit I/O ports or four multiplexed 16-bit ports for North, East, South and West communication in a two dimensional array. Control over the I/O ports is separate from the control of the data path itself, so that communication between PEs can be occurring during computational activity. Each I/O channel has associated with it two input registers and two output registers.

Simulations show that the high speed clocks should run at a maximum of 32 MHz and the system clocks at 8 MHz. Typically, the high speed clocks are expected to run four to eight times faster than the system clock. The systolic chip was built in 3μ NMOS, it is approximately 290 x 235 mils^2, and power dissipation is expected to be approximately 900 mW. It contains approximately 27,000 devices. Fixed point 32-bit multiplication and division times are estimated to be 750 nsec and 1 μsec. Data transfer rates of up to 32 MBytes/sec are possible with the chip.

5.0 SYSTEM ARCHITECTURE AND PROGRAMMING ENVIRONMENT

The computer system consists of a single-processor host machine and systolic/cellular back-end system as shown in Figure 4. The back-end system includes an array of processors, dual port data memory, program memory and a controller. Input data are loaded into the array memory formatted as data blocks. During a computation, the data blocks leave the memory row by row through one port, then "flow" through the array undergoing appropriate processing, with results returning to the data memory through the other port. The algorithms for the array are structured in such a way that they access data sequentially by rows. The addressing scheme for the dual port memory of the array is therefore based on an address queue which contains the starting addresses of the data blocks. Thus, there is no address field is in the machine instructions. When processing in the array requires access to a new block, a change-of-address instruction is executed which results in popping of a new starting address from the queue.

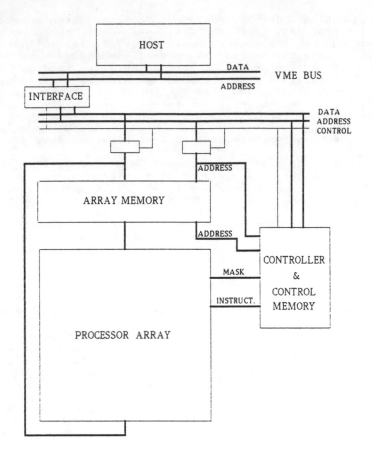

Figure 4.   System Block Diagram.

The systolic/cellular array is controlled by a simple SIMD controller. The algorithms for the array are well structured and are characterized by a deterministic control flow. Thus, as in the case of data access, the access to the instructions of the program (i.e. control flow) can be governed by a queue. The controller is a simple sequencer consisting of a program memory, in which an object code is stored, and a control queue containing destination addresses for the jump instructions of the program. All of the jumps are unconditional and are executed by popping of the next address from the control queue into the program counter.

An instruction set of the array consists of about 30 powerful, wide (approx. 100 bits) instructions. They perform arithmetic operations such as add, multiply, divide, as well as data communication operations such as memory read/write, local and nearest neighbor communication. The contents of address queue and control queue constitute part of machine level program. A mask field in the instruction allows disabling of selected processors during a given cycle. The mask encoding technique is powerful enough to implement all the masks of interest at expense of relatively small number of bits.

In the design of the programming environment the following assumptions have been made. Each program will be used many times for various sets of data. There are several special programs used most frequently (e.g. program for Faddeev algorithm or SVD algorithm) which should be available as packaged routines. In light of these assumptions, an extended assembly language has been adopted as the programming language for the array. The language gives the programmer a lot of control but provides at the same time some high level language features, which simplify the programming task.

During an execution of a computation task on the system, the following steps are performed under the control of the host. The input data is formated according to the requirements of the algorithm and the array, and is loaded into appropriate locations of the memory of the array. Part of the memory is reserved for the partial and final results. The starting addresses of the input and output data blocks are loaded into the address queue in the order in which they are used by the program of the array. The machine code is loaded into the program memory and jump addresses, appropriately ordered, are loaded into the control queue.

6.0 SYSTEM IMPLEMENTATION

The overall system organization illustrated in Figure 4 is configured physically as shown in Figure 5. It consists of a PE array unit and a Motorola 68010 based host system. The dual port RAM is used to communicate between these two basic components. Communication among various board level modules is over the VME bus (IEEE 1064 standard).

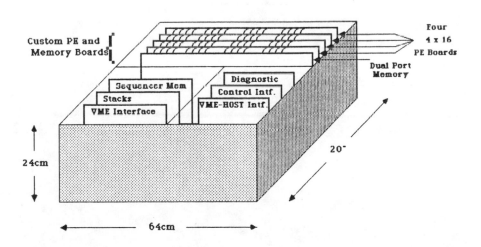

Figure 5. LAP Physical Configuration.

The 16 x 16 PE array is laid out on custom PC boards, each of which contain a 4 x 16 array of PEs. These custom boards are supplied with data by another custom board containing the dual port RAMs. Control of the PE array comes from two smaller PC boards, one containing the program memory that supplies the global instructions to the PE array and the other containing various stacks and clock generator circuitry. The chassis shown in Figure 5 contains a VME bus (lower right hand corner) which has various VME cards, including an interface to the 68010 based host. There is an interface board between the PE array bus on which the controller boards reside and this VME bus.

Maximum system performance will be in the neighborhood of 450 MOPs. This number is calculated based on two 32-bit register to register multiplies and adds in 1.1 μsec. Data transfer rate between the dual port memory and the array is approximately 90 MBytes/sec.

REFERENCES:

[1] H. C. Andrews and B. R. Hunt, Digital Image Restoration, Prentice-Hall, Inc., 1977.

[2] J. M. Speiser and H. J. Whitehouse, "Architectures for Real-Time Matrix Operations," Proc. 1980 GOMAC Conf., Houston, Texas, 19-21 Nov. 1980.

[3] T. Poggio and V. Torre, "Ill-Posed Problems and Regularization Analysis in Early Vision," Proc. Image Understanding Workshop, Oct. 1984, New Orleans, LA, pp.257-263.

[4] J. G. Nash, S. Hansen, and G. R. Nudd, "VLSI Processor Arrays for Matrix Manipulation," in VLSI Systems and Computations, Ed. by H.T. Kung, Bob Sproull and Guy Steel, Computer Science Press, 1981, pp.367-378.

[5] Frank Luk, "A Triangular Processor Array for Computing the Singular Value Decomposition," Cornell Technical Report TR 84-625, July, 1984.

[6] J. G. Nash, S. Hansen, and W. Przytula, "Systolic Partitioned and Banded Linear Algebraic Computations," Proc. SPIE Technical Symposium, San Diego, August, 1986.

[7] J. G. Nash and S. Hansen, "Modified Faddeev Algorithm for Matrix Manipulation," Proc. 1984 SPIE Conf., San Diego, CA. Aug. 1983, pp.39-46.

[8] J. G. Nash and S. Hansen, "Modified Faddeev Algorithm for Concurrent Execution of Linear Algebraic Operations," to be published in IEEE Trans. Computers.

[9] M. D. Ercegovac and J. G. Nash, "An Area-Time Efficient VLSI Design of a Radix-4 Multiplier," Proc. IEEE 1983 Int. Conf. on Computer Design, pp.684-687.

[10] J. G. Nash and C. Petrozolin, "VLSI Implementation of a Linear Systolic Array," Proc. IEEE ICASSP, March 1985, pp.779-884.

A High Performance Cascadable CMOS Transversal Filter

And Its Application To Signal Processing

M. H. Yassaie

INMOS Limited
1000 Aztec West
Almondsbury
Bristol, England BS12 4SQ

Abstract

This paper describes an optimised CMOS device with an architecture suitable for efficient execution of most signal processing algorithms such as correlation, convolution, filtering, and DFT. The device has an effective multiply-accumulation time of around 12ns for 16-bit data and is capable of 80 MOPs. The architecture of the device, and some of the related algorithms are described, together with some contrasts between this approach and conventional DSPs. The resulting ease with which high performance signal processing systems can be implemented is also illustrated, in particular how high throughput can be achieved using one or more of these devices attached to any conventional general purpose microprocessor.

1   Introduction

Many real-time signal processing systems revolve around the optimisation of data flow through a processing unit. The major functional block in this processing unit is usually a multiply-and-accumulator. For maximum efficiency the data flow is often controlled by some form of dedicated address generators and/or sequencers which add considerable complexity to the system. Also, in the limit the system performance is almost invariably dictated by the throughput of the multiply-and-accumulator unit. Conventional approaches for improving system performance have centred around either increasing the speed of the multiply-accumulator, or highly complicated pipeline and bussing architectures. However, these solutions are still ultimately constrained by the throughput of the individual multiply-accumulator(s), ie the classic Von Neumann bottleneck.

Historically, multiplication operation has been considered to be a slow and/or costly process in many processors. This has led to a severe constraint in algorithm development, as most DSP algorithms have been developed around the need for the minimisation of the number of multiplications. The cost of this multiply-optimised algorithms has often been the control complexity, irregualrity in data structures and difficulty in introducing efficient parallelism.

The advent of VLSI technology necessitates a fresh look at signal processing algorithms and their interaction with hardware architectures. Many algorithms which were not practical before can now be exploited in high performance DSP systems. The way forward is to reassess the signal processing problem by first defining the optimal algorithms (interms of memory bandwidth requirement, complexity, parallelism potential and performance) then by considering the state of the art in silicon design, refine both algorithms and hardware architecture together to arrive at the optimum combination.

This paper describes such a solution, offering an optimised architecture suitable for efficient execution of most signal processing algorithms, whilst ensuring that the ultimate performance of a system is limited by cost rather than the capabilities of a single device. The device, and some of the algorithms are described, together with some contrasts between this approach and conventional DSPs. The resulting ease with which high performance signal processing systems can be implemented is also illustrated, in particular how high throughput can be achieved using one or more of these devices attached to any conventional general purpose microprocessor.

2   Device Description

A simplified functional diagram of the device (IMS A100) is shown in Figure-1. The processing core of the device consists of 32 multiply-and-accumulation units. Data can be communicated to this multiply-and-accumulation array via either the on-board microprocessor interface or dedicated ports. Facilities are also incorporated on the chip which allow various operational modes (real/complex processing, coefficient word size, ...etc) to be set up and varied by the host machine. The 32 multiply-and-accumulation units are arranged to form a digital transversal filter. This architecture has been chosen because algorithms exist which map various DSP fuctions onto such a configuration. Figure-2 shows the architecture of the general canonical transversal filter. An alternative realisation which is functionally equivalent is shown in Figure-3. This latter configuartion has been used to implement the IMS A100 signal processor. The input sequence for the IMS A100 is fed in parallel to all 32 stages. At each stage the current input sample is multiplied by a coefficient stored in memory, and added to the output of the previous stage delayed by one processing cycle. The filter output at time $t = kT$ is therefore given by:

$$y(kT) = C(0)x(kT) + C(1)x((k-1)T) + \ldots\ldots + C(N-1)x((k-N+1)T)$$

$$y(kT) = \sum_{i=0}^{N-1} C(i)x((k-i)T) \tag{1}$$

where $x(kT)$ represents the kth input data sample, and $C(0)$ to $C(N-1)$ are the coefficients for the N stages. For the IMS A100 the input data word length is 16 bits. The coefficient word length can be programmed to be 4, 8, 12, or 16 bits. If the coefficient word length is L_C then the maximum data throughput is $\frac{40}{L_C}$ MHz. Thus for 4-bit coefficients the maximum data throughput is 10 MHz, and for 16-bit coefficients it is 2.5 MHz. The effective multiply and accumulation time (per chip) therefore lies between 3 to 12ns. To preserve complete numerical accuracy, no truncation or rounding is performed on the partial products in the multiply-accumulate array. The output of this array is calculated to a precision of 36 bits, which is sufficient to ensure that no overflow occurs. A programmable barrel shifter is located at the output of the array, which allows one of four 24 bit fields (starting at bits 7, 11, 15, or 20) within the 36 bit result to be selected for output. The selected 24 bits are always correctly rounded and are sign extended as required.

Two complete sets of coefficient memories are provided. At any instant one set of coefficients is applied to the multiply-accumulation array, whilst the other set can be accessed via a standard memory interface (capable of a 50nS access time, 100nS cycle time). Once a new set of coefficients has been loaded, the function of the two memory sets can be interchanged by invoking the "Bank Swap" facility. This facility

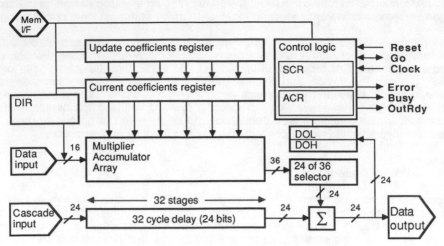

Figure-1- The block diagram of the device.

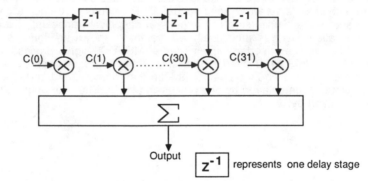

Figure 2 - Canonical transversal filter architecture.

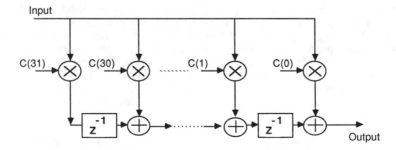

Figure 3- Modified transversal filter architecture.

is useful in adaptive filtering applications. Another feature of the device which can be activated via a control register is the "continuous Bank Swap" mode. When this mode is selected the function of the two coefficient banks will automatically be interchanged every processing cycle, allowing alternate coefficient sets to be used during successive computation cycles. This feature is particularly useful for complex data processing. To allow cascading of devices a 32 stage, 24 bit wide, shift register and a 24 bit adder are included on the chip. To cascade devices, the output of one chip is connected directly to the cascaded input of the next and same input data is applied to all cascaded devices.

Using the on-chip memory interface, the IMS A100 appears to the user as a 128 16-bit word area of fast static RAM. Of this address space, the bottom 64 words contain the two sets of coefficient registers. The remaining locations contain the control and status registers, and the data input and output registers. Input data is supplied either from the external data port or from the data input register, **DIR**, accessed via the memory interface. All 24 bits of the output can be accessed either from the external data output port, or from the two 16 bit data output registers, **DOH** and **DOL**. **DOL** contains the least significant 16 bits correctly rounded, and **DOH** contains the most significant 8 bits, which are sign extended to 16 bits. In addition to the normal memory interface signals, control is provided for functions such as interrupting the host on error, and synchronisation with other system components.

3 Algorithms

Many signal processing algorithms can be mapped onto the architecture of a transversal filter. This section briefly reviews some of these algorithms including a DFT algorithm based on Rader's Prime Number Transform-PNT [1,2 & 3]. Related index mapping techniques which involve generalised decompositions and allow calculation of long DFT sequences are also summarized. The suitability of the device for high speed correlation, convolution, digital filtering (FIR & IIR), matrix multiplication and 2-D image convolutions is also demonstrated.

3.1 A Fast DFT Algorithm For IMS A100

The algorithm described here is based around the Rader's prime radix transform and a decomposition techniques which allows partitioning of large DFT's into several shorter ones. The prime-radix algorithm has its origin in number theory and consists of three seperate operations. The first is a permutation (re-ordering) of the input data. The second operation is convolution of the permuted input data with permuted discrete cosine and sine samples. The third operation is a repermutation, which yields the DFT components in the conventional order of linear frequency. The standard DFT equation,

$$X(k) = \sum_{n=0}^{N-1} x(n) \, exp \frac{-2\pi jnk}{N} = \sum_{n=0}^{N-1} x(n) W_N^{nk} \qquad k = 0, 1, \ldots N - 1, \tag{2}$$

can be converted to a convolution between $x(n)$ and the twiddle factors W_N's. For cases where N is prime, number theory can be used to achieve this. The DFT equation can be rewritten as:

$$X(0) = \sum_{n=0}^{N-1} x(n) \tag{3a}$$

$$X(k) - x(0) = \sum_{n=1}^{N-1} x(n) W_N^{nk} \qquad k = 1, \ldots N - 1 \tag{3b}$$

i.e the expression for the zero-frequency DFT component $X(0)$ (the d.c. term) is separated. The expression for this term consists of N additions only and can be calculated directly (by the host). The bulk of the processing is, however, associated with equation (3b). To briefly demonstrate the principle of this algorithm let us consider an example for a DFT of length 7. Equation (3b) can be expressed, in matrix form, by:

$$
\begin{bmatrix}
X(1) - x(0) \\
X(2) - x(0) \\
X(3) - x(0) \\
X(4) - x(0) \\
X(5) - x(0) \\
X(6) - x(0)
\end{bmatrix}
=
\begin{bmatrix}
W^1 & W^2 & W^3 & W^4 & W^5 & W^6 \\
W^2 & W^4 & W^6 & W^1 & W^3 & W^5 \\
W^3 & W^6 & W^2 & W^5 & W^1 & W^4 \\
W^4 & W^1 & W^5 & W^2 & W^6 & W^3 \\
W^5 & W^3 & W^1 & W^6 & W^4 & W^2 \\
W^6 & W^5 & W^4 & W^3 & W^2 & W^1
\end{bmatrix}
\begin{bmatrix}
x(1) \\
x(2) \\
x(3) \\
x(4) \\
x(5) \\
x(6)
\end{bmatrix}
\tag{4}
$$

(The subscript 7 has been dropped from W_7 for convenience) By applying the permutation:

| 0 | 1 | 2 | 3 | 4 | 5 | 6 | 7 | original sequence

| 1 | 3 | 2 | 6 | 4 | 5 | 1 | 3 | reordered sequence

to both X( ) & x( ) in equation (4) we get:

$$
\begin{bmatrix}
X(1) - x(0) \\
X(3) - x(0) \\
X(2) - x(0) \\
X(6) - x(0) \\
X(4) - x(0) \\
X(5) - x(0)
\end{bmatrix}
=
\begin{bmatrix}
W^1 & W^3 & W^2 & W^6 & W^4 & W^5 \\
W^3 & W^2 & W^6 & W^4 & W^5 & W^1 \\
W^2 & W^6 & W^4 & W^5 & W^1 & W^3 \\
W^6 & W^4 & W^5 & W^1 & W^3 & W^2 \\
W^4 & W^5 & W^1 & W^3 & W^2 & W^6 \\
W^5 & W^1 & W^3 & W^2 & W^6 & W^4
\end{bmatrix}
\begin{bmatrix}
x(1) \\
x(3) \\
x(2) \\
x(6) \\
x(4) \\
x(5)
\end{bmatrix}
\tag{5}
$$

Referring to equation (5) it can be seen that the sequence $\{X(1) - x(0),\ X(3) - x(0),\ X(2) - x(0),\ X(6) - x(0),\ X(4) - x(0),\ X(5) - x(0)\}$ can be obtained by performing a circular convolution between the sequence $\{x(1), x(3), x(2), x(6), x(4), x(5)\}$ and a permuted twiddle factor set given by $\{W^1, W^3, W^2, W^6, W^4, W^5\}$. Fig.4a shows, schematically, how this circular convolution can be realised using a transversal filter structure, and Fig.4b depicts how the 7-point transform example can be implemented in **complex form** using the IMS A100 transversal filter. Details of this algorithm and its implementation can be found in reference 1.

Using prime-number algorithm there are basically two way to implement IMS A100 based DFT processors capable of handling long data blocks. The obvious approach is to cascade several devices resulting in a sufficiently large convolver capable of dealing with the whole data block size. This approach is only acceptable for moderate block sizes and becomes impractical if the data size is very large. The second approach is based on mapping technique which converts a large DFT into several short ones. This decomposition technique is particularly suitable as it also provides the basis for trade-offs between cost and speed. The essence of these mapping techniques is that, by a simple change of variable, the original complex problem is converted into several easy ones. These decomposition techniques are general mappings which encompass the well-known radix-2 FFT (for example an 8-point radix-2 FFT can be viewed as a re-ordering of the 8-data points into a 3-D array, with dimensions 2x2x2 and performing 4 2-point transforms in each one of the three dimensions, corresponding to the three passes in the normal 8-point FFT flow graph). Similar type of algorithm can be used

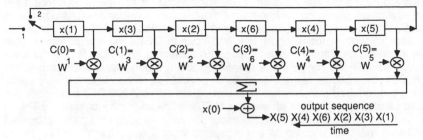

Figure 4a-Prime number transform implementation based on the transversal filter structures.

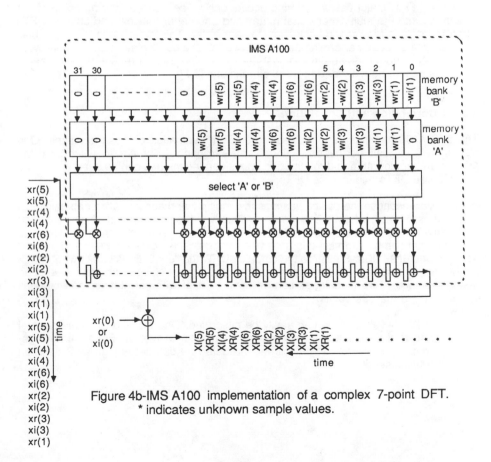

Figure 4b-IMS A100 implementation of a complex 7-point DFT.
* indicates unknown sample values.

321

with the IMS A100, with the difference that instead of using a common radix of two, a number of prime radices will be involved. For example, for a 899-point DFT, the data can be re-ordered into a 29x31 array, where the overall DFT is carried out by performing 29 31-point transforms on the rows of the array, followed by 31 29-point column transforms on the resulting array. Mathematically it can be shown that if the radices are relatively prime, no twiddle-matrix multiplications will be needed in between these short transforms [1,2,& 3]. In a practical implementation, the IMS A100 processor can be used to perform these short row and column DFT's via the prime number transform algorithm described earlier. The important fact to note here is that each set of row (or column) DFT's consists of a number of totally independent short transforms (see Figure-5). This allows various degrees of parallelism to be exploited very easily in acheiving the required performance. For example a single A100 DFT processor can be used to sequentially perform all the row DFT's followed by the column DFT's, or when extremely high processing speed is essential, several such DFT processors can be employed in parallel to compute the independent row (or column) DFT's. In the extreme case, it is possible to compute all the row and column DFT's concurrently in a pipelined system arrangement.

As the device has an on-chip memory interface, any microprocessor can be used as a host. The combination of a general purpose microprocessor and IMS A100(s) result in a high performance DFT engine. Referring to Figure-5, it can be seen that, unlike the radix-2 FFT, major data reordering occurs only twice. Once in the mapping of the data array into the multidimensional matrix and once after the row and column DFTs when the data is rearranged back into its linear order. This means that look-up tables can easily be used in performing these operations. The concepts presented here were concentrated around a two-dimensional mapping. There is no reason why the same concepts cannot be extended to more dimensions [1,3].

3.2   Correlation and Cnvolution

The IMS A100 architecture lends itself to efficient evaluation of these functions. One approach is to evaluate these functions via the frequency domain, in which case the prime-number algorithm and related decomposition techniques can be used to perform the required DFTs and inverse DFTs.

Alternatively because the device has massive multiply and accumulation capabilities, it is quite possible to perform these functions directly in the time domain. A single device is effectively a 32-tap correlator (convolver) in which the samples of the two signals to be correlated can be expressed in upto 16-bit words. This corresponds to a signal dynamic range of 96 dBs. Similar to the complex DFT, the "continuous bank swap" mode on the device can be used to perform complex (I & Q) correlation and convolutions on a single device. Two techniques can be used to deal with correlations/convolutions involving more than 32 points:

1-   As IMS A100 devices can easily be cascaded, One method is to use several cascaded devices to achieve a longer correlator/convolver. For such arrangements and with 16-bit coefficients, tha data rate can be as high as 2.5 Million samples/sec.

2- Alternatively it is possible to use various mapping techniques to decompose a long correlation/convolution into several short ones which can then be carried out via a single or a small number of devices. The host machine would merely combine the results from these short correlations/convolutions to obtain the overall result.

A simple way to decompose a long correlation/convolution of length N, between waveforms x and y, is to break up one of the waveforms, say x, into consecutive blocks of 32 sample. Each one of these blocks can then be correlated/convolved with the whole of the waveform y by loading each block into the IMS A100 coefficient registers, and using y as the input sequence. The output from these correlations/convolutions can then be combined by displacing each partial result by 32 samples, with respect to the previous one, and performing an addition operation. Note that the coefficient registers, containing blocks of waveform x, need only be updated once every time the whole of the waveform y is fed through the device, resulting in a significant saving in the memory bandwidth. More complicated decomposition technique also exist which are based on the multidimensional index mappings [4, 5]. The fact that with this decompositions, the coefficient memories need only be updated occasionally results in an impressive reduction in the memory bandwidth requirement. This is why, even with a general purpose microprocessor as the host, very impressive perfomance can be achieved.

3.3   Filtering

The application of the device for implementing Finite Impulse Response (FIR) filters is quite obvious as the device architecture is directly related to these types of filters. Implementing an FIR filter with the IMS A100 is just a matter of loading the designed filter coefficients into the device. Again the continuous swap mode can be used to implement complex (I & Q) filters if required. If the number of coefficients (filter stages) required is less than or equal to 32, a single IMS A100 would be sufficient, any unused coefficient locations being set to zero. If however, more than 32 coefficients are involved a number of IMS A100 devices can be cascaded to obtain the required filter order. Alternatively it is possible to partition a long FIR transfer function into product terms where each term has an order equal or less than 32. Then, using a single IMS A100, the data can be recirculated through the same device with different coefficients (associated with each term In the transfer function) for each circulation. In this way a very long FIR filter can be implemented with a single device at the expense of a reduction in the data rate. The programmability of the devices makes it a candidate for high speed adaptive filters. The bank swap feature can be used to achieve "on the fly" updating of the filter coefficients.

Although the IMS A100 is designed primarily for FIR type filter implementations, it can also be used in realizing Infinite Impulse response (IIR) filters. The host machine can use the continuous bank swap feature to implement IIR filters. This allows a single IMS A100 to be sufficient for the implementation of IIR filters whose order is less than or equal to 16. Figure-6 shows the coefficient memory allocations in this approach, where a's and b's are the feedback and feedforward coefficients of the IIR filter respectively and are loaded by the host processor. The procesor is also set to the continuous bank swap mode so that in one cycle the feedback coefficients (a's) and in the next cycle the feedforward coefficients (b's) are used in the calculation. If the difference between data samples and alternate output samples are written to the DIR , then the remaining output samples would correspond to the correct filter output.

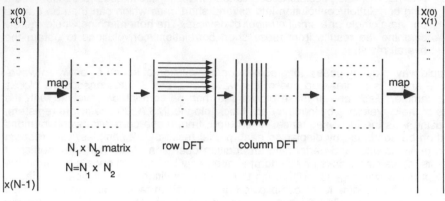

N-data

Figure-5- DFT decomposition using index mapping.

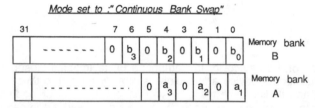

Figure 6- Coefficient memory allocation for IIR filter implementation.

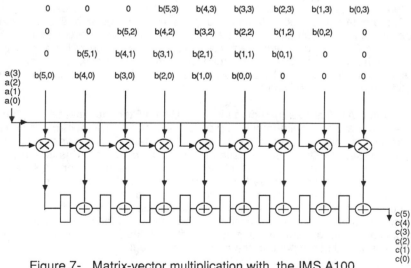

Figure 7- Matrix-vector multiplication with the IMS A100.

3.4   Matrix Multiplication

Another function that the device can perform is matrix multiplication. Figure-7 shows an example for the required data sequencing for the following matrix-vector multiplication.

$$
\begin{bmatrix} c(0) \\ c(1) \\ c(2) \\ c(3) \\ c(4) \\ c(5) \end{bmatrix} = \begin{bmatrix} b(0,0) & b(0,1) & b(0,2) & b(0,3) \\ b(1,0) & b(1,1) & b(1,2) & b(1,3) \\ b(2,0) & b(2,1) & b(2,2) & b(2,3) \\ b(3,0) & b(3,1) & b(3,2) & b(3,3) \\ b(4,0) & b(4,1) & b(4,2) & b(4,3) \\ b(5,0) & b(5,1) & b(5,2) & b(5,3) \end{bmatrix} \begin{bmatrix} a(0) \\ a(1) \\ a(2) \\ a(3) \end{bmatrix} \tag{6}
$$

Extension to matrix-matrix multiplication is simply a matter of considering the second matrix as a collection of vectors. A single device can multiply matrices of dimensions $(N \times M)$ where $N + M \leq 32$. Large matrices can be handled by partitioning techniques or by cascading several devices.

3.5   2-D image convolutions

Many applications including image processing require 2-D convolutions and correlations. Such operations are needed in image filtering, edge detection, etc. There are many ways that the IMS A100 can be used to speed up these operations. This section gives an example of how the device can be used to perform 3×3, 5×5, or larger convolutions.

Figure-8a shows part of a pixel image which is to be convolved with the 3×3 reference matrix given by Figure-8b. One way to achieve this is to load the reference matrix, as shown in Figure-8c, in one of the IMS A100 coefficient register banks, and sequence the image data through the device as shown by the arrowed path in Figure-8a. In this way every third output sample of the IMS A100 would correspond to a valid filtered pixel for the second row of the image. To proceed, the same sequence, moved down by one row, is then passed through the device which provides the filtered results for the nest row and so on. A single IMS A100 can deal with reference matrices as big as 5×5.

An alternative arrangement which gives a better throughput is one where , as shown in Figure-9a, 7 zeroes are inserted in the IMS A100 coefficient registers (between terms corresponding to the columns of the reference matrix). The data sequencing would be as shown in Figure-9c, where ten pixels from a given column are fed through the device before moving to the next column. In this scheme the first nine rows of the image are filtered in one scan, with 80% of the output data samples being valid. (Note that, using a single device, the number of inserted zeroes can be increased from 7 to 11, allowing 13 image rows to be filtered in each scan.)

The examples given here are just a small subset of possible arrangements. Remembering that the IMS A100 devices can be cascaded or used in parallel, numerous other implementations for image processing become possible.

4   Conclusions

The architecture and some of the related algorithms for a high performance signal

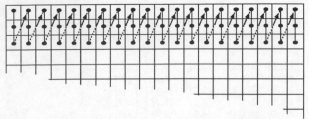

(b)- 3x3 convolution
matrix

(a)-arrows show the required data sequencing .

| 0 | 0 | 0 | 0 | ············ | 0 | 0 | a | b | c | d | e | f | g | h | i |

(c)-Coefficient register allocation for the 3x3 convolution.

Figure 8- An example of a 3x3 image convolution/correlation
with the IMS A100.

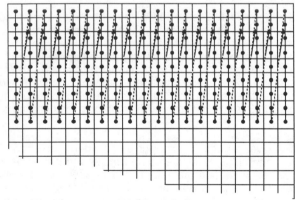

(b)- 3x3 convolution
matrix

(a)-arrows show the required data sequencing .

9 zeroes   7 zeroes   7 zeroes

| ········ | a | b | c | 0 | ········· | 0 | d | e | f | 0 | ········ | 0 | g | h | i |

(c)-Coefficient register allocation for the 3x3 convolution.

Figure 9- An improved version of the 3x3 image convolution
(correlation) with the IMS A100.

326

processing device, capable of delivering 80 MOPs, were described. The device combines high performance and very simple interfacing requirements. The on-chip standard memory interface makes the device and its related interface requirements look like those associated with a memory chip. This results in the device to be easily interfacable to most general-purpose microprocessors. Dedicated ports are also included which allow maximum performance to be achieved when required.

The algorithms described here make use of the higher level functional nature of the device and its on-chip memory to minimise the required host's memory bandwidth. For example a conventional 1024-point radix-2 FFT implemented around a multiply-accumulator would require more than 100000 memory accesses while the prime-number algorithm, for the same data size, but implemented on the IMS A100 would only involve around 16000 memory accesses. Note also that using multidimensional index mapping, the twiddle factors need to be loaded only once for each dimension. For these reasons even a slow microprocessor can achieve a respectable performance when combined with the IMS A100.

Most algorithms for the device are suitable for extension to multiprocessor environments. The advent of concurrent processors, such as the INMOS transputer and its associated language (occam), allow this potential parallelism to be exploited. Systems can be configured from arrays of transputers and A100 devices which achieve very high performance.

5 References

1   M.H. Yassaie, Discrete Fourier transform with the IMS A100, IMS A100 Application Note 2, Inmos Limited, 1986.

2   C.M. Rader, "Discrete Fourier transforms when the number of data samples is prime" *Proc. IEEE,* Vol.56, pp 1107-1109, June 1968.

3   C.S. Burrus, T.W. Parks, DFT/FFT and convolutional algorithms-theory and implementation. New York: Wiley-Interscience Publication, 1985.

4   M.H. Yassaie, Correlation and convolution with the IMS A100, IMS A100 Application Note 3, Inmos Limited, 1986.

5   C.S. Burrus, "Index mappings for multidimensional formulation of the DFT and convolution", *IEEE Trans. on ASSP*, Vol.25, pp. 239-242, June 1977.

IMS and occam are trade marks of the INMOS Group of Companies.

CHAPTER 30

A NEW FFT MAPPING ALGORITHM FOR REDUCING THE TRAFFIC IN A PROCESSOR ARRAY

Wen-Tai Lin and Chung-Yih Ho

General Electric Company/CRD
P.O. Box 8, Schenectady, New York 12301

ABSTRACT

In this paper we present a new mapping algorithm to reduce the FFT shuffling traffic by smoothing out the multiple data flows in a processor array environment. The algorithm maps multi-dimensional FFT data space into a set of memory banks to provide conflict-free access of any butterfly data pair. Only half of the memory banks is involved in inter-processor transfer; the other half remains as local memory throughout the entire FFT computation. When implementing this scheme in a mesh-connected processor array, it shows great efficiency and flexibility using LINC chip to provide the interconnection among the processor elements (PE) and their local resources such as memory banks and programmable register files.

INTRODUCTION

Various parallel FFT processing schemes have been proposed throughout the past few decades. Major efforts have been focused on reforming the intermediate data flow between stages of an FFT algorithm. The basic structural element is a butterfly processor which generates two outputs from every butterfly pair. The most considered interconnection topologies are tree, ring, mesh or direct shuffling connections. It is also shown that, in a SIMD processor array environment, the cube interconnection network provides a natural means for specifying the inter-processor data transfers required for the FFT algorithm [1]. In VLSI implementation, it is very desirable to modularize a design with local communications only. For fixed-length, one-dimensional FFT, this can be achieved by using constant geometry scheme where each FFT stage is implemented in a computation node; the address patterns in these computation nodes are identical [2]. However, it is highly desirable to be able to carry out FFT computations with variable lengths and dimensions in a general image/signal processing environment. Hence, rather than constructing specific hardware for a fixed FFT length and dimension, we are interested in providing a generic approach to implement FFT computations in a general-purpose computing environment, such as processor array.

The strategy is to smooth out the shuffling traffic by re-mapping the input/output data streams. With the mapping scheme proposed in this paper, we show that (i) the inter-processor communications are reduced by a half, (ii) the local data flows are simplified and each computation node can be identically programmed, and (iii) multiple I/O channels can be established to alleviate the traffic between the host and the processor array.

In the following we show that the FFT data shuffling can be simplified by separating the signal inputs into two halves. It is well known that for every butterfly the two indices involved differ in their parity [3]. Thus a conflict-free memory access scheme can be derived by dividing the data array into a bank of even parity (BEP) and a bank of odd parity (BOP). Following a pre-mapping procedure, the symmetrical structure of the conventional FFT data flow is preserved. It can also be easily partitioned into subtasks for multiple PEs, despite of the FFT dimensions. Moreover, it is shown that only half of the data are involved in inter-processor transfer; the other half is kept in the local memory of each PEs.

The proposed new FFT algorithm is illustrated in Fig.1 where the 16 data points are pre-mapped into a special sequence such that only 8 intermediate data points move in vertical direction between any two FFT computation stages. In other words, if the data numbered as 0,12,6,10,3,15,5,9 are stored in BEP and the rest of them are stored in BOP, then only the data stored in BOP are moved in vertical direction or involved in inter-processor transfers. The regularity shown in addressing the BOP bank also indicates that further partition can be made by evenly distributing the BEP and BOP data into multiple PEs, where each PE has a pair of local BEP and BOP banks. The address patterns in each PE can be made identical if the number of PE is a power of 2.

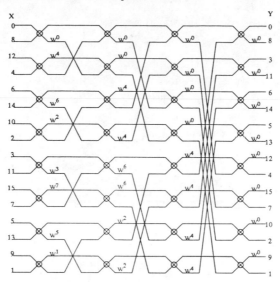

Fig.1 The smoothed 16-point FFT data flow.

INDEX MAPPING

Let $[p,q]_i$ represent the indices of a butterfly pair of an N-point FFT at stage i, i.e.

$$X_{i+1}(p) = X_i(p) + X_i(q)$$
$$X_{i+1}(q) = (X_i(p) - X_i(q))W^k,$$

where W^k is a complex multiplication constant, $i = 0, 1, ..., n-1$, and $N = 2^n$. This method is referred to as decimation-in-frequency, in-place computation. Let $<a>$ be an n-bit address variable (also called memory pointer) of the input data array MA. $<a>$ is represented in binary form as

$$<a> = (a_{n-1}, ..., a_j, ..., a_0).$$

For each $[p,q]_i$, p and q are two values of $<a>$ which differ only in bit $n-i-1$ with p smaller than q. Assume that the complex coefficients W^k are stored in a table called W-table, which holds $N/2$ distinct values. Let $<t>$ be an $(n-1)$-bit address variable of the W-table. Then for each $[p,q]_i$, k is a value of $<t>$, which can be related to $<a>$ as:

$$<t> = (a_{n-i-2}, a_{n-i-3}, ..., a_0, 0, ..., 0). \tag{1}$$

Note that (1) is equivalent to left shifting $<a>$ by filling the zeros in the least significant bit positions.

Let MA and MB be two randomly accessible memory units addressed by $< a >$ and $< b >$ respectively. Let $\oplus$ be an exclusive OR operator. Each of the following mappings defines an one-to-one correspondence relationship between $< a >$ and $< b >$:

$$< a > = P < b > \qquad (P \; mapping) \tag{2}$$
$$= (b_{n-1} \oplus b_{n-2}, \ldots, b_i \oplus b_{i-1}, \ldots, b_0).$$

$$< a > = B_n < b > \qquad (n-bits \; reversal) \tag{3}$$
$$= (b_0, \ldots, b_i, \ldots, b_{n-1}).$$

$$< a > = B_{n-1} < b > \qquad ((n-1)-bit \; reversal) \tag{4}$$
$$= (b_{n-1}, b_0, \ldots, b_i, \ldots, b_{n-2}).$$

$$< a > = L_1 < b > \qquad (rotate \; left) \tag{5}$$
$$= (b_{n-2}, \ldots, b_i, \ldots, b_0, b_{n-1}).$$

$$< a > = H < b > \qquad ((n-1)-bit \; complement) \tag{6}$$
$$= (b_{n-1}, \overline{b}_{n-2}, \ldots, \overline{b}_i, \ldots, \overline{b}_0).$$

Note that (2) can be rewritten as follows:

$$a_0 = b_0, \tag{7}$$

and

$$a_i = b_i \oplus b_{i-1} \quad for \quad i=1,\ldots,n-1.$$

Let P^{-1} be the inverse operator of P. Then the mapping of $< b > = P^{-1} < a >$ can be rewritten as

$$b_0 = a_0, \tag{8}$$

and

$$b_i = a_i \oplus b_{i-1} \quad for \quad i=1,\ldots,n-1.$$

Based on the mapping operators defined in (2) through (7), the following lemma can be easily verified.

Lemma 1
Any combination of P, B, H, and L forms an one-to-one correspondence mapping between $< a >$ and $< b >$.

Lemma 2
Let $Q \equiv B_n PH$. Then $< a > = Q < b >$ maps the data of MA into two halves in MB; the first half (i.e. $b_{n-1} = 0$) corresponds to BEP and the second half corresponds to BOP.

Let P^{-1} and Q^{-1} denote the inverse mappings of P and Q respectively. To prove the lemma, it suffices to show that the most significant bit of $Q^{-1} < a >$ is a parity bit of $< a >$. By definition,

$$< b > = Q^{-1} < a > = HP^{-1}B_n < a >$$
$$= HP^{-1}(a_0, a_1, \ldots, a_{n-1}).$$

By (6) and (8)

$$b_{n-1} = a_0 \oplus b_{n-2}$$
$$= a_0 \oplus a_1 \oplus b_{n-3}$$
$$= a_0 \oplus a_1 \oplus \cdots \oplus a_{n-1}$$

which is the parity bit of $< a >$. Thus the lemma is proved.

Lemma 3
$$B_n PH \equiv PB_n L_1.$$

Lemma 4

$$B_{n-1} \equiv B_n L_1.$$

In Fig.2 we show the implementation of Q and P mappings with the example of 16-point FFT, where Q mapping is set up for initial data distribution and P mapping is used to put the final FFT results back to their frequency order. We assume that $<a>$ is determined by $<b>$. Thus MB can be sequentially addressed and MA is randomly accessed through $<a>$. The implementation is also simplified, according to *Lemma* 3, by reducing the operation of H. Only the exclusive OR (XOR) operation involved in both mappings. The correctness of the data flow in Fig.1 can be verified through the following lemma.

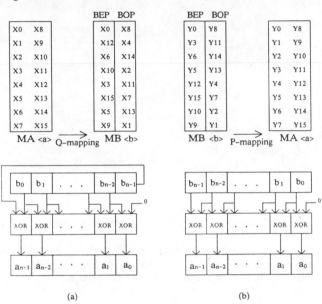

(a)   (b)

Fig.2 Two mappings involved in the I/O transfer:(a) Q mapping and (b) P mapping.

Lemma 5

Assume that $<r>$ and $<s>$ are the address variables of BEP and BOP banks respectively. If $<r>$ and $<s>$ are given as follows:

$$<r> = (0, b_{n-2}, ..., b_i, ..., b_0),$$

$$<s> = (1, b_{n-2}, ..., b_i, \overline{b}_{i-1}, ..., \overline{b}_0),$$

then $Q<r>$ and $Q<s>$ differ only in bit $n-i-1$. In other words, $<r>$ and $<s>$ generate the butterfly indices $[p,q]_i$ at the FFT computation stage i.

The proof can be attained by substituting Q with $B_n PH$ and comparing the results of the mappings over $<r>$ and $<s>$ on a bit by bit basis. From the address patterns generated by $<r>$ and $<s>$, it shows that only the addresses for BOP are changed from stage to stage. In Fig.3 we show the initialization of a 16-point FFT in a mesh-connected, 4-processor array. The weighting factors are assumed to be accessible through its local table marked as W's. However, with the decimation-in-frequency scheme it is possible to calculate a complex coefficient based on the value it used in the previous stage. Assume that the complex coefficient involved in the butterfly $[p,q]_i$ is $W^{k_i}(p,q)$, then based on (1), the complex coefficient for $[p,q]_{i+1}$ can be evaluated as

$$W^{k_{i+1}}(p,q) = -(W^{k_i}(p,q))^2.$$

In general, it is suggested that the mesh-connected PE array be equipped with wraparound connections to facilitate data shuffling via ring buses. Here we use the LINC chip to configure each PE for inter-processor communication and local data routing. With the ring buses, it is possible to program all the PEs identically. To route the local W-table output into the butterfly computation unit correctly, the LINC provides two sets of control patterns (stored in a memory unit of LINC) that define various data paths required by the butterfly operation. For each $[p,q]_i$, the control patterns are conditionally selected based on bit $n-i-1$ of $<a>$.

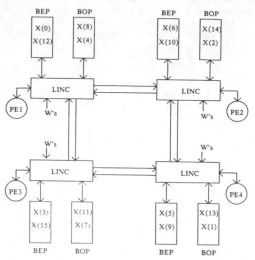

Fig.3 Partitioning the 16-point FFT task in a 2x2 processor array.

The LINC chip was designed at Carnegie Mellon University [4] and fabricated at the Corporate Research and Development Center of General Electric Company. It has a lot of nice features functioning as a glue chip in pipeline architectures. The architecture of LINC chip is shown in Fig.4 where 8 FIFO/Programmable Delay units are linked by a crossbar network; each of the 8 output channels can also be fed into a programmable register file (PRF). While the LINC-based architectures have mostly been tuned to systolic-type of processing, it is also justified that LINC may be used to construct a reconfigurable processor array for general-purpose image and signal processing [5]. Using LINC as a building block, a prototype array processor can be quickly constructed or reconfigured. Thus, rather than making unrealistic assumptions about a high performance system, the feasibility of a parallel algorithm can be justified (or bench-marked) more empirically in a LINC-based system.

PARALLEL I/O TRANSFER

In multiple PEs environment, MB is a collection of all the distributed local BEPs and BOPs. It can be shown that the I/O speed (between MA and MB) almost dominates the overall FFT computation time. Given an N-point FFT task which is further partitioned into M (with $M = 2^m$) sub-tasks. The time it takes to compute the whole FFT equals the the sum of T_c (computation time), T_i (total inter-processor transfer time) and T_m (total access time between MA and MB). Assume that all the M PEs are linearly connected, then the total inter-processor transfer time is

$$T_i = (M-1)K/2,$$

where $K = N/M$. The total processing time is expressed as follows:

$$T = T_c + T_i + T_m \qquad (9)$$

$$= (K/2) \log(KM) + (M-1)K/2 + 2MK.$$

In general, T_i depends on the connectivity of the network linking the PE array. The last term of (9) can be gracefully reduced if multiple I/O channels are installed between MA and MB.

To examine the parallel accessibility between MA and MB, the relationship between $<a>$ and $<b>$ is derived as follows:

$$
\begin{aligned}
<a> \; = \; Q<b> \; &= \; B_n PH <b> \qquad\qquad\qquad\qquad\qquad (10)\\
&= B_n P(b_{n-1}, b_{n-1} \oplus b_{n-2}, \ldots, b_{n-1} \oplus b_0)\\
&= B_n(b_{n-1} \oplus b_{n-1} \oplus {}_{n-2}, \ldots, b_{n-1} \oplus b_0)\\
&= B_n(b_{n-2}, b_{n-2} \oplus b_{n-3}, \ldots, b_{n-1} \oplus b_0)\\
&= (b_{n-1} \oplus b_0, \ldots, b_{n-2} \oplus b_{n-3}, b_{n-2})\\
&= (a_{n-1}, a_{n-2}, \ldots, a_1, a_0).
\end{aligned}
$$

Of the n bits in $<b>$, b_{n-2}, b_{n-3},..., and b_{n-m-1} are used to decode the M processors, while b_{n-1} is used to decode the local BEP and BOP banks. If MA is interleaved into M modules (with $a_{m-1}, \ldots, a_0$ as bank selection bits), then the data in MA can be moved into the PE array with M parallel channels.

The next step is to move the computed FFT data back to MA. Consider the following derivation:

A. In the conventional in-place FFT computation, the results in MA are scrambled. By applying B_n on $<a>$, the results are restored to the frequency order. According to *Lemma* 5, the address variable of BOP bank at the last FFT stage is related to $<b>$ as

$$<s> \; = \; H<b>.$$

Since $<r>$ is not affected by the H operator, we can combine $<r>$ and $<s>$ as $<b'>$ and relate it to $<b>$ as

$$<b'> \; = \; H<b>.$$

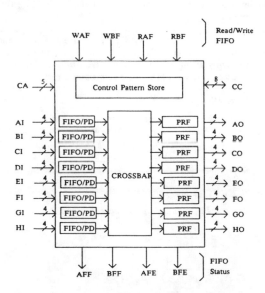

Fig.4 System overview of LINC chip.

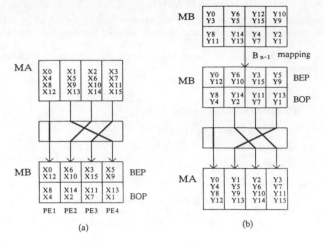

Fig.5 Parallel I/O transfer between MA and MB:(a) input phase and (b) output phase.

B. Let $< a'> = B_n < a>$. The relationship between $< a'>$ and $< b'>$ is found to be

$$< a'> - B_n < a> = B_n B_n PH < b>$$ (11)
$$= P < b'> .$$

Thus the results can be stored back to MA with M I/O channels, using the upper m bits of $< a'>$ as bank selection bits. Note that since the lower m bits of $< a>$ are used as bank selection lines at the input phase, the data sequence now stored back to MA are relatively "transposed".

C. A more desirable appraoch to output MB data is to use the same hardware configuration provided at the input phase. Let $< a'>$ and $< c'>$ be the new address variables accessing MA and MB respectively at the output phase. Given the relationship of $< a>$ and $< b>$ (which are used to access MA and MB at the input phase respectively) as

$$< a> = B_n PH < b>$$

we would like to find a mapping between $< b'>$ and $< c'>$ such that $< a'> = B_n PH < c'>$. The relationship between $< b'>$ and $< c'>$ is derived as follows:

By (11),

$$< a'> = P < b'>$$
$$= PB_{n-1}(B_{n-1} < b'>).$$

By *Lemma* 3 and 4, the PB_{n-1} mapping can be expressed as

$$PB_{n-1} = PB_n L_1 = B_n PH.$$

Therefore,

$$< a'> = PB_{n-1}(B_{n-1} < b'>) = B_n PH < c'> ,$$ (12)

where $< c'> = B_{n-1} < b'>$. (12) shows that $< a'>$ and $< c'>$ have the same relationship as that of $< a>$ and $< b>$. Hence, by pre-mapping MB with B_{n-1} operator, the I/O configuration between MA and MB remains unchanged for both the input and output phases.

The parallel access scheme is further illustrated in Fig.5 using the 16-point FFT as an example.

MAPPING MULTI-DIMENSIONAL FFT DATA SPACE

Although the previous derivations were all based upon one-dimensional FFT example, the mapping procedures may also be applied to multi-dimensional FFT problems. The main idea is to treat the multi-dimensional FFT data array as a one-dimensional array. For example given an $N \times N$ two-dimensional FFT data array MA, the data elements in MA are now indexed as a matrix with row address $< a_y >$ and column address $< a_x >$. MA may also be viewed as a one-dimensional array indexed by $< a >$, with

$$< a > = < a_y, a_x >,$$ (13)

where

$$< a_y > = (a_{2n-1}, ..., a_{n+i}, ..., a_n)$$

and

$$< a_x > = (a_{n-1}, ..., a_i, ..., a_0).$$

By *Lemma* 5, the two-dimensional FFT data flow can be formed by routing BOP bank at each computation stage. The conventional two-dimensional FFT is split into two phases: vertical and horizontal FFT computations. To treat it as a one-dimentional FFT, the first phase corresponds to the first n FFT stages and the second phase corresponds to the next n FFT stages.

In an $M \times M$ mesh-connected system, the time it takes for an $N \times N$ two-dimensional FFT computation can, in general, be estimated as follows:

Let $K = N/M$, $N = 2^n$ and $M = 2^m$, where K, n and m are integers. Let D denotes the distance of inter-processor transfer. For example, the distance between two neighbor PEs is counted as 1 and for those not involved in inter-processor transfers are counted as 0. During each phase of computation, each PE has to deal with a total distance of $M - 1$ inter-processor transfers. Since in each of these transfers, there are $K^2/2$ data elements involved. The total inter-processor transfer time spent for the entire FFT computation is $(M-1)K^2$. The total FFT processing time is given as

$$T = T_c + T_m + T_i$$ (14)

$$= K^2(\log KM + 2M^2 + M - 1).$$

The second term, which is due to single I/O traffic between MA and MB, can be gracefully reduced by employing multiple I/O channels. The straightforward approach is to divide MA into M memory banks, each communicating with a column of the $M \times M$ mesh network. Using the parallel access scheme described previously, the time complexity of the second and the third terms of (14) is reduced to $O(MK^2)$.

CONCLUSION

In this paper, we have established a unique approach to devise the FFT computation in a data flow fashion. The reduced, conflict-free data flow is derived from the combination of P, B and H mappings. The same mapping procedures can be applied to many other FFT-like data manipulations such as Bitonic sorting and Hadamard transform. In particular, we show that LINC chip supports a rapid prototyping environment for a wide variety of image/signal algorithms. Some features of LINC chip have been justified through the example of the mesh-connected FFT architecture.

References

[1] P.T. Muller, Jr., L.J. Siegel, and H.J. Siegel, "Parallel Algorithm for the Two-dimensional FFT," Proc. 5th Int'l Conf. on Pattern Recog. and Image Proc., pp.497-502, Dec. 1980.

[2] L.R. Rabiner and B. Gold, "Theory and Application of Digital Signal Processing," Prentice-Hall, Englewood Cliffs, New Jersey, 1975.

[3] M.C. Pease, "Organization of large scale Fourier processors," JAXM, vol. 16, pp.474-482, July 1969.

[4] F.H. Hsu, H.T. Kung, T. Nishizawa and A. Sussman, "LINC: The Link and Interconnection Chip," CMU internal document, May 1984.

[5] W.T. Lin and C.Y. Chin, "A Reconfigurable Array Using LINC Chip," Proceedings of the International Systolic Workshop, Oxford University, England, July, 1986.

Part IV

APPLICATION REQUIREMENTS

CHAPTER 31

32-BIT IMAGE PROCESSOR T9506 AND ITS APPLICATIONS

S. Horii, M. Kubo, C. Ohara, K. Horiguchi, Y. Kuniyasu,
E. Osaki and Y. Ohshima

TOSHIBA CORPORATION
1 Komukai, Toshiba-cho, Saiwai-ku, Kawasaki, Japan

ABSTRACT

This paper describes a newly developed 32-Bit Image Processor T9506 and its potential applications.

The T9506 is capable of carrying out 1,024-point complex FFT (Fast Fourier Transform) in 2 msec and high-precision processing of 32-bit image. It is so designed that various kinds of imagery and signal can be processed by re-writing the program stored in the memory. It will thus find a wide field of application.

By way of example, a high-performance board mounted with the T9506 and a simplified processing system interfaced with the personal computer have been developed. In addition, a large remote sensing image processing system employing a plurality of board is in the process of development.

1. INTRODUCTION

Nowadays, the demand for high-speed image processing systems is on the increase in various industrial sectors. At the same time, high-performance LSIs are drawing attention as devices for special application in the image and digital data processing fields.

As one of such devices, the T9506 is characterized first by its high-speed, high-precision operation and second by its capability of processing imagery and signals in various ways with a simple re-write of stored program. Full advantage can be taken of such characteristics in the application of the T9506 to the image processing for remote sensing, medical, FA, OA, and other purposes and to digital data processing for analyzing vibration, impact, voice, and other numerous elements.

The equipment necessary for such applications include microcomputers (small system) to general-purpose medium-to-large computers (large system) and they are multifarious in scale, form, and overall performance. As far as the boards composing each system are concerned, there are many common elements in their control bus specifications, data input/output method, kinds of processes and other aspects. Adequate consideration has been given to such factors in the production of the boards mounted with the T9506.

Construction of various systems can be achieved by use of the boards mounted with the T9506. Such systems include a simplified image processing system interfaced with the personal computer and a remote sensing image processing system employing a plurarity of boards.

In this paper, mention is made of the board mounted with the T9506, typical examples of standard system configurations, simplified image processing system as an example of the small system, remote sensing image processing system as an example of the medium-to-large system, and other products.

2. OUTLINE OF IMAGE PROCESSOR T9506

The Image Processor T9506 takes full advantage of the pipeline architecture, the latest N-well CMOS technology which provides a minimum line length of 1.2 μm, and the 2-layer aluminum wiring techniques to achieve high-speed processing and low power consumption.

This single element not only achieves a high performance of 32 bits but provides ultra high-speed operation. For example, it can perform processing of 1024-point complex FFT in 2.0 msec (the champion data being 1.0 msec).

The T9506 is designed to cover a wide range of image processing applications including remote sensing and medical image processing. It is particularly suited for FFT, affine transformation (rotation, expansion, and reduction of images), spatial filters, histrograms, and computation of sums of products.

Fig. 1 shows the block diagram of the T9506.

Since the microprogram is stored in the RAM built in PCONT which allows re-writing of its content, this processor can handle a variety of algorithms. The arithmetic units operates while data are being read from or written into the three external 1M (= 1,000,000) word x 32 bit data memories in parallel. This greatly lightens the difficulties arising from the long-pending problem of data memory and processor to achieve a high thruput. In addition, the use of a 32 x 32 bit parallel multiplier, a barrel shifter, and other parallel arithmetic circuits enables high-speed operation.

Further, this processor is capable of making an access to the chip registers and RAM or external data memories in data transfer mode through use of a 24 bit proper address. This feature permits a program to be executed concurrently with reading and writing of data into these memories and thus greatly facilitates program development.

Tables 1 and 2 summarize the main features and typical characteristics of the T9506. The external view of the T9506 is shown in Fig. 2.

3. BOARDS MOUNTED WITH IMAGE PROCESSOR T9506

Among the boards mounted with the T9506 currently are a board compatible with the multibus (IEEE 796 bus) and a board specially designed for Toshiba's purpose-built Image Processor TOSPIX-II. Although those two boards are different in size, their circuitry and processing speed are identical with each other. Fig. 3 shows the block diagram of the board mounted with the T9506. The board consists of the T9506, data memories, control bus interface, data bus interface, and peripheral circuits around the T9506. The multibus board measuring 12" x 6.75" (standard size) can be assembled into the user system.

The board mounted with the T9506 is characterized as follows.
(1) Direct interface with the user system having the multibus is capable because the control bus interface is connectable to the multibus.
(2) With 3 lines of as many as 512 x 512 x 8 bits as a high-speed data memory, this board offers the capability of superhigh-speed image and signal processing. A memory arrangement of 512 x 128 x 32 bits can be obtainable in cases where the calculation of larger bit length is required.
(3) By re-writing the program stored in the T9506, various kinds of imagery and signal processings, such as FFT, affine transformation, and inter-pixel calculation are achievable.
(4) With a high-speed data bus interface, the board permits the data transfer at a maximum of 10 M bytes/sec so that the real-time measurement system can be configured.
(5) The adoption of modularized memory and interface circuits has led to a reduction in the size and number of boards used.

Table 3 indicates the key specifications for the board. Refer to Table 2 for main processing speeds. The image or signal is processed between the T9506 and the data memory with the aid of a 10 MHz clock.

4. TYPICAL EXAMPLE OF STANDARD SYSTEM CONFIGURATION AND BASIC OPERATION FLOW

Various real-time measurement systems are configured by combining optional boards including bus conversion board, ITV interface board, and monitor interface board in addition to the board mounted with the T9506. A typical system configuration is shown in Fig. 4. This system consists of a host computer and the board mounted with the T9506, input/output board, and frame memory between the control bus and the high-speed data bus.

In this system configuration, the real-time data processing is capable because the signal input from the ITV camera or the analog signal input can be accomplished. The results of processing are connectable to the monitor TV or other outputs via the high-speed data bus without the aid of the control bus. The host computer is connectable by use of the bus conversion board in the absence of the multibus.

Fig. 5 shows the basic operation flow of the board mounted with the T9506. At first, the T9506 is initialized by the command arriving from the host computer. Next, the microprogram is transferred to the program memory built in the T9506. This microprogram can be transferred directly from the host computer. The microprogram can be also transferred to the program memory in the T9506 in such an indirect manner that it is once stored in the microprogram memory and then transferred from this memory in response to the start signal from the host computer. When various types of processes are required at one time, the latter method is advisable because all that is necessary for the transfer is to receive the start signal from the host computer. The transfer time thus can be greatly reduced. The data is sent to Data Memory 1 shown in Fig. 3 from the input board or frame memory via the high-speed data bus (see Fig. 4). The data stored on the magnetic tape or disc is sent to Data Memory 1 via the control bus and the T9506. The data is processed between the T9506 and Data Memories 1 through 3 in response to the execution command from the host computer. In the majority of cases, Data Memory 2 stores coefficients necessary for the operation. This memory, however, is designed to store imagery in the case of inter-pixel operation. The results of processing are stored in Data Memory 3. The end status signal is delivered from the T9506. Each of Data Memories 1 through 3 ordinarily has its own function as mentioned above. Depending on the process, however, each can be used for storing the input/output data or as a work memory.

The host computer transfers the results of processing to the output board or frame memory in response to the end status signal from the T9506. When the subsequent process is immediately to ensue, the microprogram is transferred anew and the results of processing are processed in the next step.

5. SOFTWARE

Necessary software include the microprogram required for the operation of the T9506, the driver software used to connect the host computer to the board mounted with the T9506 and various optional boards for microprogram entry and other controls, and the application software which is prepared according to the specific purpose. Microprograms have been successively developed for the processes shown in Table 4. Other microprograms will be available on special order.

6. SIMPLIFIED IMAGE PROCESSING SYSTEM

With the widespread need for image processing, there is an increasing demand for a compact, low-cost, and yet high-performance image processor. The system in which the T9506 is interfaced with the personal computer provides image (or signal) processing capability on a full scale. For example, such basic operations required in image processing as affine transformation, spatial filtering, and FFT, are performed in a very short time; between 0.2 to 0.5 sec including the transfer time (refer to Table 5 for details). This speed is tens to tens of thousands of times as high as that attained by only the personal computer. The throughput from each processing is on an average hundreds of times as much as that previously obtained.

Fig. 6 shows the block diagram of the simplified image processing system. The external view of the system is shown in Fig. 7. This system is composed of a personal computer connected to a processor unit. The personal computer system consists of a CPU which controls the entire system, display on which the data is indicated, hard disc on which the image data is stored, floppy disc, and printer. The processor unit is connected to the computer via the multibus/personal computer conversion board. The board mounted with the T9506 is located between the control bus and the high-speed data bus.

As options, the ITV interface board, monitor interface board, and A/D conversion board can be added to the system.

As has been described, image processing which requires complicated operations has been made possible by simply connecting the T9506 to the personal computer. In the signal processing area, vibration and voice analyses are likely to be carried out in an easy manner as is the case with the special-purpose analyzer.

7. REMOTE SENSING IMAGE PROCESSING SYSTEM

In the area of remote sensing image processing, there is a growing need for faster operation. This is mainly because of the increase in amounts of data to be processed resulting from the heightened sensor resolution and the vast amounts of computation required for synthetic aperture radar (SAR) image reproduction.

Traditional remote sensing image processing has employed all-purpose computers and dedicated image processors, and even supercomputers for SAR image reproduction. The all-purpose computer is not suited for image processing because of the difficulty in access to vast amounts of picture data having two-dimensional structure and in parallel processing. The supercomputer is also not suitable from the viewpoint of cost/performance. In terms of cost/performance, dedicated image processors are advantageous.

The latest advance in LSI technology permits considerably large-sized hardware to be designed for specific purposes. One possible approach to the development of a suitable dedicated image processor is to use parallel or pipeline configuration of one or more image processors such as T9506.

The newly developed remote sensing image processing system combines a dedicated image processor already developed with T9506 to achieve faster processing. Currently, we are checking the interface with the host computer and the processing speed and developing the microprogram for processing.

The test data of this image processor prototype promises that SEASAT SAR (100 x 100 km) image reproduction is made within 6 min when 16 T9506s are used and within about 4 min when 32 T9506s are used. This process is one which the all-purpose large computer takes several tens of hours to perform.

As stated before, the use of image processor T9506 is found to enable construction of compact, ultra high-speed image processing systems.

8. CONCLUSION

The Image Processor T9506 on the board has proved excellent in both function and performance when actually used for image processing purposes.

It has been found that the image processing system employing the T9506 is applicable to the small to the large system which is used for remote sensing purposes.

Currently, research efforts are being made to develop the board compatible with the VME bus and the one applicable to the general-purpose computer. Those boards will be new additions to the recently developed board for the multibus. Further, there are plans of improving various kinds of input/output boards and software so that the user's demand for diversified functions can be satisfied.

9. REFERENCES

1. Kubo M. et al. 1985, High speed image processing system with custom VLSI, IGARSS Digest, Amherst 7-9 October 1985, IEEE 85CH2162-6, 1097-1102.
2. Kanuma A. et al. 1986, A 200 MHz 32b pipelined CMOS image processor, ISSCC Digest, Anaheim 19-21 February 1986, IEEE 0193-6530/86/0000-0102, 102-103 and 320.
3. Kidode M. et al. 1984, Hardware implementation for robot vision, Proc. 2nd ISRR, Tokyo August 1984, Preprint 23-27.
4. Deguchi K. et al. 1983, State of the art of high speed image processor, JSICE 22(12), 1013-1020.
5. Mori K. et al. 1978, Design of local paralled pattern processor for image processing, Proc. NCC, AFIPS-47, 1025-1031.
6. Bennett J.R. et al. 1983, A fast, programmable hardware architecture for the processing of spaceborn SAR data, Proc. 17th ISRSE, Ann Arbor 9-13 May 1983, 217-232.

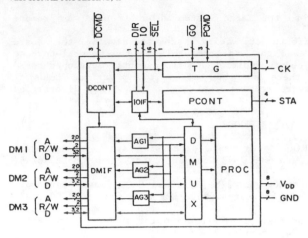

TG   Timing generator
DCONT   Transfer controller
PCONT   Arithmetic sequence unit
PROC   Arithmetic logical unit
AG1∿3   Address generator
DMUX   Data bus changeover section
DMIF   Memory interface section
IOIF   I/O interface section

Fig. 1 Block Diagram of the T9506

Fig. 2   External View of the T9506

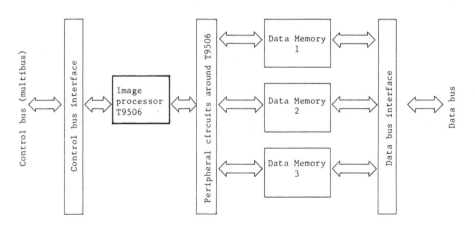

Fig. 3   Block Diagram of T9506 Board

The board mounted with the T9506 consists of the T9506, data memories, control bus interface, data bus interface, and peripheral circuits around the T9506.

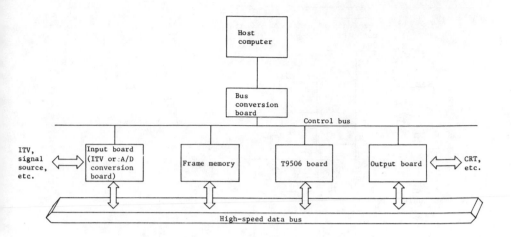

Fig. 4 Typical System Configuration

In the standard system, the board mounted with the T9506, frame memory and input/output boards are arranged between the control bus and the high-speed data bus.

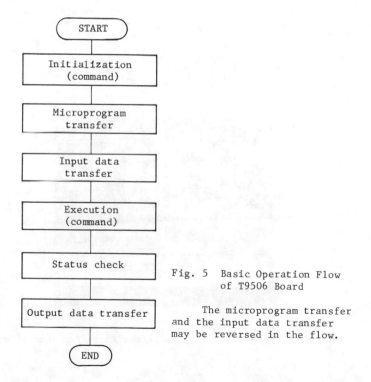

Fig. 5 Basic Operation Flow of T9506 Board

The microprogram transfer and the input data transfer may be reversed in the flow.

345

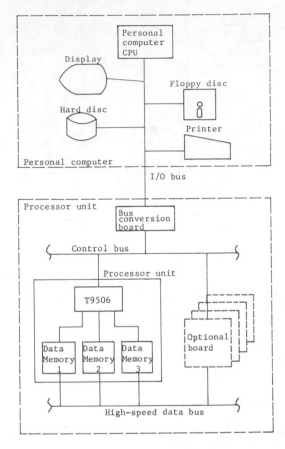

Fig. 6 Simplified Image Processing
        System

This system consists of a
personal computer interfaced with
the processor unit containing the
board mounted with the T9506.

Fig. 7 External View of Simplified Image Processing System

346

Table 1 Main Features of the T9506

DATA FORMAT	32 b FIXED POINT FORMAT
MEMORY BANDWIDTH	I MW x 32 b x 3 PORTS I 20 MBYTES/s TYPICAL
PROGRAM MEMORY (WCS)	PM : 64 W x I 6 b CM : I 6 W x 99 b
DEVICE COUNT	I 70 KTRs
CHIP SIZE	I 3.5 mm x I 3.5 mm
PACKAGE	209 PIN PGA
POWER SUPPLY	5 V , I 50mA

Table 2 Typical Characteristics of the T9506

Input clock frequency	I O MHz
FFT	2.0 ms / 1024 points, complex
Spatial filter	900 ns / pixel (@ 3 x 3 mask)
Affine transform	400 ns / pixel
Histogram	400 ns / pixel
Product sum	100 ns / term (@ 64 b sum)
Operating current	I 50 mA typical (@ 5V)

Table 3 Key Specifications for the T9506 Board

Item	Description
Control bus interface	Compatible with multibus (IEEE 796 bus)
Image memory	3 lines of 512 x 512 x 8 bits
Clock used	10 MHz
Board size	12" x 6.75" (2 boards) (for multibus) 12" x 12" (1 board) (for TOSPIX)
Number of the T9506 used	1

Table 4 List of Microprograms

Processing	Microprogram
Geometrical transformation	Affine transformation (expansion, reduction, and rotation)
Statistics	Histogram, average, and variance
Gradation transformation	Gradation transformation made referring to the table
Spatial filter	Low pass, high pass, and band pass
Orthogonal transformation	FFT and IFFT (complex and real)
Correlation	Self-correlation and inter-corrector
Absolute value	Absolute value of complex number
Inter-pixel operation	Addition, subtraction, multiplication, and division
Conjugation	Conjugation of complex number
Others	Circular shift, others

Table 5 Simplified Image Processing System Processing Speed

Process	T9506 processing time	System processing time		Remarks
		Displayed on TV monitor	Displayed on personal computer display	
FFT	471.9 msec	524.3 msec	4.2 sec	512 lines each consisting of 512 complex numbers processed
Affine transformation	105 msec	157.4 msec	3.8 sec	512 x 512 pixels
Spatial filter (3 x 3)	262 msec	314 msec	3.9 sec	Same as above
Inter-pixel operation (+, −, x)	52.4 msec	105 msec	3.7 sec	Same as above
Table look-up	210 msec	262 msec	3.9 sec	Same as above

CHAPTER 32

PRESENTATION OF W.S.I. SYSTOLIC PROCESSOR FOR IMAGE TRANSFORMATION

J.P. PETROLLI
Thomson CSF/AVG

P. GENESTIER, G. SAUCIER
Laboratoire Circuits et Systèmes
46, Avenue Félix Viallet
38031 GRENOBLE Cedex
FRANCE

J. TRILHE
Thomson EFCIS

Abstract :

This paper aims at studying a systolic processor for the geometrical transformation of digital images. Which uses a linear operator on a square domain of pixels with address and coefficients calculation. In order to execute this process on a colour video frame, the processor needs a large amount of product accumulations. This paper proposes a systolic device suitable for Wafer Scale Integration where the processing element is a 4X4 multiplier-adder.

1. INTRODUCTION

The increasing use of computer aids in avionics requires new functions to be implemented in on-board systems ; a very useful one being the real time geometrical transformation of digital images. This is used for displaying maps, which requires the display controller to follow the aeroplane movements.

Solution	Discrete components	LSI circuits	Hybrid LSI based	WSI
Nb of packages	360 LCC 52 b packages + 5 SIP of DRAM 1Mb	143 pack. PGA 200 Chips + Glue +5 SIP of DRAM 1 Mb	143 pack. +36 DRAM chips +Glue	4" Wafer + Glue +5 SIP of DRAM 1 Mb
Space required	600 X 800 mm	600 X 800 mm	250 X 250 mm	150 X 150 mm
Power consumption	40 W	17 W	17 W	16 W

Table 1 : Comparison of different possible realizations for the geometrical transformation of digital images

Added to the high performances required, the use of such systems in aeroplanes raises other constraints such as compacity and low power consumption. A predictive evaluation of the system size has been made for several technologies. Table 1 shows a comparison between several possible realizations for such a system.
The results of this comparison clearly show that the WSI implementation is the more interesting one. Moreover, a systolic architecture on a wafer provides at least two more advantages :
- Optimized memory accesses : the image pixels circulate across the circuit so that only one memory access is required for each pixel.
- Reduced pin-count which increases the system reliability.

2. GEOMETRICAL TRANSFORMATION OF DIGITAL IMAGES

Let us imagine a digital image stored in a memory. Each pixel is accessed to by a (X, Y) address. The rotation, zoom and correction (abberation) operations transform the source image into a result image which is stored in another memory. (X', Y') is the address of the output pixels.

The system must :

- first, calculate the (X, Y) address from the original (X', Y') address, (or (X', Y') from (X, Y)).

- And secondly, interpolate the resulting pixel value from the nearest samples in the source image.

The following diagram shows these different functions :

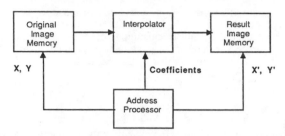

Figure 1 : General diagram of the image transformation system

Note that if (X', Y') is generated in the TV frame mode, a CRT monitor can be directly connected to the interpolator output. The address calculation can be expressed by :

$$X_i = (X' \cos q + Y' \sin q + C_x) / K_x$$
$$Y_i = (Y' \cos q - X' \sin q + C_y) / K_y$$

- (X_i, Y_i) is the original address regardless of the memory size,
- q is the rotation angle,
- C_x, C_y are the translation constants of the X and Y addresses respectively,
- K_x, K_y are the scaling factors of X and Y

2.1. Address calculation

The nearest sample address represents the N most significant bits of X_i and Y_i if N is the width of X and Y. Figure 2 shows the spatial correspondence between the resulting pixel p_i and the nearest samples in the source image.

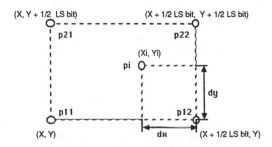

Figure 2 : Correspondance between p_i and the source image

dx and **dy** represent the relative address of p_i in the square area delimited by p11, p12, p21, p22.

In figure 3 below, the n-bits wide address represents the interpolation definition of the system.

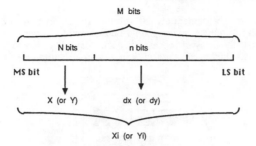

Figure 3 : Example of fields or extraction of X, dx (Y,dy) from the result of address calculation

2.2. Interpolation

Interpolation between samples is a linear operator on a square area of pixels. There are several kinds of interpolation functions. Among the different functions defined by Harry C. Andrews and Claude L. Patterson [1], the cubic-B spline and piece-wise functions are chosen because they give non-negative coefficients and use a limited square template.

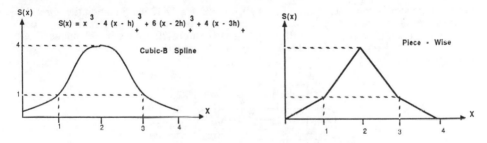

Figure 4 : Interpolation functions

The interpolation function kernel is separable, so that rows of the source image can be interpolated before columns. The following operation has therefore to be realized to perform an interpolation on a 4X4 pixels domain:

(0) : Pk = (Pk11 . Lk1 + Pk12 . Lk2 + Pk13 . Lk3 + Pk14 . Lk4) . Ck1
    + (Pk21 . Lk1 + Pk22 . Lk2 + Pk23 . Lk3 + Pk24 . Lk4) . Ck2
    + (Pk31 . Lk1 + Pk32 . Lk2 + Pk33 . Lk3 + Pk34 . Lk4) . Ck3
    + (Pk41 . Lk1 + Pk42 . Lk2 + Pk43 . Lk3 + Pk44 . Lk4) . Ck4

In the next section, a systolic processor realizing this operation is proposed.

3. SYSTOLIC PROCESSOR

The circuit specifications require a real time rotation of 1024 X 1024 pixels colour images to be performed at a refresh speed of 30 images per second. Each image pixel is coded with 4 bits for each of the 3 basic colours. The interpolation is performed on a 4X4 pixels domain. The value of each output pixel spectral component is a linear operation on the three spectral components of the operand pixels, which implies that the interpolation domain is tripled. Moreover, the simultaneous generation of the three spectral components of the output pixel triples the calculation volume. This leads to the basic specifications of the circuit which must be able to perform 200 millions interpolations per second on a 4X12 pixels domain.

In order to reach these performances, 80 interpolators (on a 4X12 pixels domain) are implemented on a wafer. Since an image pixel value is used for several interpolations, and must therefore circulate across the array of processors, each interpolator has to be able to communicate with its four neighbours, which gives the following wafer structure (see figure 5 below).

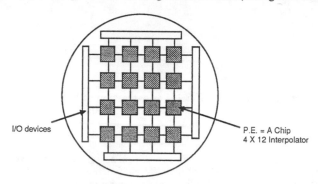

Figure 5 : General Wafer Architecture

Each interpolator is implemented as a **C H I P** performing a 4X12 interpolation (i.e. three 4X4 interpolations, one for each basic colour). The structure of these sub-interpolators, closely follows the interpolation formulae (0) by being made up of 20 processing elements able to perform the **A * B + C** operation. These processing elements (P.Es) are organized as shown in figure 6 below.

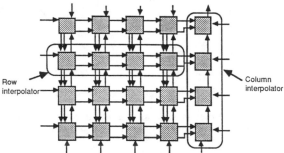

Figure 6 : 4 X 4 Interpolator Structure

The processing elements are bit-serial multiplier-adders. In the case of our image processing application the operands are 4-bits long, but their length is programmable so that 4.n bits operands can be processed (by joining several 4-bits p.Es, the only limitation being the cycle of the jointed P.Es). This limitation of the operands length to four multiples is due to some basic cells grouping required by the large number of I/Os that are necessary in the case where one-bit P.Es would have been possible. Most systolic processors use a simple carry-save adder cell. the drawback is that the P.Es array requires a bit-map organization and makes the operands circulation more difficult. The CBS serial multiplication algorithm defined by Strader and Rhyne [2] accepts full serial operands as well as a serial result. A more complex cell is required to implement this algorithm.

3.1. Multiplication algorithm

$$a = [a_n .... a_1] = \sum_{j=1}^{n} a_j . a^{j-1}$$

$$b = [b_n .... b_1] = \sum_{k=1}^{n} b_k . 2^{k-1}$$

The product is **(1)** :

$$P = \sum_{j=1}^{n} a_j . 2^{j-1} . \sum_{k=1}^{n} b_k . 2^{k-1}$$

Let P_i be the partial product at the ith iteration, so that P_m is the final product at the mth iteration

$$P_i = \sum_{j=1}^{i} a_j . 2^{j-1} . \sum_{k=1}^{i} b_k . 2^{k-1}$$

P_i can be devided into two product terms

$$P_i = \sum_{j=1}^{i-1} a_j . 2^{j-1} . \sum_{k=1}^{i} b_k . 2^{k-1} + a_i . 2^{i-1} . \sum_{k=1}^{i} b_k . 2^{k-1}$$

and then into three product terms :

$$P_i = \underbrace{\sum_{j=1}^{i-1} a_j . 2^{j-1} . \sum_{k=1}^{i-1} b_k . 2^{k-1}} + \underbrace{a_i . 2^{i-1} . \sum_{k=1}^{i} b_k . 2^{k-1}}_{B} + \underbrace{b_i . 2^{i-1} . \sum_{j=1}^{i-1} a_j . 2^{j-1}}_{A}$$

With $p_{i-1} = \sum_{j-1}^{i-1} a_j . 2^{j-1} . \sum_{j=1}^{i-1} b_k . 2^{k-1}$, we have :

(2) $p_i = p_{i-1} + A + B$

A and B are single bit operands and p_{i-1} becomes a three-bit operand after the third iteration (i=3). A five-bit adder (5/3 adder) with carries and sum registers can implement the recursive relation (2) (see figure 7 below).

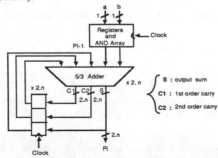

Figure 7 : basic cell for the Chen-Willoner algorithm

In [3], Chen and Willoner show that the (2) implementation requires 2.n 5/3 adders and 2.n iterations in order to get the parallel product of two n-bits operands. In order to simplify the A and B terms generation and to divide the number of adders by two, Strader and Rhyne propose the p_i terms to be right shifted. Let $q_i = p_i.2^{-i}$ be the partial product p_i left shifted at each iteration. The recursive relation (2) becomes :

$$q_i = p_{i-1}.2^{-i} + a_i.2^{-1}.[b_i........ b_1] + b_i.2^{-1}.[a_{i-1}......... a_1]$$
$$<=> \quad q_i = (q_{i-1} + a_i.[b_i........ b_1] + b_i.[a_{i-1}......... a_1]) / 2$$

$$A = b_i.[a_{i-1} .... a_1]$$
$$B = a_i.[b_i .... b_1]$$

$$(3) \quad q_i = (q_{i-1} + A + B)/2$$

Only n adders are required to implement (3) as shown in figure 8 below :

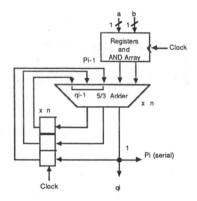

Figure 8 : Basic cell for the CBS algorithm

The (3) implementation advantages are :
- The A and B generation is simpler than in (2) implementation
- Only n 5/3 adders are necessary (instead of 2.n)
- p_i is in serial form from the LS bit to the MS bit.

Finally the CBS serial multiplier requires :
- n cascaded 5/3 adders,
- 2.N iterations or clock cycles.

The left adder can be connected to a carry save adder to achieve the add processing element function.

3.2. The processing element

Implementing the CBS algorithm described in the previous paragraph, the processing element consists of a macrocell including four basic adder cells (for 4-bits operands). The a, b operands can flow along two directions in serial mode and in two other directions in parallel mode. The result becomes the c operand in the next cell on the right and in the next higher cell.

The **FM** signal controls the macrocells (Processing elements). When it is high, the macrocell performs its first n iterations. The FM signal also controls the operand format. With n-bits wide numbers, FM is a quare signal with a period of 2.n clock cycles.

The present cell offers two solutions :

- if the same format of c must be kept across the macrocells, the LS bit of c is suppressed.

- If the LS bit of c must not be lost across then macrocells, the FM signals has to be longer in the low state than in the high   state.

Figures 9 and 10 show a timing diagram of the P.E functioning, and the P.E external connections.

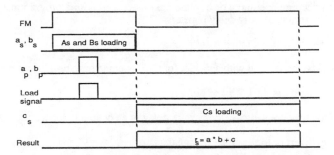

Figure 9 : Timing diagrams of the nXn multiplier macrocell

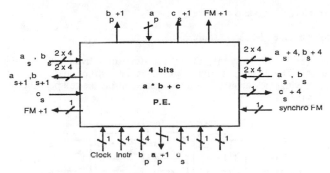

(x+n stands for x with an n clock cycles delay)

Figure 10 : External connections of a macrocell (P.E.)

3.3. Synchronization of macrocells

The synchronization between two macrocells in line is presented in the figures 11 and 12 below :

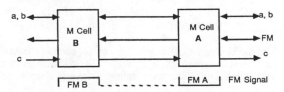

Figure 11 : Macrocells in complementary state

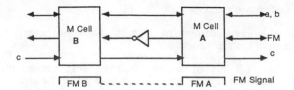

Figure 12 : Macrocells in an identical state

A result r can become operand a or b in the next macrocell with a programmable truncation of some LS bits by means of extra delay registers.

Each cell needs a 9-bits instruction word in order :

- to be the last cell with an adder for the c operand or to be a medium cell,

- to suppress the LS bit of r or not,

- to introduce a 1 to 4 clock cycles delay in an operand propagation,

- to invert or to pass the FM signal,

- to add a.b to c or to substract a.b from c,

- to select the a,b and c sources,

- to select the FM signal sources.

This instruction results from a combinatorial transformation from both a row instruction and a column instruction. The diagram of figure 13 shows the instruction distribution circuit which is based on registers with clock enable. The load signal controls the cell periodicity of the instruction affectation across the array.

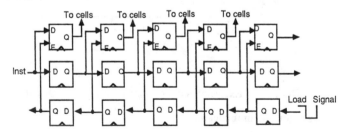

Figure 13 : Instructions distribution circuit

4. WAFER SCALE INTEGRATION

As we presented it at the beginning of section 3, 80 chips, each one containing 3X20 P.Es, must be implemented (and must be working) on the wafer. The advantages of this solution have been detailed in the introduction section. The organization of the instructions and I/Os circuits is shown in figure 14 below :

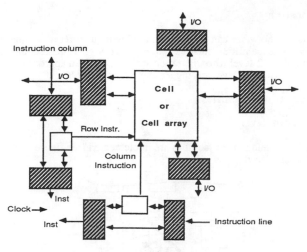

Figure 14 : processor interface general structure

A first evaluation of the basic cell complexity is about 500 transistors, which leads to :

- 2 000 transistors for a P.E. (4 basic adder cells),

- 3 X 20 X 2 000 = 120 000 transistors for a chip (4 X 12 interpolator),

- and finally to 9.6 millions transistors for the whole processor(this is only for the working part of the circuit and does not take into account the spare material that will be provided for reconfiguration (see section 4.2)).

4.1. Test strategy of the circuit

The test strategy is similar to the general one presented in [4]. It mainly includes two steps, the test of data transfer devices and the test of multipliers.

4.1.1. Test of data transfer devices

During this step, all the possible data transfer schemes will be tested :
East   West and West   East, which is achieved by shifting in either direction serial operands, which is realized by programming the cells in a scan path manner [5].

Cross transfers : this step consists in sending data in one direction and in directing them in a perpendicular direction (among those allowed by the cell possibilities).

These two steps give a first cartography of the good data paths and of the cells where both data directing devices and data storage capabilities are found working.

4.1.2. Multipliers test

After the previous step, the already declared faulty P.Es are isolated (see the next section for the bypass technique). The remaining ones are tested by propagating a wave of multiplications : the first column of P.Es performs multiplications, its results are propagated to the output, and the process is iterated column by column.
The test patterns sent as operands for the multiplications are generated in a way that limits their number by taking into account only the most probable faults that may appear in the circuit.

4.2. Reconfiguration

Since it is obvious that no wafer can be manufactured without defects (and hence without faulty P.Es), reconfiguration capabilities must be implemented so that defects can be tolerated. The circuit structure implies a 2-level reconfiguration strategy. Such an approach allows small defects to be tolerated at the P.E (chip) level and bigger ones at the wafer level.

4.2.1. Reconfiguration facilities

4.2.1.1. At the chip level

Spare multipliers (P.Es) and bypass facilities are implemented (see figure 15) so that a faulty P.E can be isolated and remplaced.

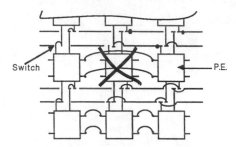

Figure 15 : Example of multiplier (i.e. P.E) bypassing in a chip

This will allow enough good chips to be found on the wafer, so that it will be possible to realize an array of the required size (80 chips).

4.2.1.2. At the wafer level

The goal here is to bypass the faulty chips (i.e. the chips that cannot be reconfigured). This requires spare chips to be implemented, and programmable data links to be provided between chips.

4.3. Test and reconfiguration process

The first step (end-of-manufacturing test) allows a first cartography of faulty nodes and data paths to be obtained. Then, the first reconfiguration attempt takes place :

- The unreconfigurable P.Es are bypassed.

- The other faulty P.Es are reconfigured using the standby multipliers.

- The faulty data paths are disconnected.

A new test of the device is performed in order to verify the reconfiguration efficiency. These steps are iterated until either a working configuration is found, or a test time limit is found, or no more reconfiguration attempt is possible.

Programming the switches used to direct data lines and to indicate whether a P.E is used or not is made with floating gate FETs during the test and reconfiguration steps. At the end of the reconfiguration process, in case of a high temperature environment of the circuit, the reconfiguration can be stored by programming fuses with a laser beam (because of the low retention capability of floating gates for temperatures exceeding 150°C).

5. APPLICATION OF THIS SYSTOLIC PROCESSOR IN IMAGE TRANSFORMATION

The present application implements the cubic-B spline function with a 4X4-coefficient kernel. The source image four samples per access. The figure 16 below shows the block of accessed samples moving along the row or the column.

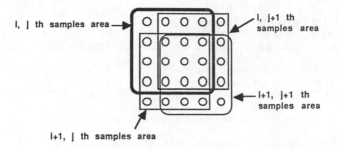

Figure 16 : Successive sample areas

The sample flow must respect the samples area moving. Figure 17 shows additions of samples and coefficients to an array macrocell interpolator.

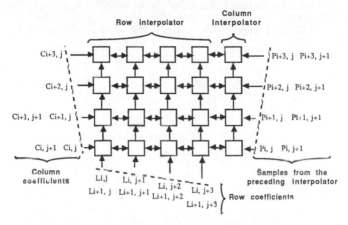

Figure 17 : Data circulation in an interpolator

The row interpolator calculates the intermediate result with the $L_{i,j}$ coefficients. The column interpolator calculates the final result p_j with the $C_{i,j}$ coefficients. Twenty macrocells constitute a systolic interpolator device. The wafer includes an array of these interpolator devices.

In case of image "zooming", the $L_{i,j}$ coefficients remain invariant per row and the $C_{i,j}$ coefficients remain invariant per column. In case of image rotation, both $L_{i,j}$ and $C_{i,j}$ are space variant.

6. CONCLUSION

The wafer is expected to integrate more than 5000 cells in the current fast CMOS technology. The address and coefficient calculation is also expected to be included in the wafer function. Any algorithm based on a bi-dimensional convolution may be applied to this WSI systolic processor.

the realization of this circuit also requires a simultaneous development of software tools for both its simulation and its programming (to which the ones required for testing and reconfiguring it must be added). The studies that led to the design of such a circuit clearly show that this kind of architectures is very a good candidate for realizing powerful specialized systems in the future.

REFERENCES

[1]   "Digital Interpolation of Discrete Images"
        Harry C. Andrew and Claude Patterson
        IEEE Transactions on Computers, vol C.25, n°2, February 1976.

[2]   "A canonical Bit-sequential Multiplier"
        Noel R. Strader and Thomas Rhyne
        IEEE Transactions on Computers, vol C-31, n°8, August 1982.

[3]   "An O(n) Parallel multiplier with Bit-sequential Input and Output"
        Chen and Willoner
        IEEE Tansactions on Computers, vol C-28, n°10, October 1979.

[4]   "Test facilities in 2-D Wafer Scale Integration"
        C. Jay, J.P. Eynard, G. Saucier
        IFIP Workshop on Wafer Scale Integration, Grenoble, March 1986.

[5]   "A logic structure for LSI testability"
        E.E. Eichelberger, T.W. Williams
        Journal of Design Automation and Fault-Tolerant Computing, vol.2, May 1978.

CHAPTER 33

ARCHITECTURAL STRATEGIES FOR DIGITAL SIGNAL PROCESSING CIRCUITS

Rajeev Jain[1], Peter A. Ruetz[2], Robert W. Brodersen[1]

[1]Department of Electrical Engineering and Computer Sciences, U.C.Berkeley, CA 94720

[2]LSI Logic, Systems Res. Lab, 1801 Page Mill Rd, P.O Box 10005 Palo Alto CA 94303-0857

ABSTRACT

The architectures used in several high-performance signal-processing chips are analyzed and the impact of various architectural strategies on the chip area and throughput are discussed. Strategies are examined (a) for on-chip signal communication, (b) for parallelism in the arithmetic and logic hardware, (c) for data storage and (d) for I/O interfaces. A set of principles are derived for designing these architectural components, which lead to efficient application-specific designs. It is shown that some of these principles are either neglected or violated by existing VLSI theories.

1. INTRODUCTION: PRACTICAL DESIGN PROBLEMS

Recent work [1-2] in the design of digital signal processing circuits shows that even in 4 μm NMOS one can integrate complex signal processing functions on a single chip with a high performance provided the right architectural trade-offs are made. For example, the 2-d 3x3 convolution algorithm for real-time image processing (10 MHz sample rate) has been integrated in 4 μm NMOS on a single chip using only 40 mm^2 die area [2]. The chip has been fabricated and used in a working image recognition system.

In this paper we will highlight the main architectural principles used in the design of the convolver and other high performance chips. Architectural strategies for lower sample rate chips such as digital audio (44 KHz) and telecommunication (8 KHz) have been discussed elsewhere [3-6].

For a given algorithm different architectural choices can be made to achieve real-time performance. The different choices will result in different chip area (or chip count) and different amounts of design effort. Moreover, they will make different demands on the performance of the circuit. These factors not only lead to different material and development costs for the signal processing system, but can also affect the feasibility of obtaining a working chip in a given technology.

For example, consider the integration of the 2-d 3x3 convolution algorithm with a 10 MHz throughput requirement. If one designs a general purpose architecture using array multipliers to allow all possible coefficient values, 9 chips may be required [7]. However, if one takes advantage of the fact that a large number of applications only require power-of-two coefficients [8] one can integrate the entire convolver on 1 chip [2]. Similarly, if the parallelism in the algorithm is exploited to construct a broadcast architecture [2], the I/O interface for the convolver can be integrated on chip using 1 MHz RAMs, whereas a pipelined systolic structure [7] requires an external 20MHz RAM chip which is more difficult to design and adds to the board size.

A major problem in the design of VLSI signal processing circuits is to make the right choice of architecture which will achieve real-time performance with a reasonable chip area and design time and can be integrated using robust circuit design techniques. This problem is more severe in high performance applications because of the limitations imposed by MOS technology on the maximum throughput/unit area for most signal processing circuits (adders, multipliers, memories).

The results obtained recently with the designs of several real-time speech and image processing circuits [9-10] indicate that to solve the architecture design problem effectively one has to devise concerted strategies for the four main parts of the architecture: (a) the on-chip signal communication, (b) the computational hardware, (b) the I/O interface, (c) the data storage elements and (d) the I/O interface. The design problems associated with these four parts are briefly described below:

(a) **On-chip signal communication**:

When several data-paths or processing elements (PEs) are used in parallel to achieve a high throughput an interconnection strategy has to be chosen to communicate signals between different PEs. The design problem in making this choice is to minimize the area overhead of any devices required solely for communication (not for any algorithmic operation) while keeping the signal propagation delay low enough to meet the constraint imposed by the desired throughput.

(b) **Computation hardware**:

There are two types of operations typically required in digital signal processing: arithmetic operations (addition and multiplication) and Boolean operations for logical computations and decision making. For the arithmetic operations two broad choices exist: bit-serial data-paths or bit-parallel data-paths constructed from basic elements such as adders, multiplexers, registers, and shifters. Boolean operations can be performed using arithmetic elements, logic elements (AND-OR gates) or logic arrays (PLAs). The choice has to be made based on the the desired sample rate, the maximum clock rate in the target technology and the complexity of the algorithmic operations. While making this choice key considerations are: the chip area, the effect on the signal communication, I/O interface and data storage requirements and the ease of circuit and layout design of the data-path.

(c) **Storage**:

The samples required by the various processing elements on the chip can either be stored in centralized storage elements such as dynamic RAMs or locally together with each PE in load/hold or pipeline registers. The advantage of a central store like a dynamic RAM is that it takes less area than registers. However, to benefit from this one has to be able to design the architecture such that the central store does not become a bottleneck. Therefore the objective in choosing between these alternatives is to lump storage of samples and data as much as possible without creating data access bottlenecks.

(d) **I/O interface**:

Usually in a signal processing system the signal source (e.g. video camera in image processing) produces the samples at a certain rate and in a certain format (e.g. raster scan). If the input interface of the processing architecture is not directly matched to this rate and format it becomes necessary to use buffering and reformatting circuits which add to the chip area and cost of the chip. In some case it even becomes necessary to use separate "glue" chips which increases the system size and production cost. The design problem is therefore to match the I/O of the processor to the output of the signal source and to the input of the post-processing or output device (e.g. display unit) while minimizing the interface circuitry so that it can be included on-chip.

In the following sections we illustrate with examples useful strategies for designing the four architectural components listed above.

2. ON-CHIP SIGNAL COMMUNICATION STRATEGY

Broadly speaking two choices exist for communicating signals between different PEs and between the PEs and storage arrays on a chip. One choice is to broadcast the output signals from different PEs on a common bus and use tri-state buffers to control which PE puts data on the bus. The data put on the bus can be simultaneously accessed by different PEs. Alternatively one can pipeline the signal transmission between PEs.

The objective in choosing between these alternatives is to communicate signals between the PEs with a minimum area overhead under the constraint of the desired transmission rate.

2.1 Broadcast communication

Broadcast communication only requires the space for metal or poly wires to implement the buses and does not use any active devices. The total signal propagation delay in this

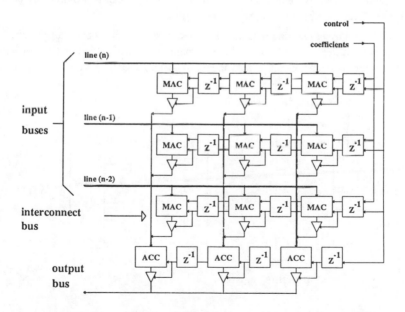

Fig.1 Broadcast architecture for computing the 2-d 3×3 convolution algorithm.

approach is the RC delay of the bus and the set-up time for the input nodes of PEs connected to the bus. Fig.1 shows a broadcast architecture for 2-d convolution which has been integrated in 4 μm NMOS. There are 3 input buses, 3 interconnect buses and 1 output bus. Each input bus feeds data from the I/O interface to the 3 multiplier-accumulator (MAC) elements in a row. Each interconnect bus connects the outputs of the 3 MACs in a column to the input of an accu-mulator element (ACC). Finally the output bus connects the output of the ACCs to the I/O interface. In each case the data is clocked onto the bus at a 10 MHz rate and clocked into a PE from the bus at a 10 MHz rate. Calculations using the MOSIS 4 μm NMOS process parameters showed that this data rate is feasible and it has also been verified on a working chip. Hence the propagation delay with broadcast is sufficiently low to allow real-time image processing at a 10 MHz sample rate in 4 μm NMOS technology. In the chip implementation of the broadcast architecture [2] some of the buses actually run the entire length of the chip.

2.2 Pipeline communication

Pipelining breaks up the propagation delay along an interconnection path by using addi-tional clocked registers so that the throughput is increased at the expense of a higher latency and chip area. The latency is often acceptable. However, circuit design experience shows that each pipeline register consumes about $\frac{1}{3}$ the chip area of an adder, a fact often overlooked in VLSI architecture theories. Fig.2 shows a pipelined systolic architecture which performs the same convolution algorithm as the broadcast architecture in fig.1. The systolic architecture requires 45 extra registers which are only used for pipelining the interconnection between the 9 PEs i.e. they perform no algorithmic operations. These registers consume the equivalent of 15 adders in chip area. Observe that since the 10 MHz throughput can be achieved by broadcast the extra investment for pipelining is unnecessary in this case.

Although the throughput in a pipelined interconnection is independent of the number of PEs to be interconnected, it is limited by the maximum rate at which the pipeline registers can be clocked (assuming the pipeline is synchronous). This limitation arises due to the problem of

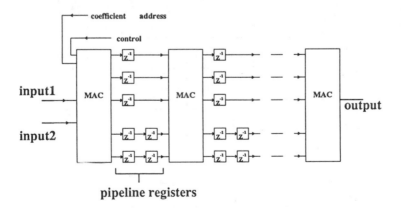

Fig.2 Pipelined architecture for 2-d 3×3 convolution algorithm.

on-chip clock generation and distribution. In fact due to this problem it is not always obvious whether synchronous pipeline communication can actually offer a higher throughput than broadcast communication in a given technology. An interesting topic of research would be to investigate whether asynchronous pipeline communication would offer a higher throughput and at what additional cost in chip area [11].

We therefore conclude that broadcast communication minimizes the area overhead for on-chip signal communication and is to be preferred over pipelining whenever it can satisfy the throughput requirements. Practical designs show that broadcast can be used up to data rates of at least 10 MHz in 4 μm NMOS.

2.3 Determination of communication strategy

The communication strategy for the chip architecture is determined by the manner in which the parallelism in the algorithm is exploited. This is illustrated for the example of the 2-d 3×3 convolution algorithm, in fig.6. For each position of the 3×3 window one can simultaneously compute 9 products with 9 different samples and accumulate them in each sample interval. This choice of parallel computation has been used to design the pipelined array of fig.2. The same algorithm can also be implemented by computing in each sample interval the 3 different products with a common input sample, which contribute to 3 successive output samples. This interpretation leads to a broadcast array.

$$y(i,j) = \sum_{k=1}^{3} \sum_{l=1}^{3} a_{kl} \, x(i-k+2, j-l+2)$$

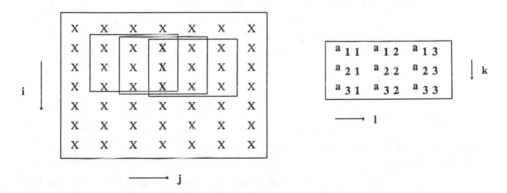

Fig.6 Graphical illustration of the 2-d 3×3 convolution algorithm showing that each input pixel contributes to three successive output pixels.

3. COMPUTATIONAL HARDWARE DESIGN

In a signal processing circuit the processing elements or data-paths must be designed to execute the algorithmic operations in the desired sequence and at the required sample rate. For low sample rate systems one can usually employ general-purpose microprogrammable data-paths [4]. The execution of an algorithm on such a data-path takes several clock cycles.

For high performance circuits where the desired throughput is close to the clock rate it becomes more efficient to use data-paths which are dedicated to the required algorithmic operations. In a dedicated data-path a separate pipelined operator (e.g. adder, multiplier, z^{-1} register) is provided for each algorithmic operation so that the algorithm can be executed in one clock cycle. Furthermore, if the algorithmic operations on different input samples can be performed simultaneously several data-paths can be used in parallel to speed up the processing. Three design choices that have to be made are: the hardware for the operators in the data-path, the structure of the data-path, and the parallel configuration when multiple data-paths are used.

3.1 Bit-serial vs. Bit-parallel data-paths

The basic hardware choices for the arithmetic elements in the data-path are bit-serial and bit-parallel. Let us consider the adder and the z^{-1} elements.

The registers required for z^{-1} elements consume the *same* chip area in bit-serial and bit-parallel hardware. In both cases N-bit registers are required. In the bit-serial case one requires an N-bit shift register and in the bit-parallel case an N-bit load/hold or clocked register is needed.

A bit-serial adder occupies less chip area than a bit-parallel adder. However, bit-serial addition is slower on two accounts. First, because each bit of the sample is processed sequentially in time, it takes N clock cycles to process a sample where N is the wordlength. Secondly the overhead associated with clock rise and fall times and register delay has a relatively greater effect upon the clock cycle in the bit-serial adder. The total time required to perform a bit-serial addition on N-bit samples is approximately given by:

$$t_{serial} = N \ \{\max(t_{carry}, t_{sum}) + t_{reg} + t_{clk}\} \tag{1}$$

where t_{carry} and t_{sum} are the propagation delays through the carry and sum parts of the adder respectively, t_{reg} is the register delay and t_{clk} is the total time lost in a clock cycle due to finite clock rise and fall times and clock separation. The time required to add two N-bit samples with a bit-parallel ripple-carry adder (other structures exist which are faster) is given by:

$$t_{parallel} = N \times t_{carry} + t_{sum} + t_{reg} + t_{clk} \tag{2}$$

Observe that the effects of register delay and clock delay are N times larger for the bit-serial addition than for the bit-parallel addition. Furthermore, the bit-serial delay is determined by N times the maximum of t_{carry} and t_{sum} whereas the bit-parallel delay is always affected by N times t_{carry} only. It is always possible to make t_{carry} smaller than t_{sum} by using single-gate delay carry-stages.

To understand how the above factors affect the trade-offs between bit-serial and bit-parallel let us look at the bit-serial systolic architecture for 2-d 3×3 convolution shown in fig.3 [12]. Since the bit-serial clock rate is $\frac{1}{3}$ of the sample rate, three identical 3×3 arrays of bit-serial processors are used. In a bit-parallel implementation only one 3×3 array would suffice since the clock rate would then be equal to the sample rate. The 27 bit-serial adders in the 3 bit-serial arrays occupy less chip area than the 9 bit-parallel adders required in a single bit-parallel array. However, the registers for z^{-1} elements in the bit-serial arrays occupy 3 times as much chip area as required in the bit-parallel array since in both cases registers are equally big. Apart from this area disadvantage the bit-serial arrays also complicate the I/O interface.

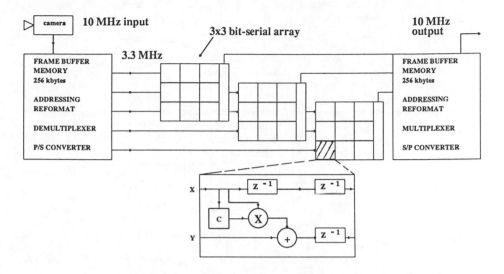

Fig.3 Bit-serial systolic architecture for 2-d 3×3 convolution.

Therefore in high throughput applications the use of multiple bit-serial processors to compensate for the low throughput of bit-serial processing is counterproductive.

We conclude that for high performance circuits where the sample rate is close to or equal to the maximum clock rate in the target technology, bit-parallel hardware is more area efficient than bit-serial hardware. It is advantageous to use bit-serial hardware if the sample rate is close or equal to $\dfrac{f_c}{N}$ where f_c is the maximum clock rate for the bit-serial adder.

3.2 Structure of dedicated bit-parallel data-paths

The different ways to exploit the parallelism in the algorithm not only affect the interconnection strategy for the multiple data-paths as shown in sec.2.3, but also determine the structure of the data-paths and the possible floorplans for the chip. In deciding on a particular choice the major factors to consider are the regularity, modularity and the circuit and layout design time for the data-paths.

To cut the design time an approach which has been used successfully for several high performance circuits [10,13] is to construct data-paths that can be automatically layed out from a structural-level input using a cell-library based bit-slice data-path compiler [14]. The structure of some data-paths for image processing which have been automatically layed out are shown in figs.4-5. These data-paths are internally pipelined and can process samples at a rate of 10 MHz. It can be seen these data-paths are made from a common set of basic cells.

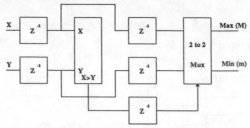

(a) 2-way sort element built from comparator, multiplexer and register cells

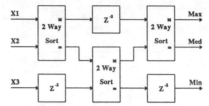

(b) 3-way sort elements built from 2-way sort elements

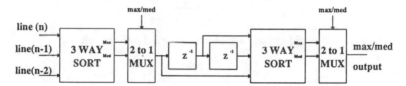

(c) data path for computing maximum or medium in a 3x3 window

Fig.4. Data path for finding the maximum and separable median used in a 3×3 sorting filter for noise rejection.

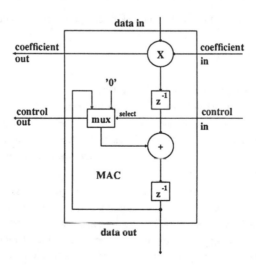

Fig.5 Data path structure for MAC element in the convolver of fig.1

3.3 Logic hardware

When the algorithm requires logical computations such as in dilation and erosion of binary images [8] it is much more area efficient to use logical processing elements rather than arithmetic elements. Fig.7 shows the architecture of a logical convolver [10] using logical PEs which are much smaller than the MAC elements in the linear convolver of fig.1.

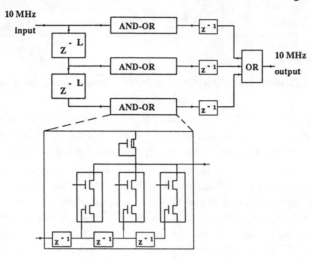

Fig.7 Logical convolver architecture using logic operators

For decision-making operations and for computations which can be efficiently performed by table-look up techniques (such as calculating tangents of angles) the use of PLAs leads to low design time and efficient silicon utilization. PLAs are broadcast logical arrays where each PE is simply a transistor. It is interesting to note that conventional techniques such as systolic arrays do not allow the use of PLAs. If a PLA is systolized by introducing pipelining between the PEs, the estimated increase in area of the PLA would be by a factor 10. PLAs have been used extensively in several high performance chips for both control hardware as well as decision-making operations in the algorithm [9-10]. Fig.8 shows an example of a contour tracer architecture for image recognition. The entire trace algorithm is executed by a single PLA in real-time. The PLA performs logical comparisons, tangent calculations and determines the pixel to be tested in the next sample interval.

4. DATA STORAGE MECHANISM

The way in which the parallelism in the algorithm is exploited also affects the kind of storage elements required on the chip.

In the contour tracer (fig.8), in each clock cycle, several operations are performed in parallel on a single pixel value by the logical elements in the PLA. A central RAM is used to store all the pixels and the values are read out sequentially and transferred to the PLA.

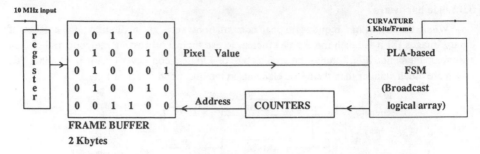

Fig.8 Contour tracer architecture for an image recognition system

In a systolic implementation each PE would operate on separate data in each clock cycle so that a central store can not be used. Such an approach requires distributed storage elements which consume more area and may not necessarily offer any throughput advantage.

The broadcast convolver architecture of fig.1 is designed such that the input sample (pixel) values of only the past two image lines need to be stored. Since the same sample value in an image line is broadcast to a row of PEs in each clock cycle, the past sample values for each image line can be stored centrally in a dynamic RAM. Circuit design experience shows that this provides a very efficient means of storage since the RAMs are very compact. In this case the dynamic RAMs also interface the processor directly to the video signal source (fig.9) so that no external memories are required and no time is lost in transferring the data from the signal source to the PEs (see next section).

In general purpose array machines such as the MPP [15] the storage is distributed and provided locally with each PE. In this case the input data has to be first stored in an external memory device and then transferred to the local memories of the PEs before it can be processed. This requires additional hardware and can create an I/O bottleneck. To avoid this bottleneck and maintain real-time capability expensive hardware has to be used to provide a very high data rate between the external memory and the local memories. Thus distributing the storage substantially increases the size and cost of the array as well as of the I/O hardware.

5. I/O INTERFACE STRATEGY

While designing the processor architecture it is important to take into account the input signal source and output or post-processing device with which the processor has to be interfaced. This point is illustrated by the two different architecture designs for the 2-d convolution architecture shown in figs.1 and 2.

The broadcast architecture in fig.1 [2] is designed to match the rate and format of the 10 MHz raster scan video signal obtained from a standard video camera. It has 3 inputs which require pixels (samples) from 3 successive lines of the image in each sample interval. Further, each input requires the sample sequence to be the same as the line scan. These 3 inputs can therefore be simply obtained from the video input by using 2 line buffers as shown in fig.9. The line buffers can be implemented on-chip using 2×512 byte RAMs. The output sequence from the processor is also a raster scan format so it can be connected directly to a standard video display device or another processor with a similar input interface as in fig.9.

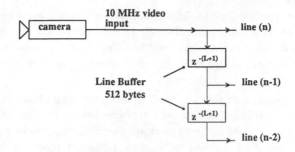

Fig.9 Input interface for broadcast convolver architecture shown in fig.1

The pipeline architecture design in fig.2 on the other hand can not directly accept a raster scan signal. It requires two 10 MHz input sequences which have to be derived from the raster scan signal by reformatting it. This results in the need for a number of additional chips, including a frame buffer memory of at least 256 Kbytes and addressing hardware to reformat the pixel sequence. This extra hardware is only required to interface the pipeline architecture to the video source. As shown by the broadcast design (fig.1) this extra expense can be completely avoided while providing the same system performance.

6. SUMMARY

VLSI architecture design methods for monolithic signal processing circuits must take into account four major factors (a) the on-chip signal communication, (b) the choice of computational hardware, (c) the data storage mechanism and (d) the I/O interface. Optimizing any of these factors without considering the others can be counterproductive.

Guiding principles for optimizing these factors have been illustrated based on practical chip designs. These principles indicate that an effective VLSI architecture design methodology for high performance signal processing circuits should try to give preference a) to broadcast communication over pipelining, b) to bit-parallel hardware over bit-serial, c) centralized storage over distributed storage and should allow the designer to match the processor I/O to system sample rate and format.

As shown by several examples existing VLSI architecture theories actually contradict or neglect the above principles. This is because these theories are based on abstractions of VLSI circuit properties which are oversimplified and sometimes incorrect.

Based on simplified estimations of circuit propagation delays, it has traditionally been considered impossible to allow broadcast of data on shared buses and pipelined communication has been widely advocated as a necessary basis for VLSI architectures. Practical designs show that broadcast communication not only works satisfactorily at sample rates as high as 10 MHz even in a relatively mature technology like 4 μm NMOS, but it is also more area efficient than pipelined communication.

The use of bit-serial processing elements which is supposed to lead to low routing area and smaller chips actually turn out to consume much more area than bit-parallel processors due to the relatively low throughput of bit-serial processing.

In conventional theories it has been considered important to avoid centralized storage due to expected memory access bottlenecks. We have found that the use of central storage elements such as dynamic RAM is not only feasible in many designs but is also more area efficient than distributed storage elements such as load/hold registers. Furthermore, distributing the storage actually creates I/O bottlenecks.

Finally, our experience shows that with a careful matching of the processor architecture to the external data input/output devices one can often eliminate the need for external memory or I/O control chips. Most designs based on existing architectural theories neglect this point altogether so that even if the resulting architecture is compact enough to fit on a single chip a number of glue chips are required for interfacing it with the rest of the system.

(Research supported by GE, MICRO, DARPA)

REFERENCES

[1] R. A. Kavaler, T. G. Noll, M. Lowy, and R. W. Brodersen, "A Dynamic Time Warp IC for a One Thousand Word Recognition System", *Proc. of ICASSP'84*, San Diego, March 1984, pp.25B6.1-4.

[2] P.Ruetz and R.W.Brodersen, "A Realtime Image Processing Chip Set", *Digest IEEE Int. Solidstate Circ. Conf.*, Anaheim, Feb. 1986, pp. 148-149.

[3] J.Rabaey, R.W.Brodersen, "Experiences with automatic generation of audio band digital signal processing circuits", *Proc. of ICASSP'86*, Tokyo, 1986, pp.1541-1544.

[4] S.Pope, J.Rabaey, R.W.Brodersen, "Automated design of signal processors using macrocells", *VLSI signal processing*, IEEE press, 1984, pp. 239-251.

[5] J.Vanginderdeuren, H. De Man, A. Delaruelle, H.V. Wyngaert, "A digital audio filter using semi-automated design", *Digest IEEE Int. Solidstate Circ. Conf.*, Anaheim, Feb. 1984, pp. 88-89.

[6] R.Jain et. al., "Custom Design of a VLSI PCM-FDM Transmultiplexer from system specifications to circuit layout using a computer-aided design system", *IEEE J. Solidstate Circuits*, vol. SC-19, Feb. 1986, pp. 73-85.

[7] H.T.Kung and R.L.Picard, "One dimensional systolic arrays for multidimensional convolution", in *VLSI for Pattern Recognition and Image Processing*, Ed. King-sun Fu, Springer-Verlag, 1984, pp.9-24.

[8] W.K.Pratt, *Digital Image Processing*, Johan Wiley, 1978.

[9] R.A.Kavaler, *The design and evaluation of a speech recognition system for engineering workstations*, Ph.D Thesis, U.C.Berkeley, Memo No. UCB/ERL M86/39.

[10] P.A.Ruetz, *Architectures and design techniques for real-time image processing ICs*, Ph.D Thesis, U.C.Berkeley, Memo No. UCB/ERL M86/37.

[11] S.Y.Kung, "An algorithm basis for systolic/wavefront array software" *Proc. ICASSP'84*, San Diego, March 1984, pp. 25A.2.1-4.

[12] H.T.Kung, "A systolic 2-d convolution chip", in *Multicomputers and Image Processing*, Eds. K.Preston,Jr. and L.Uhr,.R Academic Press, 1982, pp. 373-384.

[13] B.Richards, *Design of a video histogrammer using automated layout tools*, Ph.D Thesis, U.C.Berkeley, Memo No. UCB/ERL M86/38.

[14] P.Ruetz, R.Jain, C.S.Shung, J.M.Rabaey, G.M.Jacobs, and R.W.Brodersen, "Automatic layout generation of Real-Time digital image processing circuits", *Proc. CICC'86*, Rochester, May 1986, pp. 111-115.

[15] J.L.Potter, *The Massively Parallel Processor*, The MIT Press, 1985.

SYNCHRONOUS DATA FLOW:
DESCRIBING DSP ALGORITHMS FOR PARALLEL COMPUTATION[1]

Edward A. Lee
David G. Messerschmitt

Department of Electrical Engineering and Computer Science
University of California, Berkeley, CA. 94720

ABSTRACT

Data flow is a natural paradigm for describing DSP applications for concurrent implementation on parallel hardware. Data flow programs for signal processing are directed graphs where each node represents a function and each arc represents a signal path. Synchronous data flow (SDF) is a special case of data flow (either atomic or large grain) in which the number of data samples produced or consumed by each node on each invocation is specified a-priori. Nodes can be scheduled statically (at compile time) onto single or parallel programmable processors so the run-time overhead usually associated with data flow evaporates. Conditions for correctness of SDF graph are explained and scheduling algorithms are described for homogeneous parallel processors sharing memory. A voiceband data modem example illustrates the efficacy of the technique. Also, the use of SDF for automatic VLSI circuit generation from a functional description is proposed.

1. THE DATA FLOW PARADIGM

For concurrent implementation, a signal processing task is broken into subtasks which are then automatically, semiautomatically, or manually scheduled onto parallel processors, either at compile time (*statically*) or at run-time (*dynamically*). Automatic breakdown of an ordinary sequential computer program is an appealing concept [1], but the success of existing techniques is limited because sequential programs do not often exhibit the concurrency available in the algorithm. If the programmer provides the breakdown as a natural consequence of the programming methodology, we should expect more efficient use of concurrent resources.

DSP systems are often described using *block diagrams* consisting of functional blocks connected by signal paths. An example is a voiceband data modem, illustrated in figure 1. The figure illustrates an implementation of a 2400 bit per second, 600 baud, frequency-division-multiplexed, full-duplex data modem with bandsplitting filters and a fractionally-spaced passband adaptive equalizer [2-4]. Block diagram descriptions are modular, meaning that once a block is defined it is easily re-used. They can also be hierarchical, where a block may itself represent another block–diagram, yielding descriptions with much of the elegance of structured programming. For example, the PLL block in figure 1 might be expanded as shown in figure 2. Such diagrams exhibit much of the concurrency inherent in the algorithm.

The block diagrams of figures 1 and 2 are a *data flow graphs*. The fundamental premise behind data flow graphs is that each node represents a function that can be invoked whenever input data is available [5-7]. Functions may be elemental (addition, multiplication, etc.) or non–elemental (digital filters, FFT units, modulators, phase locked loops, etc.), and the directed arcs represent paths taken by successive data samples. The

[1] This research was sponsored in part by the National Science Foundation Grant ECS-8211071, an IBM Fellowship, and a grant from Shell Development Corporation.

complexity of the functions (or the "granularity") will affect the amount of parallelism available. If the granularity is at the level of signal processing subsystems (second order sections, butterfly units, etc.), the paradigm is called *large grain data flow* (LGDF) [8–11]. Otherwise, it is called *atomic* data flow, from the Greek word *atomos*, meaning indivisible. Atomic data flow is obviously a special case of LGDF. Because the program execution is controlled by the availability of data, data flow programs are said to be *data driven* [12].

Synchronous data flow (SDF) is a special case of data flow (either atomic or large grain) in which the number of samples produced or consumed each time a node is invoked is specified *a-priori* as part of the node definition. These numbers are shown in figures 1 and 2 adjacent to the input and output arcs of each node.

LGDF is ideally suited for signal processing, and has been adopted in simulators in the past [13]. Other signal processing systems use a data-driven paradigm to partition a task among cooperating processors [14], and many so called "block diagram languages" have been developed to permit programmers to describe signal processing systems more naturally. Some examples are BLODI [15], PATSI [16], BLODIB [17], LOTUS [18], DARE [19], MITSYN [20], Circus [21], and TOPSIM [22]. The technique we propose in this paper shares the appropriateness of block diagram languages, but is more flexible and involves less overhead. In particular, multiple sample rates are managed efficiently, and the mapping of computations onto computing resources is done *statically* at compile time.

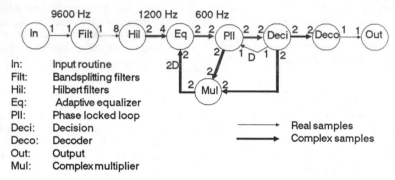

Figure 1. A block diagram of a 2400 bit per second, 600 baud modem. Three sample rates are evident. The numbers beside each input and output indicate the number of samples produced or consumed each time the node is invoked. The bold arcs indicate complex signals.

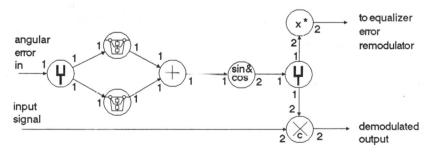

Figure 2. Detail of the PLL of figure 1. Represented with icons, the nodes are (from left to right) a *fork* (which replicates each input sample on two output paths), a pair of *biquads*, an *adder*, a *sine and cosine* computation (or table lookup), a complex *conjugator*, another fork, and a complex *multiplier*.

One way to achieve flexibility and retain the convenience of expressing algorithms as block diagrams is to use LGDF, as done in BLOSIM [13, 23], a single-processor digital signal processing simulator that naturally accommodates multiple sample rates and asynchronous systems. The control is *dynamic* because the order of invocation of nodes is determined at run-time. Such dynamic control mechanisms are traditional in data flow implementations, but they involve considerable overhead. The amount of memory required for buffering is difficult to determine, and even for fully synchronous systems may depend on the order in which the blocks are invoked. In programmable DSPs or VLSI implementations, memory may be a scarce commodity, and such inefficiencies must be avoided. Secondly, the supervisory overhead can be substantial. Such overhead is not required for implementations of SDF graphs.

Signal flow graphs, special cases of SDF graphs, have been used to describe linear, single-sample-rate systems for parallel implementation [24]. The method in [24] does not consider the repetitive nature of a desired schedule, and therefore does not always identify the best schedules. It also does not support multiple sample rates. In spite of these deficiencies, this representation has been endorsed by others [25, 26].

The term "signal flow graph" is often used to described single-sample-rate data flow graphs, regardless of whether the the system is linear. We define *homogeneous* SDF graphs to be SDF graphs where each time a node is invoked it consumes or produces exactly one sample on each input or output arc. Multiprocessor implementations of algorithms specified this way have been explored in [27-29]. This admirable work has some deficiencies is our application, however. Primarily, it has no provision for multiple sample rates, thus restricting the range of applications. Also, the complexity of the scheduling technique may become unmanageable for some important target architectures.

Reduced dependence graphs are specifications of systems in terms of periodic acyclic precedence graphs, where only one period is illustrated, and its dependence on previous periods is done by indexing [30, 31]. The resulting description is similar to homogeneous data flow graphs. Reduced dependence graphs are used to describe *regular iterative algorithms*, which can then be mapped onto processor arrays. This approach is suitable for descriptions of well structured algorithms to be implemented in systolic arrays. The range of applications is again excessively limited for our objectives.

Computation graphs were introduced in 1966 by Karp and Miller [32] and were further explored by Reiter [33]. They are essentially equivalent to SDF graphs, but our use of the model differs significantly. Karp and Miller concentrate on fundamental theoretical considerations, for example proving that computation graphs are *determinate*, meaning that any admissible execution yields the same result. Such a theorem, of course, also underlies the validity of data flow. Other early analysis using the general computation graph model concentrates on graphs that *terminate*, or deadlock after some time. Most DSP applications, however, do not terminate, so these results are not useful in this application. Simplified versions of the model have been explored by Commoner and Holt [34] and Reiter [35], but the restrictions imposed on the model are excessive. Computation graphs have been shown to be a special case of *Petri nets* [36-38] or *vector addition systems* [39]. These more general models can be used to describe asynchronous systems, but implementations generally require expensive dynamic flow control.

Common objections leveled against traditional data flow center around the overhead required to dynamically schedule and synchronize the nodes. However, synchronous DSP systems, in which sample rates are known and are rational multiples of one another, can be specified as SDF graphs, and can be statically scheduled onto parallel processors.

2. SYNCHRONOUS DATA FLOW GRAPHS

A node in a data flow graph is said to be *synchronous* if we can specify *a-priori* the number of input samples consumed on each input and the number of output samples produced on each output each time the node is invoked. A synchronous node is shown in figure 3(a) with a number associated with each input or output specifying the number of inputs consumed or the number of outputs produced. These numbers are part of the node definition. A large grain example, a digital filter node, would have one input and one output, and the number of input samples consumed and output samples produced would be unity. A 2:1 decimator node would also have one input and one output, but would consume two samples for every sample produced. A *synchronous data flow* (SDF) graph is a network of synchronous nodes, as in figure 3(b).

2.1. Implementation Architectures

An obvious implementation would have separate hardware for each node in the graph. In this case, SDF is a description of hardware as well as a description of the computation; the SDF graph is isomorphic with the implementation architecture. In some ways, SDF is useful as hardware description, but our intent is that SDF be used primarily for functional description. In this case, an isomorphic implementation architecture is not likely to be efficient, in that much of the hardware is likely to be idle much of the time.

If SDF is used primarily as a functional description, then an algorithm designer can specify a system graphically using a graphics workstation, and a compiler is required to translate the functional description into an efficient parallel implementation. The nodes of the graph must be scheduled onto a single or parallel processor architecture. One class of parallel architectures that we consider is homogeneous parallel processors sharing memory without contention. A practical programmable DSP of this type uses extensive pipelining with interleaved concurrent programs [41].

Given a SDF graph, we may want to achieve a given computation rate with a minimal amount of hardware. In this case, scheduling alone is not sufficient because the number of processors and their interconnection must be determined. An approach suitable for VLSI is based on parametrized macrocell silicon compilers such as LAGER [42]. A chip consists of multiple relatively simple programmable processors with parameters (such as wordsize, memory size, and the presence or absence of hardware features) determined by the computation required. Such designs can be automatically generated from SDF functional descriptions.

We have mentioned three possible target architectures: isomorphic hardware mappings, homogeneous parallel processors sharing memory, and custom VLSI designed with parametrized programmable processors.

2.2. Inconsistent Sample Rates

We assume that a SDF graph describes a repetitive computation to be performed on an infinite stream of input data. Hence, the desired schedule is periodic. It is not always possible to construct a practical periodic schedule for a SDF graph, however. Consider the SDF graph of figure 4(a). To start the computation, node 1 can be invoked because it has no input paths and hence needs no data samples. After invoking node 1, node 2 can be invoked, after which node 3 can be invoked. This sequence can be repeated. But node 1 produces twice as many samples on arc 3 as node 3 consumes. An infinite repetition of this schedule therefore causes an infinite accumulation of samples in the buffer associated

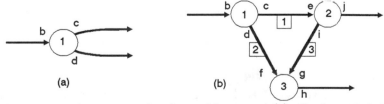

Figure 3. A synchronous data flow (SDF) node (a) and graph (b). The flags attached to the arcs simply identify them with a number. The input and output arcs are ignored for now.

with arc 3. This implies an unbounded memory requirement, which is clearly not practical.

In a DSP sense, the SDF graph has *inconsistent sample rates*. Node 3 expects as inputs two signals with the same sample rate but gets two signals with different sample rates. The SDF graph of figure 4(b) does not have this problem. A periodic admissible sequential schedule repeats the invocations {1,2,3,3}. Node 3 is invoked twice as often as the other two. It is possible to automatically check for consistent sample rates and simultaneously determine the relative frequency with which each node must be invoked. To do this, we need a little formalism.

2.3. Characterizing SDF Graphs

Consider the SDF graph of figure 3(b). The connections to the outside world are not considered, for now. Thus, a node with only inputs from the outside is considered a node with no inputs, which can be scheduled at any time. The limitations of this approximation are discussed in [43]. A SDF graph can be characterized by a matrix similar to the incidence matrix associated with directed graphs in graph theory. It is constructed by first numbering each node and arc, as in figure 3(b), and assigning a column to each node and a row to each arc. The $(i,j)^{th}$ entry in the matrix is the amount of data produced by node j on arc i each time it is invoked. If node j consumes data from arc i, the number is negative, and if it is not connected to arc i, then the number is zero. For the graph in figure 3(b) we get

$$\Gamma = \begin{vmatrix} c & -e & 0 \\ d & 0 & -f \\ 0 & i & -g \end{vmatrix} \tag{1}$$

This matrix can be called a *topology matrix*, and need not be square, in general.

If a node has a connection to itself (a *self loop*), then only one entry in Γ describes this link. This entry gives the net difference between the amount of data produced on this link and the amount consumed each time the node is invoked. This difference should clearly be zero for a correctly constructed graph, so the Γ entry describing a self loop should be a zero row.

We can replace each arc with a FIFO queue (buffer) to pass data from one node to another. The size of the queue will vary at different times in the execution. Define the vector $b(n)$ to contain the queue sizes of all the buffers at time n.

For the sequential (single processor) schedule, only one node can be invoked at a time, and for the purposes of scheduling it does not matter how long each node runs. Thus, the time index n can simply be incremented each time a node finishes and a new node is begun. We specify the node invoked at time n with a vector $v(n)$, which has a one in the position corresponding to the number of the node that is invoked at time n and zeros for each node that is not invoked. For the system in figure 3(b), $v(n)$ can take one of three values for a sequential schedule,

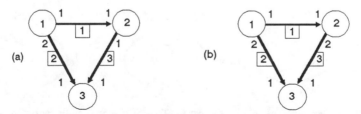

Figure 4. (a) An example of a defective SDF graph with sample rate inconsistencies. (b) A corrected SDF graph with consistent sample rates.

$$v(n) = \begin{vmatrix} 1 \\ 0 \\ 0 \end{vmatrix} OR \begin{vmatrix} 0 \\ 1 \\ 0 \end{vmatrix} OR \begin{vmatrix} 0 \\ 0 \\ 1 \end{vmatrix} \qquad (2)$$

depending on which of the three nodes is invoked. Each time a node is invoked, it will consume data from zero or more input arcs and produce data on zero or more output arcs. The change in the size of the buffer queues caused by invoking a node is given by

$$b(n+1) = b(n) + \Gamma v(n) \qquad (3)$$

The topology matrix Γ characterizes the effect on the buffers of running a node program.

This simple computation model is powerful. First we note that the computation model handles delays. The term *delay* is used in the signal processing sense, corresponding to a sample offset between the input and the output. We define a *unit delay* on an arc from node A to node B to mean that the nth sample consumed by B will be the $(n-1)^{th}$ sample produced by A. This implies that the first sample the destination node consumes is not produced by the source node at all, but is part of the initial state of the arc buffer. Indeed, a delay of d samples on an arc is implemented in our model simply by setting an initial condition for equation (3). Specifically, the initial buffer state, $b(0)$, should have a d in the position corresponding to the arc with the delay of d units.

To make this idea firm, consider the example system in figure 5. The symbol "D" on an arc means a single sample delay, while "2D" means a two sample delay. The initial condition for the buffers is thus

$$b(0) = \begin{vmatrix} 1 \\ 2 \end{vmatrix}. \qquad (4)$$

Because of these initial conditions, node 2 can be invoked once and node 3 twice before node 1 is invoked at all. Delays, therefore, affect the way the system starts up. Clearly, every directed loop in an SDF graph requires at least one delay, or the system cannot be started.

2.4. Identifying Inconsistent Sample Rates

Inconsistent sample rates preclude construction of a periodic sequential schedule with bounded memory requirements. A necessary condition for the existence of such a schedule is that $rank(\Gamma) = s-1$, where s is the number of nodes. This is proven in [43], so we merely give the intuition here. In figure 4(a), the topology matrix is

$$\Gamma = \begin{vmatrix} 1 & -1 & 0 \\ 0 & 1 & -1 \\ 2 & 0 & -1 \end{vmatrix} \qquad rank(\Gamma) = s = 3,$$

so no periodic admissible sequential schedule can be constructed. The SDF graph of figure 4(b) has the topology matrix

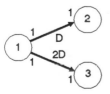

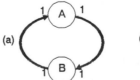

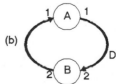

Figure 5. An example of a SDF graph with delays on the arcs.

Figure 6. Two SDF graphs with consistent sample rates but no admissible schedule.

$$\Gamma = \begin{bmatrix} 1 & -1 & 0 \\ 0 & 2 & -1 \\ 2 & 0 & -1 \end{bmatrix}, \qquad rank\ (\Gamma) = s = 2.$$

It is also proved in [43] that a topology matrix with the proper rank has a strictly positive (element-wise) integer vector $\mathbf{q}$ in its nullspace, meaning that $\Gamma\mathbf{q}$ is the zero vector. For figure 4(b), a set of such vectors is

$$\mathbf{q} = J \begin{bmatrix} 1 \\ 1 \\ 2 \end{bmatrix},$$

for any positive integer J. Notice that the length of $\mathbf{q}$ is s, the number of nodes. Notice further that $\mathbf{q}$ specifies the number of times we should invoke each node in one cycle of a periodic schedule. Node 3 gets invoked twice as often as the other two nodes, for any positive integer J. This is proved using equation (3) by observing that if each node is invoked the number of times specified by $\mathbf{q}$, the amount of data $b(n)$ left in each buffer ends up equal to the amount before the invocations. Hence, the schedule can be repeated infinitely often with finite memory.

Valuable information is obtained from the topology matrix. Its rank can be used to verify consistent sample rates, and its nullspace gives the relative frequency with which nodes must be invoked.

2.5. Insufficient Delays

Even with consistent sample rates, it may not be possible to construct a periodic admissible sequential schedule. Two examples of SDF graphs with consistent sample rates but no such schedules are shown in figure 6. Directed loops with insufficient delays are an error in the construction of the SDF graph and must be identified to the user. It is shown in [43] that a large class of scheduling algorithms will always run to completion if a periodic admissible sequential schedule exists, and will fail otherwise. Running such an algorithm is a simple way of verifying the correctness of the SDF graph. The class of algorithms is described in the next subsection.

2.6. Scheduling for a Single Processor

Given a positive integer vector $\mathbf{q}$ in the nullspace of Γ, one cycle of a periodic schedule invokes each node the number of times specified by $\mathbf{q}$. A sequential schedule can be constructed by selecting a *runnable* node, using (3) to determine its effect on the buffer sizes, and continuing until all nodes have been invoked the number of times given by $\mathbf{q}$. We define a class of algorithms.

DEFINITION (*CLASS S ALGORITHMS*): Given a positive integer vector q s.t. $\Gamma\mathbf{q} = 0$ and an initial state for the buffers $\mathbf{b}(0)$, the ith node is *runnable* at a given time if it has not been run q_i times and running it will not cause a buffer size to go negative. A *class S algorithm* is any algorithm that schedules a node if it is runnable, updates $\mathbf{b}(n)$ and stops (*terminates*) only when no more nodes are runnable. If a class S algorithms terminates before it has scheduled each node the number of times specified in the q vector, then it is said to be *deadlocked*.

Class S algorithms ("S" for Sequential) construct static schedules by simulating the effects on the buffers of an actual run for one cycle of a periodic schedule. That is, the nodes need not actually run. Any dynamic (run time) scheduling algorithm becomes a class S algorithm simply by specifying a stopping condition, which depends on the vector q. It is proven in [43] that any class S algorithm will run to completion if a periodic admissible sequential schedule exists for a given SDF graph. Hence, successful completion of the algorithm guarantees that there are no directed loops with insufficient delay. A suitable class S algorithm for sequential scheduling is

1. Solve for the smallest positive integer vector $q \in \eta(\Gamma)$.
2. Form an arbitrarily ordered list L of all nodes in the system.
3. For each $\alpha \in L$, schedule α if it is runnable, trying each node once.
4. If each node α has been scheduled q_α times, STOP.
5. If no node in L can be scheduled, indicate a deadlock (an error in the graph).
6. Else, go to 3 and repeat.

A synchronous data flow programming methodology offers concrete advantages for single processor implementations. The ability to interconnect modular blocks of code (nodes) in a natural way could considerably ease the task of programming high performance signal processors, even if the blocks of code themselves are programmed in assembly language. The gain would be somewhat analogous to that experienced in VLSI design through the use of standard cells. For synchronous systems, the penalty in run time overhead is minimal. But a single processor implementation cannot take advantage of the explicit concurrency in a SDF description. The next section is dedicated to explaining how the concurrency in the description can be used to improve the throughput of a multiprocessor implementation.

2.7. Scheduling for Parallel Processors

Clearly, if a workable schedule for a single processor can be generated, then a workable schedule for a multiprocessor system can also be generated. Trivially, all the computation could be scheduled onto only one of the processors. Usually, however, the throughput can be increased substantially by distributing the load more evenly. We show in this section how the multiprocessor scheduling problem can be reduced to a familiar problem in operations research for which good heuristic methods are available.

We assume a tightly coupled parallel architecture, so that communication costs are not the overriding concern. Furthermore, we assume homogeneity; all processors are the same, so they process a node in a SDF graph in the same amount of time. It is not necessary that the processors be synchronous, although the implementation will be simpler if they are.

A periodic admissible *parallel* schedule is a set of lists $\{\psi_i \; ; i = 1, \cdots, M\}$ where M is the number of processors, and ψ_i specifies a periodic schedule for processor i. We discuss methods for constructing *blocked* schedules, in which all processors finish one cycle before the next cycle begins (cf. [27, 29]). If p is the *smallest* positive integer vector in the nullspace of Γ then a cycle of a schedule must invoke each node the number of times given by $q = J p$ for some positive integer J. J is called the *blocking factor*, and for blocked schedules, there is sometimes a speed advantage to using J greater than unity. If the "best" blocking factor is known, then construction of a good parallel schedule is not too hard.

For a sequential schedule, precedences are enforced by the schedule. For a multiprocessor schedule, the situation is not so simple. We will assume that some method enforces the integrity of the parallel schedules. That is, if a schedule on a given processor dictates that a node should be invoked, but there is no input data for that node, then the processor halts until this input data is available. The task of the scheduler is thus to construct a schedule that avoids deadlocks and minimizes the *iteration period*, defined to be the run time for one cycle of the schedule divided by J. The mechanism to enforce the integrity of the communication between nodes on different processors could use semaphores in shared memory or simple "instruction-count" synchronization, where no-ops are executed as necessary to maintain synchrony among processors.

The first step is to construct a graph describing the precedences in $q = J p$ invocations of each node. The graph will be acyclic. A precise class S algorithm accomplishing this construction is given in [43] so we merely illustrate it with the example in figure 7(a). The SDF graph in figure 7(a) is neither acyclic nor a precedence graph. Node 1 should be invoked twice as often as the other two nodes, so $p = [2\ 1\ 1]^T$. Further, given the delays on two of the arcs, we note that there are three periodic admissible sequential schedules with unity blocking factor, $\phi_1 = \{1,3,1,2\}$, $\phi_2 = \{3,1,1,2\}$, or $\phi_1 = \{1,1,3,2\}$. A schedule that is not admissible is $\phi_1 = \{2,1,3,1\}$, because node 2 is not immediately runnable. Figure 7(b) shows the precedences involved in all three schedules. Figure 7(c) shows the precedences using a blocking factor of two ($J=2$). The graph in figure 7(a) shows self loops for each node. This means that successive invocations of the same node cannot overlap in time. Some practical SDF implementations have such precedences in order to preserve the integrity of the buffers between nodes. In other words, two processors accessing the same buffer at the same time may not be tolerable, depending on how the buffers are implemented. The self-loops are also required, of course, if the node has a *state* that is updated when it is invoked. We will henceforth assume that all nodes have self loops, thus avoiding the potential implementation difficulties.

If we have two processors available, a schedule for $J=1$ is

$$\psi_1 = \{3\}$$

$$\psi_2 = \{1,1,2\}.$$

When this system starts up, nodes 3 and 1 will run concurrently. The precise timing of the run depends on the run time of the nodes. If we assume that the run time of node 1 is a single time unit, the run time of node 2 is two time units, and the run time of node 3 is three time units, then the timing is shown in figure 8(a). The shaded region represents idle time. A schedule constructed for $J=2$, using the precedence graph of figure 7(c) will perform better. An example is

$$\psi_1 = \{3,1,3\}$$

$$\psi_2 = \{1,1,2,1,2\}$$

and its timing is shown in figure 8(b). There is no idle time, so no faster schedule exists.

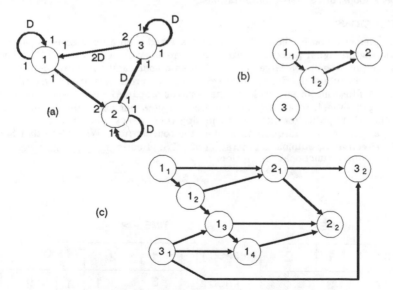

(a)

(b)

(c)

Figure 7. (a) A synchronous data flow graph with self-loops
(b) An acyclic precedence graph for $J=1$.
(c) An acyclic precedence graph for $J=2$.

The construction of the acyclic precedence graph is handled by the class S algorithm given in [43]. The remaining problem of constructing a parallel schedule given an acyclic precedence graph is a familiar one. It is identical with assembly line problems in operations research, and can be solved for the optimal schedule, but the problem is NP complete [44]. This may not be a problem for small SDF graphs, and for large ones we can use well studied heuristic methods, the best being members of a family of "critical path" methods [45]. An early example, known as the Hu-level-scheduling algorithm [46], closely approximates an optimal solution for most graphs [45, 47], and is simple.

2.8. Silicon Compilation

To automatically generate a chip from a functional description, we must solve a different scheduling problem. A throughput constraint must be satisfied with a minimum amount of hardware, rather than attempting to maximize the throughput for a given amount of hardware. A direct solution is to begin with a single processor schedule and determine the throughput. The number of processors required to achieve the desired throughput can then be estimated. The scheduling problem can then be solved for successively increasing numbers of processors. If the Hu level scheduling algorithm is used, this is a practical approach.

2.9. Iteration Bound

Recall that the iteration period is defined to be the run time for one cycle of a blocked schedule divided by J, the blocking factor. For the special case of homogeneous SDF graphs, the iteration period is bounded from below by the maximum over all loops of the total computation time in the loop divided by the number of delays in the loop [27, 48]. A method for computing the iteration bound for general SDF graphs is given in [40]. The iteration period is a measure of the speed (or throughput) of a schedule, and the iteration bound gives the best possible throughput for all blocking factors and for any number of parallel processors. As expected, the maximum throughput is determined by the feedback loops, or recursive computations.

3. CONCLUSIONS

An implementation of the modem of figures 1 and 2 on a homogeneous parallel programmable digital signal processor with shared memory is described in [41]. The complete SDF graph contains 28 large grain and atomic nodes and three sample rates. Using the Hu-level scheduling algorithm, up to seven parallel processors are fully utilized (have no significant idle time in one cycle of the periodic schedule). For seven processors, the iteration bound is met, so adding more processors does not speed up the schedule. We conclude that this application exhibits surprising concurrency, and that relatively simple scheduling algorithms are adequate to exploit the concurrency. We believe that SDF will be equally effective for automatic generation of VLSI circuits using parametrized processing elements from a functional description.

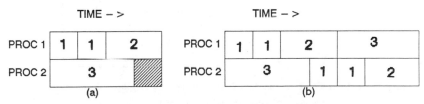

Figure 8. One period of each of two periodic schedules for the SDF graph of figure 8. In (a), $J=1$, while in (b), $J=2$.

References

1. Padua, D. A., Kuck, D. J., and Lawrie, D. H., "High-Speed Multiprocessors and Compilation Techniques," *IEEE Trans. on Computers* C-29(9) pp. 763-776 (Sept. 1980).

2. Ungerboeck, Gottfried, "Fractional Tap-Spacing and Consequences for Clock Recovery in Data Modems," *IEEE Transactions on Communications*, (August, 1976).

3. Gitlin, R. D. and Weinstein, S. B., "Fractionally-Spaced Equalization: An Improved Digital Transversal Equalizer," *Bell System Technical Journal* 60(2)(February, 1981).

4. Falconer, D. D., "Jointly Adaptive Equalization and Carrier Recovery in Two-Dimensional Digital Communication Systems," *Bell System Technical Journal* 55(3)(March 1976).

5. Dennis, J. B., "Data Flow Supercomputers," *Computer* 13(11)(Nov., 1980).

6. Dennis, J. B. and Misunas, D. P., "A Computer Architecture for Highly Parallel Signal Processing," *Proc. of the 1974 Nat. Comp. Conf.*, pp. 402-409 (1984).

7. Watson, I. and Gurd, J., "A Practical Data Flow Computer," *Computer* 15(2)(Feb. 1982).

8. Davis, A. L., "The Architecture and System Method of DDM1: A Recursively Structured Data Driven Machine," *Proc. Fifth Ann. Symp. Computer Architecture*, pp. 210-215 (April, 1978).

9. Rumbaugh, J., "A Data Flow Multiprocessor," *IEEE Trans. on Computers* C-26(2) p. 138 (Feb. 1977).

10. Babb, Robert G., "Parallel Processing with Large Grain Data Flow Techniques," *Computer* 17(7)(July, 1984).

11. Ackerman, William B., "Data Flow Languages," *Computer* 15(2)(Feb., 1982).

12. Treleaven, P. C., Brownbridge, D. R., and Hopkins, R. P., *Data Driven and Demand Driven Computer Architecture*, University of Newcastle upon Tyne, Newcastle upon Tyne, England (1981). Technical Report

13. Messerschmitt, D. G., "Structured Interconnection of Signal Processing Programs," *Proceedings of Globecom 84*, (Dec., 1984).

14. Snyder, Lawrence, "Parallel Programming and the Poker Programming Environment," *Computer* 17(7)(July, 1984).

15. Kelly,, Lochbaum,, and Vyssotsky,, "A Block Diagram Compiler," *BSTJ* 40(3)(May, 1961).

16. Gold, B. and Rader, C., *Digital Processing of Signals*, McGraw-Hill (1969).

17. Karafin, B., "The new block diagram compiler for simulation of sampled-data systems," *AFIPS Conference Proceedings* 27 pp. 55-61 (1965). Spartan Books

18. Dertouzous, M., Kaliske, M., and Polzen, K., "On line simulation of block-diagram systems," *IEEE Trans. on Computers* C-18(4)(April, 1969).

19. Korn, G., "High-speed block-diagram languages for microprocessors and minicomputers in instrumentation, control, and simulation," *Computers in Electrical Engineering* 4 pp. 143-159 (1977).

20. Henke, W., "MITSYN - An Interactive Dialogue Language for Time Signal Processing," *MIT Research Laboratory of Electronics memo. no. RLE-TM-1*, (Feb. 1975).

21. Crystal, T. and Kulsrud, L., *Circus*, Institute for Defense Analysis, Princeton, NJ (Dec., 1974). CRD Working Paper

22. Dipartimento di Elettronica, Politecnico di Torino,, *TOPSIM III - Simulation Package for Communication Systems - User's Manual.*

23. Messerschmitt, D. G., "A Tool for Structured Functional Simulation," *IEEE Journal on Selected Areas in Communications* SAC-2(1)(January, 1984).

24. Crochiere, R. E. and Oppenheim, A. V., "Analysis of Linear Digital Networks," *Proceedings of the IEEE* 63(4) pp. 581-595 (April, 1975).

25. Brafman, J. P., Szczupak, J., and Mitra, S. K., "An Approach to the Implementation of Digital Filters using Microprocessors," *IEEE Trans. on Acoustics, Speech, and Signal Processing* ASSP-26(5) pp. 442-446 (Oct. 1978).

26. Zeman, J. and Moschytz, G. S., "Systematic Design and Programming of Signal Processors, Using Project Managment Techniques," *IEEE Trans. on Acoustics Speech and Signal Processing* ASSP-31(6)(December 1983).

27. Schwartz, David A., "Synchronous Multiprocessor Realizations of Shift-Invariant Flow Graphs," *Georgia Institute of Technology Technical Report DSPL-85-2*, (July 1985). PhD Dissertation

28. Barnwell, Thomas P., Hodges, C. J. M., and Randolf, Mark, "Optimum Implementation of Single Time Index Signal Flow Graphs on Synchronous Multiprocessor Machines," *Proceedings of the Int. Conf. on Acoustics, Speech, and Signal Processing*, (May 3-5, 1982).

29. Barnwell, Thomas P. and Schwartz, D. A., "Optimal Implementation of Flow Graphs on Synchronous Multiprocessors," *Proc. 1983 Asilomar Conf. on Circuits and Systems*, (Nov., 1983).

30. Karp, R. M., Miller, R. E., and Winograd, S., "The Organization of Computations for Uniform Recurrence Equations," *Journal of the ACM* **14** pp. 563-590 (1967).

31. Rao, Sailesh K., *Regular Iterative Algorithms and their Implementations on Processor Arrays*, Information Systems Laboratory, Stanford University (October, 1985). PhD Dissertation

32. Karp, R. M. and Miller, R. E., "Properties of a Model for Parallel Computations: Determinacy, Termination, Queueing," *SIAM Journal* **14** pp. 1390-1411 (November, 1966).

33. Reiter, Raymond, *A Study of a model for Parallel Computations*, University of Michigan (1967). Doctoral Dissertation

34. Commoner, F. and Holt, A. W., "Marked Directed Graphs," *Journal of Computer and System Sciences* **5** pp. 511-523 (1971).

35. Reiter, Raymond, "Scheduling Parallel Computations," *JACM*, (14) pp. 590-599 (1968).

36. Peterson, James L., "Petri Nets," *Computing Surveys* **9**(3)(September, 1977).

37. Peterson, James L., *Petri Net Theory and the Modeling of Systems*, Prentice-Hall Inc., Englewood Cliffs, NJ (1981).

38. Agerwala, Tilak, "Putting Petri Nets to Work," *Computer*, p. 85 (December, 1979).

39. Karp, Richard M. and Miller, Raymond E., "Parallel Program Schemata," *Journal of Computer and System Sciences* **3** pp. 147-195 (1969).

40. Lee, E. A., "A Coupled Hardware and Software Architecture for Programmable Digital Signal Processors," *Memorandum No. UCB/ERL M86/54, EECS Dept., UC Berkeley*, (1986). PhD Dissertation

41. Lee, E. A. and Messerschmitt, D. G., "A Coupled Hardware and Software Architecture for Programmable Digital Signal Processors: Parts I and II," *IEEE Transaction on ASSP (submitted)*, (1986).

42. Pope, S., Rabaey, J., and Brodersen, R. W., "An Integrated Automatic Layout Generation System for DSP Circuits," *IEEE Trans. on Computer-aided Design* **CAD-4**(3) pp. 285-296 (July 1985).

43. Lee, E. A. and Messerschmitt, D. G., "Static Scheduling of Data Flow Programs for Synchronous Digital Signal Processing," *To appear in IEEE Trans. on Computers*, (1986).

44. Coffman, E. G. Jr., *Computer and Job Scheduling Theory*, Wiley, New York (1976).

45. Adam, T. L., Chandy, K. M., and Dickson, J. R., "A Comparison of List Schedules for Parallel Processing Systems," *Comm. ACM* **17**(12) pp. 685-690 (Dec., 1974).

46. Hu, T. C., "Parallel Sequencing and Assembly Line Problems," *Operations Research* **9**(6) pp. 841-848 (1961).

47. Kohler, W. H., "A Preliminary Evaluation of the Critical Path Method for Scheduling Tasks on Multiprocessor Systems," *IEEE Trans. on Computers*, pp. 1235-1238 (Dec., 1975).

48. Renfors, Markku and Neuvo, Yrjo, "The Maximum Sampling Rate of Digital Filters Under Hardware Speed Constraints," *IEEE Trans. on Circuits and Systems* **CAS-28**(3)(March 1981).

A Reconfigurable Concurrent
VLSI Architecture for Sound Synthesis

John Wawrzynek

Computer Science
California Institute of Technology
Pasadena, California

Introduction

Past attempts at modeling the dynamics of musical instruments to produce musical sounds have met with limited success. Physically based models are necessarily complicated because they must include subtle effects, and all physical models are computationally expensive. Even simple models are not computable in real time on conventional computers. The inability to perform extensive listening experiments has hindered refinement of models of musical instruments.

We have developed a computing system that outperforms conventional computers several hundreds of times for sound synthesis and related applications. The architecture is simple and can be extended easily. It is based on reconfigurable concurrent computation and VLSI. A linear array of simple processing elements is employed, with an interconnection matrix between processors that can be reconfigured. The realization of a new instrument model involves a configuration of the connection matrix between the processing elements and a configuration of connections to the outside world for updates of parameters and inter-chip communication. Our system has been implemented in CMOS technology, and can generate realistic musical sounds in real time. Although it was developed for sound synthesis, our architecture is capable of solving a wide range of problems very efficiently.

Sound Synthesis

Traditionally, because of computational complexity, systems designed to generate musical sounds have focused on mechanisms to reproduce the *harmonic content* of a sound, with little concern for the dynamics of the musical instrument or sound source. The way a sound evolves in time is as important as harmonic content to human audio perception. Sound sources in nature evolve in complex ways and, except in isolated cases cannot be modeled as simple oscillators with amplitude envelopes; Such systems produce a characteristic "electronic" sound. In struck sound sources, the harmonics of the sound evolve from an initially noisy signal as the natural modes of the system "filter" the excitation function. The human audio perception system is particularly sensitive to the evolution of harmonics from noise.

Nonlinearities in the driving mechanism of wind instruments aid human perception of evolving musical sounds by (1) providing harmonic content to otherwise pure resonance tones of the instrument, and (2) providing a mode-mixing mechanism for the transfer of energy from one resonance mode to another. Although, these effects are well known, existing electronic musical instruments do not include nonlinearities. A nonlinear element in a feedback loop provides a novel and efficient means to model wind instruments.

Struck Instrument Model

We have developed a simple model that produces the sound of struck and plucked instruments, for example, a marimba or a guitar string. Furthermore, it provides a parameterization that is natural to a musician and useful in a musical context. This model and all others presented here are directly executable on our computing engine to generate sounds in realtime.

The model comprises two pieces; the *attack section* and the *resonator bank* (Figure 1). The attack section models the impact of the striking or plucking device on the actual instrument. An impulse is fed to a second-order section that is tuned with a Q value close to critical damping. The output of the *attack resonator* is fed to the input of the *noise modulation section*. This computation adds to the input signal an amount of random signal, or noise, proportional to the level of the input.

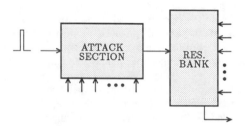

Figure 1: Struck Instrument Model

The output of the noise modulation section is used to drive a parallel connection of second-order sections that serve as resonators. The latter are tuned to the major resonances of the sound source being modeled. The parameters of the attack section—attack resonator frequency and Q value, signal-to-noise ratio, and attack level—are adjusted to produce a variety of musical timbres.

In a typical application, a pianolike keyboard is used to control the instrument. Pressing a key generates an impulse that is sent to the attack resonator. The key position determines the coefficients loaded into the resonator bank, and the key-press velocity controls the level of the noise modulation coefficient in the attack section (higher key velocities correspond to more noise being introduced into the the system and hence a higher attack level).

Instrument Model with Nonlinearity

Figure 2 shows the configuration of a model employing a nonlinear element, that has been used successfully to generate flutelike tones and sounds of other woodwind instruments [3]. It is composed of three pieces: (1) a nonlinear element that computes a third-order polynomial, (2) a noise modulation section that adds an amount of noise proportional to the size of the signal at its input, and (3) a resonator bank that has second-order resonators tuned to frequencies corresponding to the resonances of the musical instrument. These elements are connected in a cascade arrangement, forming a closed loop. When the loop gain is sufficiently high, and the system is disturbed, it oscillates with modes governed by the tuning of the resonator bank. Typically, the loop gain is controlled by the gain of the nonlinear element, G. For small values of G, the feedback is too small and the system does not oscillate. If G is just large enough, the system will oscillate with a very pure tone as it operates in the nearly linear range of the nonlinear element. If the nonlinear gain G is set to an even higher value, the signal is increased in amplitude and is forced into the nonlinear region. The nonlinearity shifts energy into higher frequencies, generating a harsher, louder tone.

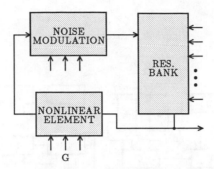

Figure 2: Instrument Model with Nonlinearity

In a typical application, the loop gain is set by controlling G according to the velocity of a key-press. A slowly pressed key sets a small G value, and thus generates a soft, pure tone. A quickly pressed key sets a larger G value, and hence produces a louder, harsher tone. When the key is released, the G value returns to just slightly less than the point at which the loop gain is large enough to sustain oscillation. G is not returned to zero, so the signal dies out exponentially with time; the time constant is controlled by the value of G.

Composite Model

Because both models contain several parts in common, they can be combined into one structure, with the addition of an extra coefficient to control the feedback, as shown in Figure 3. A detailed view of the composite model is shown in the form of a computation graph for our computing engine in Figure 5. Each rectangle in the graph represents the operation of add/multiply/delay and, optionally, mod 2^{32}, $(A + B \times M)z^{-1}$, as illustrated in Figure 4.

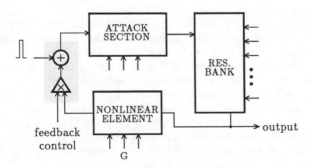

Figure 3: Composite Instrument Model

$$Y = (A + BM)z^{-1}$$

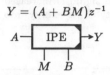

Figure 4: Processing Node

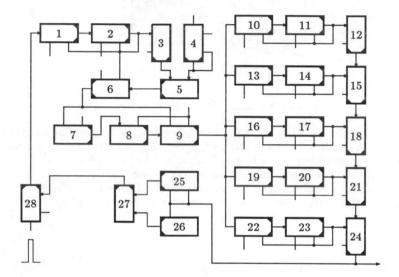

Figure 5: Computation Graph for Composite Model

COMPUTING ENGINE

In this section, we present a computer architecture designed specifically for the finite-difference computations used in generating musical sounds. Our computational task is the realtime evaluation of the fixed computation graphs of the variety presented in the previous section. In these graphs, each computation node is one or more members of the set of operations: plus, times, mod 2^{32}, and delay. Input to each node is either the output of another node or an externally supplied coefficient. Samples and coefficients flow to computation nodes across the arcs of the graph. Each processor in our computer is mapped to one and only one node and each communication channel in our machine is mapped to one and only one arc. This concept of a one-to-one mapping is a deviation from the traditional approach, in which a single processor is time-multiplexed to perform the function of each node sequentially, and memory is used to form "interconnect."

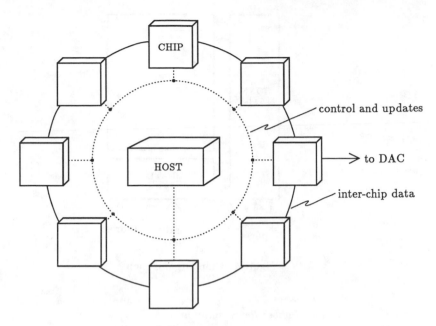

Figure 6: Typical System Configuration

Architectural Overview

Our machine is structured as a number of inner-communicating chips, each responsible for computing a piece of a computation graph. We assume that our task may be organized such that somewhat independent subgraphs may be split off and solved fairly independently in a small number of clustered chips (possibly just one), alleviating the need for very high bandwidth between chips and between clusters of chips. Musical sound synthesis has this locality property.

Figure 6 shows a typical system configuration (many others are possible). The chips are organized in a ring structure with each chip communicating to its nearest neighbors. They are controlled by a global master, or *host*, that provides initialization information and coefficient updates during the computation. The host also provides an interface either to an external controlling device, such as a pianolike keyboard, or to a disk file containing musical-score information.

Each chip comprises three major pieces, as illustrated in Figure 7. An array of identical processing elements responsible for arithmetic and delay operations forms the first piece. The second piece is a buffer for holding coefficients supplied by the host; these coefficients, with the outputs of other processing elements, serve as operands for the processing elements. The third piece is an reconfigurable interconnection matrx that serves all the chips communication needs; it connects processing elements to one another, to the output of the coefficient update buffer, and to input and output connectors of the chip. The exact patterns of communication are determined by setting switches in the matrix prior to the computation. Throughout a computation, these switches remain constant and thus the topology of the computation graph is fixed; it can be changed, however, between computations.

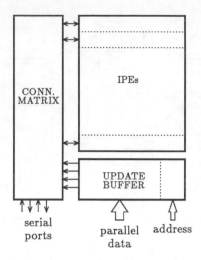

Figure 7: Chip Organization

Quantization Errors and Number Representation

Errors exist in our system due to the quantization of numbers used to represent signals and coefficients. Of course, it is the goal of every system designer to minimize these errors. We needed to select a number representation (fixed versus floating, number of bits) and a method of arithmetic, that would reduce the effects of quantization to a tolerable level.

Quantization of Coefficients. In signal processing systems, using a finite number of bits for the representation of coefficients results in imprecise pole and zero placement. In essence, coefficient quantization causes a system to exhibit a *finite* set of behaviors. A problem arises because the behaviors are not evenly distributed. Consider the second-order difference equation we used as a resonator:

$$y_n = 2R\cos\theta_c y_{n-1} - R^2 y_{n-2} + x_n$$

This system has a pair of poles that appear in the z-plane at $z = Re^{j\theta_c}$ and $z = Re^{-j\theta_c}$. The system impulse response is a damped sine wave with frequency and damping related to θ and R. With $R = 1$, the impulse response is a sine wave of constant amplitude; the system is an oscillator. The coefficients R^2 and $2R\cos\theta_c$ may take on only a finite set of values; therefore, the poles may take on only a finite set of positions in the z-plane. In a musical context this means that only a particular set of frequencies and damping values are attainable. At the extremes of frequency and damping the attainable frequencies are very sparse. To make matters worse, pitch is perceived as the log of frequency, further spreading out the attainable frequencies in the low range. A small number of bits is not enough to generate a frequency range with enough resolution to be musically useful. In fact, human frequency discrimination has been measured at about 0.2 or 0.3% [6]. To satisfy this tolerance for the fundamental frequencies of a piano, generated using the second-order resonator as a test case, at least 24 bit fractional coefficients are necessary.

Of possibly greater concern is *relative* frequency. Two tones that are meant to have a exact ratio in their frequencies may generate beat frequencies because of errors in the ratio due to quantization of coefficients.

Quantization Noise. Each time a digital multiplication is performed on two N bit numbers a 2N bit result is formed, and the result must be rounded (or truncated) to fit within the N bit number representation. Rounding introduces a small error signal (less than one LSB). Some researchers have approximated this effect by modeling it as a noise signal added to the signal at each multiplier in a system [5]. Although this model assumes that the error signal is a white-noise sequence, is uniformly distributed, and is uncorrelated with the signal, it is probably fairly accurate for signals as complex as those in music. Using this model, the effect of quantization may be derived for various systems or filter forms. Oppenheim & Shaffer computed the variance of the output noise due to arithmetic rounding in a second-order two-pole system. Using their result we have computed the output noise for several values of center frequency and R, as shown in Figure 8. The values of R were chosen to cover a wide range of values for the damping constant τ. The sampling frequency is the digital audio standard of 44057 samples/second. The output noise level was computed in terms of the *number of bits* necessary to represent such a level. From Figure 8 it is apparent that, for systems with long time constants, at least 16 bits of each sample may be in error due to quantization, therefore a number representation with much more than 16 bits is necessary.

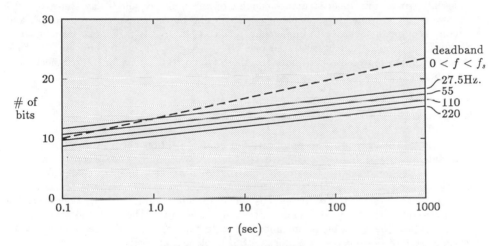

Figure 8: Quantization Effects

Limit Cycles. One type of limit cycle in digital systems results in so called deadbands—or intervals of signal values around zero in which a system can experience self-sustaining oscillations due to rounding of state values [1]. As previously, we computed the effect in the *number of bits* for a second-order two-pole system. The amplitude of the limit cycle in the second-order system is only a function of the da mping coefficient R. This amplitude (in bits) is plotted, with the curves showing the quantization "noise" in Figure 8. Limit cycles are potentially a more harmful problem than is quantization "noise" because limit cycles concentrate their energy at a particular frequency unlike quantization "noise" that is more uniformly distributed. These curves may be interpreted as bounds on the quantization noise and limit cycle amplitude and used as a guide for choosing the number of bits to use to represent signals. From the curves, 16 or even 24 bits clearly are not enough. We have chosen a two's-complement representation that employs a sign bit, two integer bits, and 29 bits of fraction.

This analysis says nothing about composition of second-order sections or even other filter forms, but does treat an important case gives a flavor for how bad quantization effects can be in general.

The Processors

Bit serial processing offers two attractive features for our application. First, the processing elements are physically small, so large numbers of them can be integrated on a single chip. Bit serial processing also facilitates bit serial communication, simplifying communication channels; single wires can be used to interconnect processing elements. One potential drawback is the *latency* incurred with each operation—from the operands arrival until the total answer's arrival at the output. In our application, however, we *want* a delay at each processing step, so the latency is an advantage.

Various bit serial multiplication schemes have been implemented and presented in the literature [2]. We wished to provide maximum processing power per unit chip area as was possible with current technology. Therefore, we chose the simplest multiplication scheme that met the constraints placed by standard digital audio rates. Our processor is a serial/parallel multiplier structure capable of one multiply/add/delay step per word time, we call it an inner product element (IPE). The multiplier structure is simple and therefore requires little space to implement in silicon [4]. Inputs arrive one bit at a time, LSB first, and the output is generated one bit at a time. All inputs and outputs have the same number representation; therefore, there are no restrictions for interconnection of processors or the connection of coefficients.

The Connection Matrix

The connection matrix provides point-to-point communication between processors, from the update buffer and, bidirectionally with the outside world. The matrix is *programmable*; the interconnection patterns within the matrix are not fixed, but are changeable from external control, made possible by a storage cell located at each cross point in the matrix and logic to set the state of the storage cells. In addition to programmability the connection matrix also takes advantage of the inherent locality in sound synthesis computations and is discretionary in the allowable interconnection patterns, saving in chip area and providing for the growth of the processing power of VLSI implementations.

Figure 9 shows the basic structure of the connection matrix and its interface to the other components. Note that the horizontal wires, or *tracks*, are used to bring in signals from off-chip and to send signals off-chip, as well as to provide communication between processing elements. One possible configuration is to dedicate one track per processing element output, which guarantees that any IPE can communicate with any other IPE. Such a configuration, however, grows as the square of the number of processing elements; in musical sound synthesis applications it is a waste of chip area. In Figure 10, we have mapped the computation graph in Figure 5 onto the processor array by assigning nodes of the graph to IPEs and routing the interconnections, assuming tracks could be broken arbitrarily. The assignment of processors to nodes in the graph was ordered from left to right across the array for consecutive number nodes. Clearly, all tracks have many breaks and there is a large number of small links and a relatively smaller number of larger links, and so on. In the modification shown in Figure 11; tracks no longer span the entire array of IPEs, but rather are split at one or more points along their lengths. There is one track of links for length 2, one for length 4, and so on, doubling the length of the links for each track until the entire length of the array is spanned in one link. The breaks in the tracks are arranged to avoid any two breaks lining up vertically, and consequently to maximize the potential communication between pairs of processors. This matrix grows as $Nlog(N)$—rather that N^2, as does the earlier version—thus saving area. The computation graph of Figure 5 has been mapped into the new structure, in Figure 12. The ordering of the nodes in the graph has been perturbed to make a better match. All but one network is routed in the modified matrix; one additional track is used to handle that network.

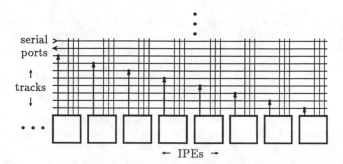

Figure 9: Basic Structure of Connection Matrix

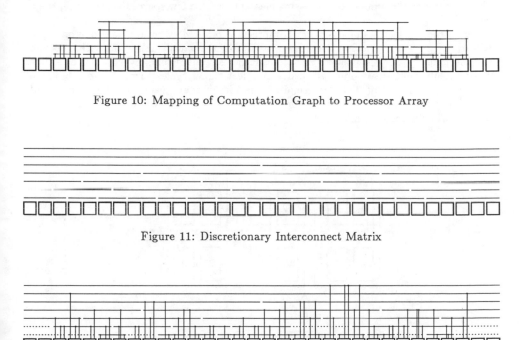

Figure 10: Mapping of Computation Graph to Processor Array

Figure 11: Discretionary Interconnect Matrix

Figure 12: Mapping of Computation Graph to Modified Matrix

We are currently establishing the best configuration for the connection matrix and are developing algorithms for assignment of computational nodes to processors in the array. Although the assignment problem is NP-complete, heuristic algorithms that find the inherit localities in our applications perform very well and are aided through the addition of a few extra tracks in the matrix and few extra processing elements in the array.

The Update Buffer

The update buffer is simply a register bank to hold coefficients (that is, inputs to the processing elements supplied from the host computer). Input to the update buffer is a parallel connection to a standard computer memory bus. The outputs of the update buffer are bit serial lines that run through the connection matrix to the processing elements. For maximum flexibility in the assignment of processing elements, no a priori correspondence is made between update buffer registers and processing elements; this assignment is made by programming the connection matrix.

One important feature of the update buffer is that it is double buffered. Coefficients can be sent from the host computer to the update buffer without effecting the ongoing computation. Only, after all the coefficients of a new set of updates have arrived in the buffer is a signal sent to update them simultaneously. If some coefficients were allowed to change before others did, instability could result.

Empirically, we have found that our applications average about one coefficient per processing element. This fact defines the nominal number of registers in the update buffer to be the same as the number of processing elements, with a few spares to cover exceptional cases.

The structure of the update buffer comprises two RAM structures laid one on top of the other (Figure 13). The first RAM is writable from the parallel input bus with a decoder that selects one coefficient (row). The second RAM is readable one bit (column) at a time, all coefficients being read simultaneously. A select signal cycles through the columns of the second RAM one bit at a time sending the bits of the registers to the output, LSB first. Under control from the host computer, a transfer signal copies the contents of the first RAM to the second one.

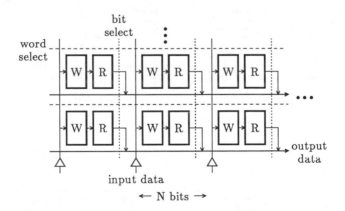

Figure 13: Dual Ram Structure of Update Buffer

CONCLUSION

We have used VLSI technology to tailor an architecture to the computations required for modeling the dynamics of musical instruments. The key to the efficiency of our machine differentiates it from other concurrent architectures: No processing cycles are used for communication—the processors are dedicated to arithmetic operations and the connection matrix is programmed to provide the communication for a specific task. The computations required for modeling musical instruments are representative of a larger class of problems that can be formulated using systems of finite difference equations. Are architecture should be equally efficiently for these related problems.

Figure 14 shows a photo of a prototype of a CMOS implementation of one chip from our system, fabricated using a 3 μ feature size. This version has 20 IPEs and an interconnection matrix. New versions contain more that 30 processors per chip—roughly enough to synthesize a single musical voice per chip in real time. Details of the circuits may be found in [4].

Figure 14: Chip Photograph

ACKNOWLEDGMENT

This work was supported by the System Development Foundation.

REFERENCES

[1]   Jackson, L. B., "An Analysis of Limit Cycles due to Multiplication Rounding in Recursive Digital (Sub) Filters," *Proc. 7th Annu. Allerton Conf. Circuit System Theory*, pp. 69–78, 1969.

[2]   Lyon, R. F., "Two's Complement Pipeline Multipliers," *IEEE Transactions on Communications*, April 1976, pp. 418–425.

[3]   McIntyre M. E., Schumacher R. T., Woodhouse, J., "On the Oscillations of Musical Instruments," *Journal of the Acoustical Society of America*, November 1983, pp. 1325–1344.

[4]   Mead, C. A. and Wawrzynek J. C., "A New Discipline for CMOS Design: an Architecture for Sound Synthesis," *1985 Chapel Hill Conference on Very Large Scale Integration*, Edited by Henry Fuchs, Computer Science Press, 1985.

[5]   Oppenheim A. V., and Schafer R., *Digital Signal Processing*, Prentice-Hall, 1975.

[6]   Pickles J. O., *An Introduction to the Physiology of Hearing*, Academic Press, 1982.

[7]   Wawrzynek, J. C. and Mead, C. A., "A VLSI Architecture for Sound Synthesis," *VLSI Signal Processing: A Bit-Serial Approach*, Edited by Peter Denyer and David Renshaw, 1985 Prentice-Hall, pp. 277–297.

Author Index